Intelligent Systems Reference Library

Volume 264

The aim of this series is to publish a Reference Library, including novel advances and developments in all aspects of Intelligent Systems in an easily accessible and well structured form. The series includes reference works, handbooks, compendia, textbooks, well-structured monographs, dictionaries, and encyclopedias. It contains well integrated knowledge and current information in the field of Intelligent Systems. The series covers the theory, applications, and design methods of Intelligent Systems. Virtually all disciplines such as engineering, computer science, avionics, business, e-commerce, environment, healthcare, physics and life science are included. The list of topics spans all the areas of modern intelligent systems such as: Ambient intelligence, Computational intelligence, Social intelligence, Computational neuroscience, Artificial life, Virtual society, Cognitive systems, DNA and immunity-based systems, e-Learning and teaching, Human-centred computing and Machine ethics, Intelligent control, Intelligent data analysis, Knowledge-based paradigms, Knowledge management, Intelligent agents, Intelligent decision making, Intelligent network security, Interactive entertainment, Learning paradigms, Recommender systems, Robotics and Mechatronics including human-machine teaming, Self-organizing and adaptive systems, Soft computing including Neural systems, Fuzzy systems, Evolutionary computing and the Fusion of these paradigms, Perception and Vision, Web intelligence and Multimedia.

Indexed by SCOPUS, DBLP, zbMATH, SCImago.

All books published in the series are submitted for consideration in Web of Science.

Zoltán Gellér · Vladimir Kurbalija ·
Miloš Radovanović · Mirjana Ivanović

Recent Advances in Time-Series Classification—Methodology and Applications

Springer

Zoltán Gellér
Faculty of Philosophy
Department of Media Studies
University of Novi Sad
Novi Sad, Serbia

Miloš Radovanović
Faculty of Sciences
Department of Mathematics
and Informatics
University of Novi Sad
Novi Sad, Serbia

Vladimir Kurbalija
Faculty of Sciences
Department of Mathematics
and Informatics
University of Novi Sad
Novi Sad, Serbia

Mirjana Ivanović
Faculty of Sciences
Department of Mathematics
and Informatics
University of Novi Sad
Novi Sad, Serbia

Corresponding Member of Serbian
Academy of Sciences and Arts (SASA)
Belgrade, Serbia

ISSN 1868-4394 ISSN 1868-4408 (electronic)
Intelligent Systems Reference Library
ISBN 978-3-031-77529-1 ISBN 978-3-031-77527-7 (eBook)
https://doi.org/10.1007/978-3-031-77527-7

This Springer imprint is published by the registered company Springer Nature Switzerland AG
The registered company address is: Gewerbestrasse 11, 6330 Cham, Switzerland

Preface

Time series are used to store, analyze and visualize continuous numerical data points collected in various fields. With the increasing volume of data, the interest in researching tasks related to mining time series data and particularly in classification has become increasingly prominent. A critical aspect of time-series, distance-based classification is choosing an appropriate similarity/distance measure to compare time series effectively. Numerous similarity measures have been proposed and utilized in this field, many of which rely on dynamic programming. However, the computational requirements of dynamic programming can be a limiting factor in practical applications. One approach to mitigate this issue is to apply global constraints to restrict the search area in the warping (editing) matrix. On the other hand, these constraints may change the behavior of original measures and impact the accuracy of classification, making it crucial to understand their implications. Therefore, the selection of a suitable similarity measure and the consideration of global constraints are essential factors in the successful realization of distance-based classification tasks of time series.

This book, through several aspects, investigates the impact of constraints on elastic time-series similarity measures and how it can assist practitioners and researchers in selecting suitable measures. The findings can contribute to the development of effective indexing techniques for accelerating nearest-neighbor calculations, where some papers suggest that measures with constraints around 5% yield similar classification accuracies as their unconstrained counterparts. At the same time, computational time is reduced by two and in some cases even three orders of magnitude.

Many studies highlight the relevance of the nearest neighbor in time-series classification, with the 1NN classifier performing well compared to other, more complex classifiers. Additionally, the utilization of the kNN classifier yields to more stable results. The special status of the most similar neighbor can significantly influence the diversity of class labels in k-neighborhoods. Therefore, the usage of weighted kNN classifier (the kNN classifier where closer neighbors contribute more to the final selection of the class) can merge the positive effects of these two aspects. The experimental evaluation of weighted classification is the second important contribution of our work.

This book consists of the following chapters:

1. Introduction
2. Time Series and Similarity Measures
3. Time Series Classification
4. The Impact of Global Constraints on the Accuracy of Elastic Similarity Measures
5. The Influence of Weighting Schemes on Classification Accuracy
6. Combined Influence of Weighting Schemes and Global Constraints on Classification Accuracy
7. Conclusion

In the introductory chapter, a brief outline of the considered scientific field is provided and the motivation and goals of our research are highlighted.

All necessary concepts in time-series analysis are introduced in Chap. 2. This chapter provides the definition of time series, presentation of the examined distance measures and the concept of global constraints. The description of the data repository and the framework for time-series analysis which is used throughout this book is also presented here.

Chapter 3 introduces concepts needed for time-series classification and performance evaluation. Here, a comprehensive overview of the nearest neighbor classifier and its variants is given. In addition, numerous weighting schemes for kNN classifier are explained in detail. At the end of the chapter, an overview of the basic methods for evaluation of classifier performance is given.

The first chapter which presents the results of the experiments (Chap. 4) explores the impact of global constraints on the four basic elastic measures presented in Chap. 2: Dynamic Time Warping (DTW), Longest Common Subsequence (LCS), Edit distance with Real Penalty (ERP) and Edit Distance on Real sequence (EDR). In three phases of experiments, the impact of constraining elasticity on the intensity of changes in the NN graph was investigated and the findings were verified by comparing unconstrained measures with their constrained counterparts using optimal warping-window widths.

Chapter 5 compares the performances of 1NN and variants of the kNN classifier employing dynamic programming-based distance measures and several versions of L_p-based distances (Euclidean, Manhattan and Chebyshev distance) with the aim to comprehend the importance of the most similar neighbor and to enhance the accuracy of kNN by focusing on close neighbors by applying various weighting approaches.

Finally, the last chapter which presents the results of the experiments (Chap. 6) combines findings from the previous two chapters and empirically elaborates the cooperative effects of applying global constraints on elastic distances and weights in kNN classification of time series. The findings from this research underscore the significance of the nearest neighbor in time-series classification and suggest ways to boost the performance of the kNN classifier using different weighting techniques and global constraints in distance measure computation.

In the end, Chap. 7 summarizes the findings presented in this book, gives relevant conclusions and outlines the possibilities for further research.

The authors would like to thank the Faculty of Philosophy and the Faculty of Sciences of the University of Novi Sad for making the computers of their computer classrooms and computation centers available for generating distance matrices.

The authors would like to express special thanks to Eamonn Keogh for compiling and sharing the UCR time-series datasets and to all data contributors whose efforts were indispensable to this research.

Novi Sad, Serbia

Zoltán Gellér
Vladimir Kurbalija
Miloš Radovanović
Mirjana Ivanović

Contents

1 Introduction 1
References 4

2 Time Series and Similarity Measures 9
2.1 Time Series 9
2.1.1 Data Mining Tasks 12
2.1.2 Time-Series Preprocessing 16
2.2 The UCR Time Series Classification Archive 19
2.3 Similarity Measures 25
2.3.1 Euclidean Distance (L_2) 27
2.3.2 Dynamic Time Warping (DTW) 31
2.3.3 Longest Common Subsequence (LCS) 39
2.3.4 Edit Distance with Real Penalty (ERP) 47
2.3.5 Edit Distance on Real Sequence (EDR) 59
2.3.6 A Comparative Overview of the Properties of Distance Measures 65
2.4 Global Constraints 68
2.5 Distance Matrices 79
2.5.1 Framework for Analysis and Prediction (FAP) 80
2.5.2 Distributed Distance Matrix Generation 82
References 86

3 Time Series Classification 95
3.1 The Nearest Neighbor Rule 97
3.1.1 Dudani's Weighting Functions 99
3.1.2 Macleod's Weighting Function 100
3.1.3 The Fibonacci Weighting Function 100
3.1.4 The Uniform and Dual-Uniform Weighting Functions 101

3.1.5 Zavrel's Weighting Function ... 104
3.1.6 The Dual Distance-Weighted Function ... 106
3.2 Classifier Evaluation Methods ... 108
References ... 109

4 The Impact of Global Constraints on the Accuracy of Elastic Similarity Measures ... 115
4.1 Evaluation Methodology ... 116
4.1.1 Changes in the NN Graph ... 116
4.1.2 Classification Accuracy—Individual Warping-Window Widths ... 119
4.1.3 Classification Accuracy—Optimal Warping-Window Widths ... 120
4.2 Dynamic Time Warping (DTW) ... 123
4.2.1 Changes in the NN Graph ... 123
4.2.2 Classification Accuracy—Individual Warping-Window Widths ... 124
4.2.3 Classification Accuracy—Optimal Warping-Window Widths ... 125
4.3 Longest Common Subsequence (LCS) ... 131
4.3.1 Changes in the NN Graph ... 131
4.3.2 Classification Accuracy—Individual Warping-Window Widths ... 133
4.3.3 Classification Accuracy—Optimal Warping-Window Widths ... 134
4.4 Edit Distance with Real Penalty (ERP) ... 139
4.4.1 Changes in the NN Graph ... 139
4.4.2 Classification Accuracy—Individual Warping-Window Widths ... 141
4.4.3 Classification Accuracy—Optimal Warping-Window Widths ... 142
4.5 Edit Distance on Real Sequence (EDR) ... 148
4.5.1 Changes in the NN Graph ... 148
4.5.2 Classification Accuracy—Individual Warping-Window Widths ... 149
4.5.3 Classification Accuracy—Optimal Warping-Window Widths ... 151
4.6 Highlights ... 156
References ... 157

5 The Influence of Weighting Schemes on Classification Accuracy 161
5.1 Euclidean Distance (L_2) 166
5.2 Manhattan Distance (L_1) 176
5.3 Chebyshev Distance (L_∞) 185
5.4 Dynamic Time Warping (DTW) 195
5.5 Longest Common Subsequence (LCS) 205
5.6 Edit Distance with Real Penalty (ERP) 213
5.7 Edit Distance on Real Sequence (EDR) 226
5.8 Highlights 236
References 239

6 Combined Influence of Weighting Schemes and Global Constraints on Classification Accuracy 241
6.1 Dynamic Time Warping (DTW) 245
6.2 Longest Common Subsequence (LCS) 261
6.3 Edit Distance with Real Penalty (ERP) 275
6.4 Edit Distance on Real Sequence (EDR) 290
6.5 Highlights 305
References 307

7 Conclusion 311
References 316

Appendix A: Similarity Measures 319

Appendix B: Weighting Functions 323

Appendix C: The Relationship Between the Average Difference in Performance and the Average Distance from the Diagonal 325

Appendix D: Explanation of the Behavior of kNN Classifier Variants with $k = 2$ and $k = 3$ 327

Abbreviations

1N	1NN classifier paired with the corresponding unconstrained measure
1NN	Nearest neighbor rule
AV	Average
DD	Dual distance-weighted function
DDIT	Combination of the dual distance-weighted function and the Itakura parallelogram
DDSC	Combination of the dual distance-weighted function and the Sakoe-Chiba band
DN	Dudani's weighting scheme
DNIT	Combination of Dudani's weighting scheme and the Itakura parallelogram
DNSC	Combination of Dudani's weighting scheme and the Sakoe-Chiba band
DTW	Dynamic Time Warping
DU	Dual-uniform weighting function
DUIT	Combination of the dual-uniform weighting function and the Itakura parallelogram
DUSC	Combination of the dual-uniform weighting function and the Sakoe-Chiba band
EDR	Edit Distance on Real sequence
ERP	Edit distance with Real Penalty
FAP	Framework for Analysis and Prediction
FB	Fibonacci weighting function
IN	Inverse distance weight
INIT	Combination of the inverse distance weight and the Itakura parallelogram
INSC	Combination of the inverse distance weight and the Sakoe-Chiba band
IS	Inverse of the squared distance weight
ISIT	Combination of the inverse of the squared distance weight and the Itakura parallelogram

ISSC	Combination of the inverse of the squared distance weight and the Sakoe-Chiba band
IT	Itakura
kNIT	Combination of the majority-voting *k*-nearest neighbor rule and the Itakura parallelogram
kNN, kN	Majority-voting *k*-nearest neighbor rule
kNSC	Combination of the majority-voting *k*-nearest neighbor rule and the Sakoe-Chiba band
LCS	Longest Common Subsequence
MC	Macleod's weighting function
MR	Mean rank
pp	Percentage point
RK	Rank weight
SC	Sakoe-Chiba
SCV	Stratified cross-validation
SCV10x10	10 times repeated tenfold stratified cross-validation
SCV1x9	Ninefold stratified cross-validation
SD	Standard deviation
UCR	UCR Time Series Classification Archive
UN	Uniform weighting function
ZA	Zavrel's weighting function

Chapter 1
Introduction

In today's society, data, information and knowledge play crucial role in numerous aspects of human activities. Data refers to individual units of information that are raw and unorganized and are rarely useful alone. Information is derived from data through processing and organizing it in a context, while knowledge is a more abstract concept which represents understanding based on extensive experience with information. Data is present in all human activities which are closely related to usage of computers, involving activities like collecting, storing and analyzing it. By combining different interpretations, data mining can be outlined as the systematic exploration of extensive data repositories through various analytical techniques with the aim to discover and extract useful information [1–4].

Temporal data, specifically time series, represents data that is collected over time and is used in a wide range of domains such as finance, science, ecology, telecommunications, medicine and meteorology [2, 5, 6]. Time series analysis utilizes various methodologies, taken mostly from artificial intelligence and data science fields, and involves tasks like prediction, anomaly detection, clustering and classification. Moreover, the manipulation with this kind of data has led to the development of numerous methodologies in research community.

Time series, the most prevalent form of temporal data, are usually represented as sequences of ordered pairs, where each value describes some parameter of the observed phenomenon at a specific point in time. Time series can be continuous or discrete, multivariate or univariate, deterministic or stochastic and may contain missing values. In this book, one-dimensional, discrete, equidistant (measured at equally spaced time intervals) time series without missing values, are considered. Time series analysis aims to identify patterns in data, filter out noise, detect outliers and make predictions about future values. To accomplish these tasks, researchers and practitioners rely on statistical, data mining and machine learning approaches [2]. Understanding important components of time series analysis is essential for drawing trustworthy and meaningful conclusions and making informed decisions.

Z. Gellér et al., *Recent Advances in Time-Series Classification—Methodology and Applications*, Intelligent Systems Reference Library 264,
https://doi.org/10.1007/978-3-031-77527-7_1

The application of statistical modeling to time series data has a rich and extensive history [2], especially in economics and econometrics applications. However, the challenge of managing growing amounts of data has led to increased attention on temporal data mining, encompassing tasks like anomaly detection, indexing, clustering, classification, segmentation and prediction [1, 7]. This book focuses on time-series classification, which faces a constantly growing interest in recent years.

Various machine learning techniques have been explored for time-series classification [8–15], including decision trees [16], neural networks [17], support vector machines [18], Bayesian classifiers [19] and more. Surprisingly, the straightforward nearest-neighbor (1NN) method proved to be very competitive compared to these more complex approaches [13, 20–22].

The nearest neighbor (1NN) method is highly esteemed in data mining community [23]. The idea behind this simple classifier is self-explanatory: it assigns, to an unknown sample, the class of its closest neighbor from the training data [24]. The k-nearest neighbor (kNN) algorithm extends this principle by examining the k closest data points and selecting the class that appears most frequently within this group [25]. Another variation, the distance-weighted k-nearest neighbor method [26], assigns weights to neighbors' classes based on their proximity to the unclassified instance (closer neighbors are awarded with higher weights) and by summing the weights assigned to each class among the k nearest neighbors, it selects the class with the largest total weight.

A critical aspect of implementing the 1NN or kNN rule is selecting an appropriate distance measure to find the nearest neighbors. To compare time series data points, various similarity-based distance measures have been proposed [27–30]. Among them, the most widely employed are the Euclidean distance [31] and Dynamic Time Warping (DTW) [32]. The Euclidean distance is very intuitive and extremely fast to compute. However, it is sensitive to shiftings and scalings along the time axis as well as to noise and outliers. Unfortunately, these kinds of distortions occur commonly in the real-world data and cannot be easily eliminated using pre-processing techniques. DTW is probably the most researched and used among the elastic measures proposed to alleviate these deficiencies, while other common approaches include: Longest Common Subsequence (LCS) [33], Edit Distance with Real Penalty (ERP) [34] and Edit Distance on Real sequence (EDR) [35].

Elastic measures [36] are highly robust to various distortions but at the cost of lower efficiency compared to Euclidean distance. The implementation of all these measures is based on dynamic programming, but its quadratic computational complexity can be a limiting factor for larger problems. To address this issue, global constraints (such as the Sakoe-Chiba band [37] and the Itakura parallelogram [38]) can be used to restrict the search area and speed up computation [39, 40]. Additionally, global constraints have been shown to improve the accuracy of classification of distance-based classifiers compared to unconstrained measures [13, 41, 42]. The impact of global constraints on various similarity measures needs to be carefully investigated to understand their advantages and limitations [40, 43, 44].

In a 1NN graph, each time series is connected by a directed edge to the time series it most closely resembles. Given that the 1NN classifier assign the class of the most similar data point to unclassified instances, alterations to the 1NN graph can substantially impact the classification accuracy. To gain insights into the effects of global constraints, Chap. 4 investigates the changes in the 1NN graph as the size of the constraint changes. This analysis will provide a deeper understanding of the advantages and limitations of global constraints and how they impact the accuracy of the 1NN classifier.

The 1NN classifier was chosen due to its reported strong performance in classifying time-series data compared to other methods [7, 20–22, 45]. Moreover, its accuracy serves as a direct indicator of the underlying similarity measure's quality, making it a suitable method for investigating the impact of global constraints on classification accuracy. By examining how changes in the 1NN graph affect the classes of nearest neighbors and the overall accuracy of the classifier, a deeper understanding of the benefits of global constraints can be gained.

In the literature, various functions have been proposed for computing the weights of nearest neighbors in kNN classification tasks [24–26, 46–51], where each method usually claimed superiority over previous solutions. These papers often evaluated the newly introduced schemes on a small number of UCI machine learning repository datasets [52]. One of our goals was to compare these weighting schemes in a fair manner for time series data. The analyzed weighting methods are presented in Chap. 3, while Chap. 5 aims to compare them specifically within the time series domain considering unconstrained elastic similarity measures and distances based on the L_p norm. The motivation for this study is the belief that the 1NN rule may outperform the majority voting kNN rule. The research focuses on publicly available labeled time-series datasets from the UCR Time Series Repository [53], relying on three variants of the Minkowski distance (Euclidean, Manhattan and Chebyshev) and the most common elastic similarity measures used to compare time series (DTW, LCS, ERP and EDR). Statistical analysis is used to support the findings.

Finally, Chap. 6 investigates joint influence of global constraints and the effect of weights and neighborhood size on classification accuracy. It combines the five most promising weighting schemes (selected according to the results of Chap. 5) and global constraints of the four fundamental elastic similarity measures. All the results are compared with the performance of the 1NN classifier combined with the corresponding unconstrained elastic measure in order to make the results comparable with the previous chapters. This chapter actually examines the combined effect of two important parameters in distance-based classification of time series: r—the width of the warping window of the constrained elastic measures and k—the number of nearest neighbors considered by the kNN variants. The considered values of r and k are the same as in Chaps. 4 and 5 respectively.

The significance of time series data across different domains has resulted in a variety of solutions, often independently implemented and described in the literature. A freely available, open-source software package providing implementations of core time series algorithms is crucial to support research in this domain. Therefore, we developed the Framework for Analysis and Prediction (FAP) [54], a freely available library used for conducting experiments presented in this book. The library's basic features are outlined in Chap. 2.

Fair comparison of new and existing methodologies is crucial in every academic work. Therefore, all experiments within this book are performed on the largest publicly available time series repository of labeled data—the UCR Time Series Classification Archive [53]—which is also described in Chap. 2. The UCR Time Series Classification Archive is a valuable resource for researchers in the field of time-series data mining, containing 128 labeled datasets from a variety of research domains. These datasets include data from astronomy, sport, medicine, robotics, biology, seismology, food science, physics and more. However, the raw datasets in the archive differ in various properties such as: existence of missing values, normalization, inconsistent class enumeration and different time series length (within the same dataset). To address these inconsistencies, all datasets in the repository are preprocessed using Paparizzos' script [55] to ensure that all of them are z-normalized, have no missing values, and that time series belonging to the same dataset have the same length. This standardization process allows for fair comparisons between different methods when benchmarking on the UCR datasets.

Since time-series classification has gathered significant attention in research, this book will try to give a comprehensive overview of the behavior of the most common distance-based classifier(s) combined with elastic similarity measures which are more robust than the commonly accepted Euclidean distance. Global constraints offer a way to improve the speed and accuracy of similarity measures in time-series data analysis. By using techniques like the Sakoe-Chiba band and the Itakura parallelogram, the computational complexity of dynamic programming can be reduced without sacrificing accuracy. On the other hand, the usage of weighted kNN classifier can provide more stable classification results, yet favoring the first neighbor which has undoubtedly a special role in time-series data. Therefore, we believe that our efforts will provide valuable insights into the benefits of using global constraints and weights in distance measure calculations and time-series classification. This comprehensive analysis contributes to the ongoing development and refinement of techniques for analyzing and mining time series data. Going forward, continued research in this area is expected to further optimize time-series classification methods for handling increasing volumes of data.

References

1. Han, J., Pei, J., Tong, H.: Data mining: concepts and techniques, 4th edn. Morgan Kaufmann (2022)
2. Laxman, S., Sastry, P.S.: A survey of temporal data mining. Sadhana **31**, 173–198 (2006). https://doi.org/10.1007/BF02719780
3. Sumathi, S., Sivanandam, S.N.: Introduction to data mining and its applications. Springer (2006). http://books.google.rs/books?id=QwveL0jFk_gC
4. Witten, I.H., Frank, E., Hall, M.A., Pal, C.J.: Data mining: practical machine learning tools and techniques, 4th edn. Morgan Kaufmann (2017). https://doi.org/10.1016/C2015-0-02071-8
5. Esling, P., Agon, C.: Time-series data mining. ACM Comput. Surv. **45**, 12:1–12:34(2012). https://doi.org/10.1145/2379776.2379788

6. Hand, D.J., Mannila, H., Smyth, P.: Principles of data mining, MIT Press, Cambridge, MA, USA, 2001. http://books.google.rs/books?id=SdZ-bhVhZGYC.
7. Ding, H., Trajcevski, G., Scheuermann, P., Wang, X., Keogh, E.: Querying and mining of time series data: experimental comparison of representations and distance measures. In: Proc. VLDB Endow., VLDB Endowment, pp. 1542–1552 (2008). https://doi.org/10.14778/1454159.1454226
8. Grabockal, J., Bedallig, E., Schmidt-Thieme, L.: Efficient classification of long time-series. ICT Innov Secur. Intell. Syst. **207**(2013), 47–57 (2012)
9. Górecki, T., Łuczak, M.: Using derivatives in time series classification. Data Min. Knowl. Discov. **26**, 310–331 (2013). https://doi.org/10.1007/s10618-012-0251-4
10. Jeong, Y.-S., Jeong, M.K., Omitaomu, O.A.: Weighted dynamic time warping for time series classification. Pattern Recognit. **44**, 2231–2240 (2011). https://doi.org/10.1016/j.patcog.2010.09.022
11. Radovanović, M., Nanopoulos, A., Ivanović, M.: Time-series classification in many intrinsic dimensions. In: Proc. 2010 SIAM Int. Conf. Data Min., Society for Industrial and Applied Mathematics, Philadelphia, PA, pp. 677–688 (2010). https://doi.org/10.1137/1.9781611972801.59
12. Spiegel, S., Jain, B.-J., Albayrak, S.: Fast time series classification under lucky time warping distance. In: Proc. 29th Annu. ACM Symp. Appl. Comput., ACM, New York, NY, USA pp. 71–78 (2014). https://doi.org/10.1145/2554850.2554885
13. Xi, X., Keogh, E., Shelton, C., Wei, L., Ratanamahatana, C.A.: Fast time series classification using numerosity reduction. In: Proc. 23rd Int. Conf. Mach. Learn.—ICML '06, pp. 1033–1040. ACM Press, New York, USA (2006). https://doi.org/10.1145/1143844.1143974
14. Ye, L., Keogh, E.: Time series shapelets: a new primitive for data mining. In: Proc. 15th ACM SIGKDD Int. Conf. Knowl. Discov. Data Min., pp. 947–956. ACM, New York, NY, USA (2009). https://doi.org/10.1145/1557019.1557122
15. Middlehurst, M., Schäfer, P., Bagnall, A.: Bake off redux: a review and experimental evaluation of recent time series classification algorithms. Data Min. Knowl. Discov. (2024). https://doi.org/10.1007/s10618-024-01022-1
16. Rodríguez, J.J., Alonso, C.J.: Interval and dynamic time warping-based decision trees. In: Proc. 2004 ACM Symp. Appl. Comput., pp. 548–552. ACM, New York, NY, USA (2004). https://doi.org/10.1145/967900.968015
17. Nanopoulos, A., Alcock, R., Manolopoulos, Y.: Feature-based classification of time-series data. Int. J. Comput. Res. **10**, 49–61 (2001)
18. Wu, Y., Chang, E.Y.: Distance-function design and fusion for sequence data. In: Proc. Thirteen. ACM Conf. Inf. Knowl. Manag.—CIKM '04, pp. 324–333. ACM Press, New York, USA (2004). https://doi.org/10.1145/1031171.1031238
19. Pavlovic, V., Frey, B.J., Huang, T.S.: Time-series classification using mixed-state dynamic Bayesian networks. In: Proceedings. 1999 IEEE Comput. Soc. Conf. Comput. Vis. Pattern Recognit. (Cat. No PR00149), pp. 609–615. IEEE Comput. Soc. (1999). https://doi.org/10.1109/CVPR.1999.784983
20. Ruiz, A.P., Flynn, M., Large, J., Middlehurst, M., Bagnall, A.: The great multivariate time series classification bake off: a review and experimental evaluation of recent algorithmic advances. Data Min. Knowl. Discov. **35**, 401–449 (2021). https://doi.org/10.1007/s10618-020-00727-3
21. Bagnall, A., Lines, J., Bostrom, A., Large, J., Keogh, E.: The great time series classification bake off: a review and experimental evaluation of recent algorithmic advances. Data Min. Knowl. Discov. **31**, 606–660 (2017). https://doi.org/10.1007/s10618-016-0483-9
22. Jiang, W.: Time series classification: nearest neighbor versus deep learning models. SN Appl. Sci. **2**, 721 (2020). https://doi.org/10.1007/s42452-020-2506-9
23. Wu, X., Kumar, V., Ross Quinlan, J., Ghosh, J., Yang, Q., Motoda, H., McLachlan, G.J., Ng, A., Liu, B., Yu, P.S., Zhou, Z.-H., Steinbach, M., Hand, D.J., Steinberg, D.: Top 10 algorithms in data mining. Knowl. Inf. Syst. **14**, 1–37 (2008). https://doi.org/10.1007/s10115-007-0114-2
24. Cover, T., Hart, P.: Nearest neighbor pattern classification. IEEE Trans. Inf. Theory **13**, 21–27 (1967). https://doi.org/10.1109/TIT.1967.1053964

25. Fix, E., Hodges, J.L.: Discriminatory analysis. Nonparametric discrimination: consistency properties. Int. Stat. Rev. / Rev. Int. Stat. **57**, 238–247 (1989). https://doi.org/10.2307/1403797
26. Dudani, S.A.: The distance-weighted k-nearest-neighbor rule. IEEE Trans. Syst. Man. Cybern. SMC-6, 325–327 (1976). https://doi.org/10.1109/TSMC.1976.5408784
27. Giusti, R., Batista, G.E.A.P.A.: An empirical comparison of dissimilarity measures for time series classification. In: 2013 Brazilian Conf. Intell. Syst., pp. 82–88. IEEE (2013). https://doi.org/10.1109/BRACIS.2013.22
28. Lines, J., Bagnall, A.: Time series classification with ensembles of elastic distance measures. Data Min. Knowl. Discov. 1–28 (2014)
29. Serrà, J., Arcos, J.L.: An empirical evaluation of similarity measures for time series classification. Knowl.-Based Syst. **67**, 305–314 (2014). https://doi.org/10.1016/j.knosys.2014.04.035
30. Górecki, T., Łuczak, M., Piasecki, P.: An exhaustive comparison of distance measures in the classification of time series with 1NN method. J. Comput. Sci. **76**, 102235 (2024). https://doi.org/10.1016/j.jocs.2024.102235
31. Faloutsos, C., Ranganathan, M., Manolopoulos, Y.: Fast subsequence matching in time-series databases. ACM SIGMOD Rec. **23**, 419–429 (1994). https://doi.org/10.1145/191843.191925
32. Berndt, D., Clifford, J.: Using dynamic time warping to find patterns in time series. In: Usama, R.U., Fayyad, M. (ed.) Knowl. Discov. Databases Pap. from 1994 AAAI Work., pp. 359–370. AAAI Press, Seattle, Washington (1994). http://dblp.uni-trier.de/rec/bib/conf/kdd/BerndtC94
33. Vlachos, M., Kollios, G., Gunopulos, D.: Discovering similar multidimensional trajectories. In: Proc. 18th Int. Conf. Data Eng., pp. 673–684. IEEE Comput. Soc. (2002). https://doi.org/10.1109/ICDE.2002.994784
34. Chen, L., Ng, R.: On the marriage of Lp-norms and edit distance. In: Nascimento, M.A., Özsu, M.T., Kossmann, D., Miller, R.J., Blakeley, J.A., Schiefer, K.B. (eds.) Proc. 2004 VLDB Conf., pp. 792–803. Elsevier (2004). https://doi.org/10.1016/B978-012088469-8.50070-X
35. Chen, L., Özsu, M.T., Oria, V.: Robust and fast similarity search for moving object trajectories. In: Proc. 2005 ACM SIGMOD Int. Conf. Manag. Data—SIGMOD '05, pp. 491–502. ACM Press, New York, USA (2005). https://doi.org/10.1145/1066157.1066213
36. Holder, C., Middlehurst, M., Bagnall, A.: A review and evaluation of elastic distance functions for time series clustering. Knowl. Inf. Syst. **66**, 765–809 (2024). https://doi.org/10.1007/s10115-023-01952-0
37. Sakoe, H., Chiba, S.: Dynamic programming algorithm optimization for spoken word recognition. IEEE Trans. Acoust. **26**, 43–49 (1978). https://doi.org/10.1109/TASSP.1978.1163055
38. Itakura, F.: Minimum prediction residual principle applied to speech recognition. IEEE Trans. Acoust. **23**, 67–72 (1975). https://doi.org/10.1109/TASSP.1975.1162641
39. Kurbalija, V., Radovanović, M., Geler, Z., Ivanović, M.: The influence of global constraints on DTW and LCS similarity measures for time-series databases. In: Dicheva, D., Markov, Z., Stefanova, E. (eds.) Third Int. Conf. Software, Serv. Semant. Technol. S3T 2011 SE-10, pp. 67–74. Springer, Berlin, Heidelberg (2011). https://doi.org/10.1007/978-3-642-23163-6_10
40. Geler, Z.: Role of similarity measures in time series analysis. University of Novi Sad, Serbia (2015). (PhD thesis). https://nardus.mpn.gov.rs/handle/123456789/1703
41. Ratanamahatana, C.A., Keogh, E.: Three myths about dynamic time warping data mining. In: Proc. 2005 SIAM Int. Conf. Data Min., pp. 506–510. Society for Industrial and Applied Mathematics, Philadelphia, PA (2005). https://doi.org/10.1137/1.9781611972757.50
42. Geler, Z., Kurbalija, V., Ivanovic, M., Radovanovic, M., Dai, W.: Dynamic time warping: Itakura vs Sakoe-Chiba. In: 2019 IEEE Int. Symp. Innov. Intell. Syst. Appl., pp. 1–6. IEEE (2019). https://doi.org/10.1109/INISTA.2019.8778300
43. Kurbalija, V., Radovanović, M., Geler, Z., Ivanović, M.: The influence of global constraints on similarity measures for time-series databases. Knowl.-Based Syst. **56**, 49–67 (2014). https://doi.org/10.1016/j.knosys.2013.10.021

44. Geler, Z., Kurbalija, V., Ivanović, M., Radovanović, M.: Elastic distances for time-series classification: Itakura versus Sakoe-Chiba constraints. Knowl. Inf. Syst. **64**, 2797–2832 (2022). https://doi.org/10.1007/s10115-022-01725-1
45. Lima, F.T., Souza, V.M.A.: A large comparison of normalization methods on time series. Big Data Res. **34**, 100407 (2023). https://doi.org/10.1016/j.bdr.2023.100407
46. Macleod, J., Luk, A., Titterington, D.: A re-examination of the distance-weighted k-nearest neighbor classification rule. IEEE Trans. Syst. Man Cybern. **17**, 689–696 (1987). https://doi.org/10.1109/TSMC.1987.289362
47. Mitchell, T.M.: Machine learning. McGraw-Hill Inc., New York, NY, USA (1997)
48. Zavrel, J.: An empirical re-examination of weighted voting for k-NN. In: Proc. 7th Belgian-Dutch Conf. Mach. Learn., pp. 139–148 (1997)
49. Pao, T.-L., Chen, Y.-T., Yeh, J.-H., Cheng, Y.-M., Lin, Y.-Y.: A comparative study of different weighting schemes on KNN-based emotion recognition in mandarin speech. In: Huang, D.-S., Heutte, L., Loog, M. (eds.) Adv. Intell. Comput. Theor. Appl. With Asp. Theor. Methodol. Issues, pp. 997–1005. Springer, Berlin, Heidelberg (2007). https://doi.org/10.1007/978-3-540-74171-8_101
50. Gou, J., Xiong, T., Kuang, Y.: A novel weighted voting for K-nearest neighbor rule. J. Comput. **6**, 833–840 (2011). https://doi.org/10.4304/jcp.6.5.833-840
51. Gou, J., Du, L., Zhang, Y., Xiong, T.: A new distance-weighted k-nearest neighbor classifier. J. Inf. Comput. Sci. **9**, 1429–1436 (2012)
52. Kelly, M., Longjohn, R., Nottingham, K.: UCI machine learning repository (n.d.). https://archive.ics.uci.edu. Accessed 7 Mar 2024
53. Anh Dau, H., Keogh, E., Kamgar, K., Michael Yeh, C.-C., Zhu, Y., Gharghabi, S., Ann Ratanamahatana, C., Chen, Y., Hu, B., Begum, N., Bagnall, A., Mueen, A., Batista, G.: Hexagon-ML, the UCR time series classification archive (2019). https://www.cs.ucr.edu/~eamonn/time_series_data_2018/
54. Kurbalija, V., Radovanović, M., Geler, Z., Ivanović, M.: A framework for time-series analysis. In: Dicheva, D., Dochev, D. (eds.) Artif. Intell. Methodol. Syst. Appl. SE-5, pp. 42–51. Springer, Heidelberg (2010). https://doi.org/10.1007/978-3-642-15431-7_5
55. Paparrizos, J.: 2018 UCR time-series archive: backward compatibility, missing values, and varying lengths (2019). https://github.com/johnpaparrizos/UCRArchiveFixes

Chapter 2
Time Series and Similarity Measures

2.1 Time Series

Nowadays human population and society is based on various varieties of data, information and knowledge. Usually people mix these concepts in everyday use. However in particular areas and domains, professional and scientific environments there is necessity to precisely separate these categories.

According to Shannon [1], *data* (singular *datum*) are individual units of information and they are rarely useful alone. In popular publications usually *data* are seen as the facts or details, needed to be put into some context and represent the source from which *information* is derived. Data is raw, unorganized facts that need to be analysed and processed in order to offer something useful for people who will use it. Consequently information is "something useful", obtained from data by processing, organizing, structuring or presenting it in a given context. Knowledge is the most abstract concept among these three and it represents understanding based on extensive experience dealing with information on a subject. The fundamental concept of data refers to the *representation* or *encoding* of existing information or knowledge in a structured format that facilitates efficient manipulation and analysis.

In academic environments data simply represents units of information. Data is present in all human activities starting from businesses, finance, governance, science and continuing in virtually every other form of human organizational activity.

In last several decades data has become closely related to computers and activities like collecting, storing, analysing and drawing some important conclusions and patterns of behavior from it. Various types of data exist within different human activities: Natural, Cultural, Geographical, Meteorological, Financial, Transport, Scientific and a lot of other. In this monograph we will concentrate on a specific type of temporal data, so called "time series".

Temporal data is one among numerous data types. They may refer to a wide range of discrete events, moving objects, repeated observations and so on. Generally speaking, temporal data addresses the data in which temporal aspects and changes

Z. Gellér et al., *Recent Advances in Time-Series Classification—Methodology and Applications*, Intelligent Systems Reference Library 264,
https://doi.org/10.1007/978-3-031-77527-7_2

over time play an essential role. Temporal data is simply data that represents a state in time. Numerous application domains influence the collection and processing of large amounts of temporal data for various purposes like: financial data analysis, scientific applications, medical applications, manufacturing systems, geological and meteorological domains and so on.

In its simplest form, temporal data represents an array of values that describes how the observed phenomenon changes over time. Each of these values characterize the examined phenomenon at distinct time points, and such a series is called a time series [2, 3]. For example, time series appears when measuring daily temperature at a particular place or observing the value of the stocks of a particular company over the time. The defining characteristic of these series is the chronological order of the observations, meaning they are arranged based on the time they were collected. Time series are suitable for representing and analysing the results of observations in different areas of science, including biology, economics, medicine, meteorology, ecology and others [4, 5].

Large amounts of time series data are collected in different domains with the goal of using the results of their analysis in meaningful ways. Characteristic tasks related to time series analysis include prediction, anomaly detection, clustering and classification. The applicability of time series in a wide range of different domains has increased the interest of the academic community in examining different areas of time series analysis leading to the introduction of a large number of new methodologies [6–14].

The *representation* of time series depends on the particular domain of application. The most common one is probably in the form of a finite list of ordered pairs. Formally, a *time series* A of length n is defined as a sequence of ordered pairs (called *points*) as follows [15–18]:

$$A = ((a_1, t_1), (a_2, t_2), \ldots, (a_n, t_n))$$

where $a_i \in R^d$ is the value that quantifies the observed phenomenon's behavior at time $t_i \in T$. The core premise is that $t_i < t_{i+1}$ for every $i < n$, i.e. that the ordered pairs are sorted chronologically—in ascending order with respect to the time component. The length of the sequence (n) is usually referred as the *dimension* of the time series and is denoted by $|A|$.

Let $1 \leq i \leq j \leq n$. If we denote the ith element of A by $A[i]$, then $A\left[i{:}j\right] = \left((a_i, t_i), (a_{i+1}, t_{i+1}), \ldots, \left(a_j, t_j\right)\right)$ refers to a *contiguous* (or *continuous*) *subsequence* of A. In the literature, instead of the term contiguous subsequence, only the term *subsequence* is often used [19, 20], or alternatively, the terms *subseries* [21] and *segment* [22].

Based on the frequency of observations, time series can be *continuous* or *discrete*. A time series is *multivariate* if more than one value is involved in the observations (i.e. when $d > 1$). When the values of a time series can be generated using some mathematical function, it is said to be *deterministic*. If no such function exists, it is

Table 2.1 Examples of time series: annual coffee production in Angola[1] (*A*); exchange rates of euro[2] (*B*); blood pressure change (*C*)

A	4.26	3.30	1.26	2.04	1.98	3.50	5.00	11.52	4.00	11.98	9.95	10.19
B	0.58	0.54	0.50	0.60	0.59	0.58	0.59	0.59	0.58	0.57	0.58	0.58
C	106	104	103	105	110	108	110	102	114	106	106	103

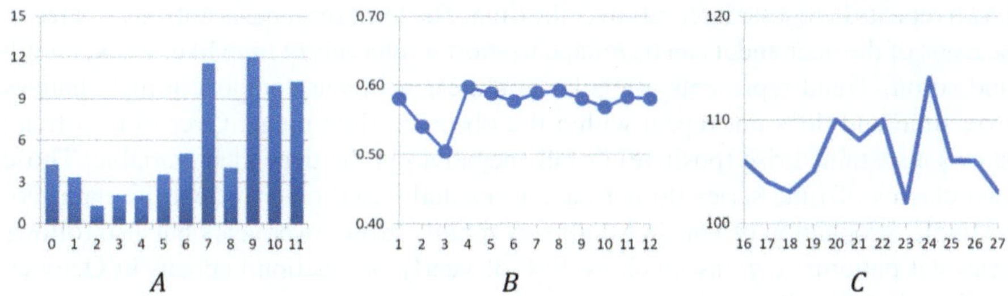

Fig. 2.1 Visualization of the time series listed in Table 2.1

said to be a *stochastic* or *nondeterministic*. Time series may contain *missing values* [23]: ordered pairs (a, t) within which a has no value.

In the rest of this book, we will consider only one-dimensional (*univariate*, $a_i \in R$) discrete time series without missing values, measured at equally spaced time-intervals, i.e. $t_{i+1} - t_i = c$, for every $i < n$, where c is a (positive) constant. Under these assumptions, the time component can be excluded, thus time series can be represented as simple sequences of real numbers: $A = (a_1, a_2, \ldots, a_n)$. Three examples of such time series are given in Table 2.1. The first series (A) describes the annual coffee production (in thousands of tons, rounded to two decimal places) in Angola in the period from 2000 to 2011. The second one (B) reports the exchange rates of euro during 2020 from January to December at the National Bank of Serbia (NBS). Since all the values are between 117 and 118, for the sake of readability, only the decimal part (rounded to two digits) of the corresponding values are presented. The third example (C) represents changes in the systolic blood pressure of a patient measured over 12 consecutive days (from the 16th to the 26th in the observed month). A visual representation of these time series is given in Fig. 2.1.

Time series analysis is an essential tool in the process of drawing meaningful conclusions based on the collected data so that the gained knowledge could be used for the realization of various tasks like decision support, predicting future behavior of some observed phenomenon and so on. There are two important goals of time series analysis: (i) using a sequence of observations to identify the nature of the phenomenon and (ii) predicting/forecasting the future values of the time series variable.

---

[1] http://world-statistics.org/index-res.php?code=FAO-PROD-656-5510?name=Coffee,%20green%20-%20Production%20(Tonnes).

[2] https://www.nbs.rs/sr/drugi-nivo-navigacije/statistika/ks_stat/.

A core assumption in time series analysis is that the data exhibits a structure or logical pattern. However in almost all real-world situations data contains some kind of random noise (usually called error) which makes process of identifying the pattern more difficult. Accordingly, almost all time series analysis techniques must include filtering out noise to identify the pattern more relevant.

Most time series patterns are characterized by two essential components: *seasonality* and *trend*. Seasonality represents the phenomena that may have similar nature, and it repeats in regular intervals over the time. The term seasonal is not only related to seasons of the year and it can be related to shorter intervals of time like: week, month and so on. Trend represents a regular nonlinear or linear component that changes over time and does not repeat within the observed time range. Over time, a trend shows a sustained rise (positive) or fall (negative) in the dependent variable. Those two classes of time series do not have to exclude each other in real-life data. For example, production of vine in a company rapidly grows over years but also follows seasonal patterns (e.g., as much as 30% of yearly production happens in October, while only 6% in September).

Two additional components connected to time series are: *Cyclic Pattern* and *Random Pattern*. Cyclic behavior is like seasonal behaviors and shows fluctuations upwards and downwards that are not connected to season. The occurrence of peaks and troughs in fact does not show any regularity. A characteristic example could be the annual count of large earthquakes in some particular region. Substantial peaks and troughs appear but at irregular intervals. There will be no significant peaks or troughs at unpredictable times nor will random pattern show predictable peaks and troughs. Instead, the data exhibits random variation around a relatively stable average value.

2.1.1 Data Mining Tasks

In the last several decades, with rapid development of ICT (Information and Communications Technology), the Internet, the World Wide Web and globalization of human society, wide variety of new application domains are recognized and established. Generating huge amounts of different data from heterogeneous sources led towards new era of science, technology and functioning of human society. Challenging disciplines and research areas in ICT have been appearing constantly and at the moment we are witnesses of a highly networked world based on Sensor networks, Internet of Things (IoT), Smart Environments, Big Data that generate enormous amounts of different kinds of data including time series. The intensifying reliance on the utilization of time series in solving different problems has induced an increased interest in researching various tasks related to their analysis and mining [3, 4, 24–32]:

- **Indexing** (*query by content, similarity search, discovery of known patterns*) deals with the problem of fast and efficient retrieving of time series from a database that are the most similar to a given query series (pattern) Q relying on a given (di)similarity measure. The task can be to find the k most similar series (*k-Nearest Neighbors query*, Fig. 2.2a) or to find all the series that do not differ from Q more than some threshold value δ (*δ-range query*, Fig. 2.2b). In the case of *whole series matching*, the query sequence Q and the elements of the database are considered as equal objects in the sense of comparing complete series. *Subsequence matching* is a variation of query by content with the aim of finding time series that contain subseries similar to Q. Indexing usually involves *dimensionality reduction* and using simpler *representations* of time series.

The simplest way to find the time series of a database that are the most similar to Q is using *linear (sequential) scanning* which involves computing the similarity between Q and each individual element of the database. However, in the case of large databases and/or long time series, this brute-force approach may be very time consuming and computationally inefficient. Sequential scanning can be avoided by indexing the time series included in the database. This is achieved by building an auxiliary data structure called the *index structure* or simply the *index*. The aim of the index structure is to reduce the number of similarity calculations by clustering the time series and thus providing faster access to the most promising ones.

- **Clustering** (*unsupervised learning*, *grouping*) represents the process of grouping together similar time series of a database based on their degree of (dis)similarity (Fig. 2.3). The groups of similar series are called *clusters*. The goal is to find homogeneous clusters: series belonging to the same cluster should be as similar as possible, and they should differ as much as possible from the elements of the other clusters. While *whole time-series clustering* deals with complete time series, in *subsequence clustering* clusters are derived from subseries that are extracted from a single or multiple longer time series.
- **Classification** (*supervised learning*) assigns new, yet unclassified time series to one of several predefined classes. Unlike clustering where the groups are not known and the task is their discovery, in the case of classification they are known

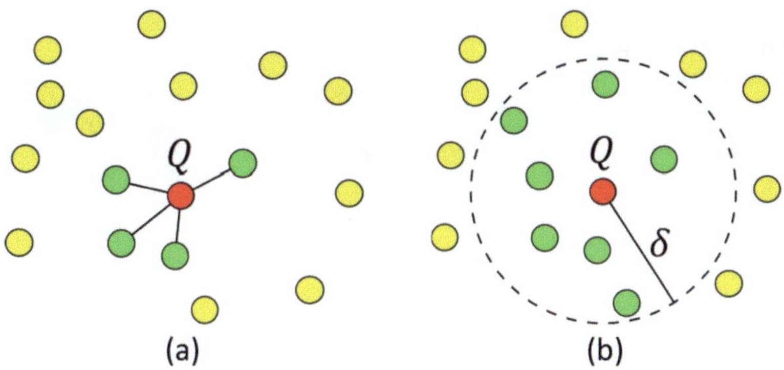

Fig. 2.2 Searching for the k most similar time series (**a**), and for all of the series that do not differ from A more than the threshold value δ (**b**)

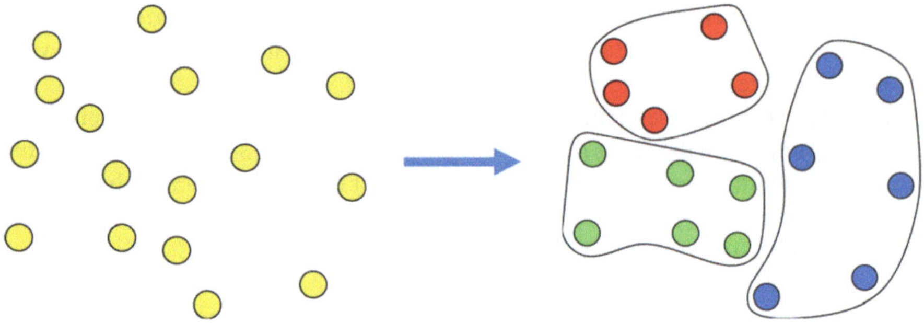

Fig. 2.3 Grouping similar time series into homogeneous clusters

in advance. The algorithm that performs the classification is called the *classifier*. In order for a classifier to be able to classify new time series, it must first learn how to distinguish the classes. This is achieved by *training* the classifier on an example dataset called the *training set*. For each time series of the training set, it is known which class it belongs to, and the classifier uses these examples to learn how to recognize the classes based on the characteristics of the series that belong to them. In Fig. 2.4, the time series whose class is to be determined is marked in yellow, and the training set contains three classes: red, green and blue.

- **Summarization** explores techniques for creating compact approximate *representations* of individual time series or entire datasets/clusters so as to obtain an informative form, cleared of irrelevant details but retaining the most important characteristics. When it comes to individual time series, one approach is to describe them using descriptive statistics: mean value, median, mode and variance. Other possibilities includes *dimensionality reduction, visualization, linguistic characterization* (using natural language expressions) and others.
- **Segmentation** (*dimensionality reduction, approximation, representation*) divides the time series into a set of non-overlapping intervals called segments (Fig. 2.5).

Fig. 2.4 Classifying unknown time series

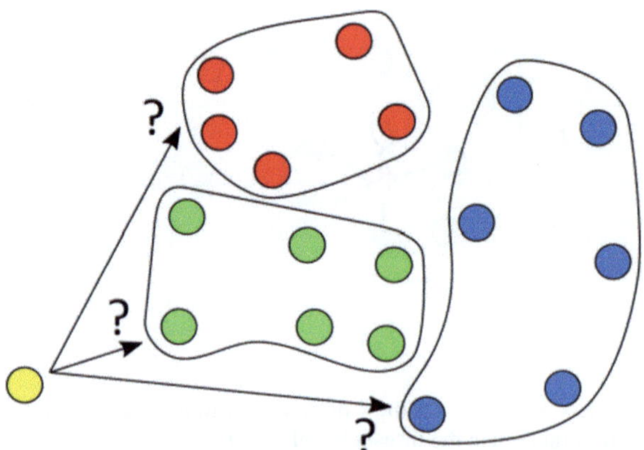

The segments are constructed in such a way that they represent a good approximation of the initial time series and to achieve dimensionality reduction, i.e. so that their number is considerably less than the length of the time series. Different segmentation objectives can be distinguished, for example:

- to make an approximation containing a specific number of segments,
- to limit the approximation error of individual segments with a threshold,
- to limit the total approximation error of all the segments with a threshold,
- to divide the time series into homogeneous sections.

- **Prediction** (*forecasting*) focuses on forecasting, estimating future values of time series based on previously observed data (Fig. 2.6). Forecasting techniques can be broadly classified into two categories: *statistical* methods and *machine learning* approaches.

  To build a predictive *model* (a simplified representation of the observed phenomenon), statistical methods require a prior knowledge about the data distribution. Statistical methods are *parametric*: the accuracy of the models depends on how optimized are their parameters to fit the data.

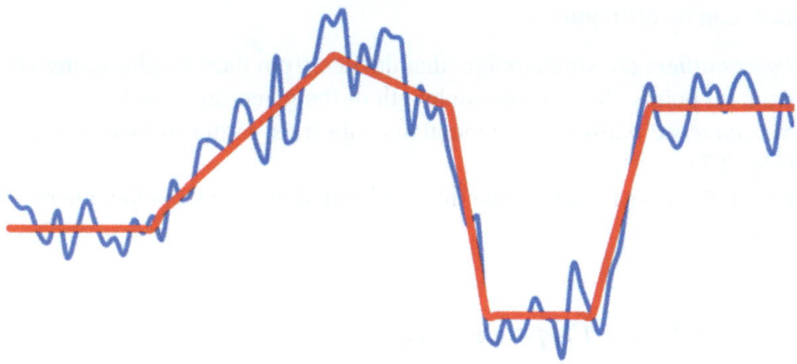

Fig. 2.5 Approximation of a time series (blue line) by segmentation (red line)

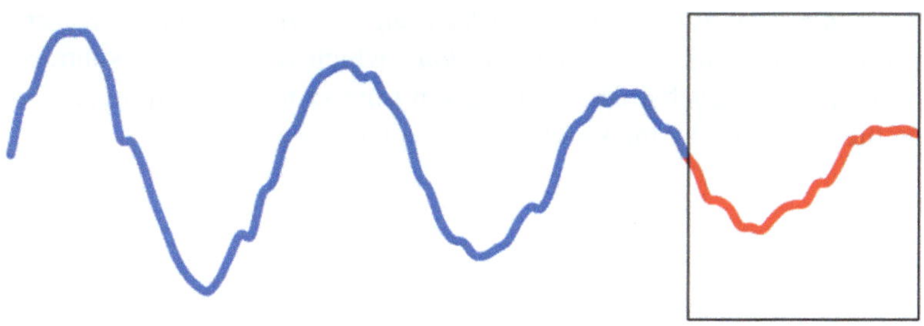

Fig. 2.6 Predicting future values (red line) based on previous values (blue line)

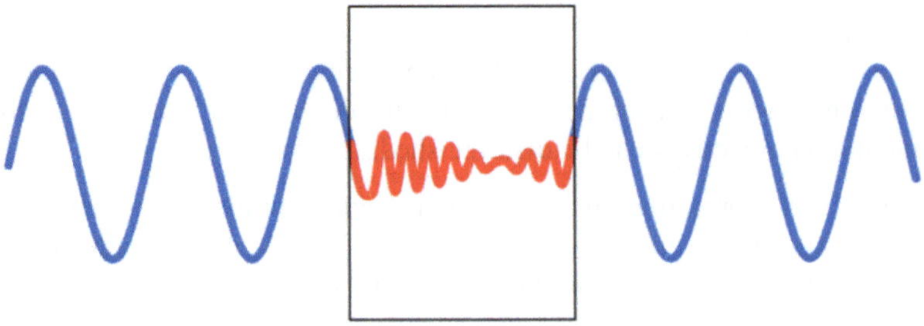

Fig. 2.7 Anomaly (red line) detection in time series

Machine learning models describe time series without a prior knowledge about their distribution. As these models do not depend explicitly on parameters to model the observed phenomenon, they are known as *non-parametric* models.

- **Anomaly detection** (*outlier detection*) techniques seek to find anomalies in time series: abnormal, surprising, interesting or unexpected subseries. According to the classic definition given by Hawkins [33], *outliers* represents observations that deviate from the expected behavior. When it comes to time series, three types of outliers can be distinguished:
  - *Point outliers* are single points that deviate from their neighbouring points or from all points along the entire length of the given time series.
  - *Subsequence outliers* are anomalous subseries within individual time series (Fig. 2.7).
  - *Outlier time series* are whole time series that deviate from the others.

2.1.2 *Time-Series Preprocessing*

For time series to be meaningfully comparable, they must often be prepared using different *preprocessing transformations.* One of the most commonly used such transformations is *z-normalisation* (*standardisation, z-score normalisation, zero-mean normalisation, standard score normalisation*), which is applied by subtracting the mean value μ_A from each element of a given time series $A = (a_1, a_2, \ldots, a_n)$ and dividing the difference by the standard deviation σ_A [6, 34, 35]:

$$A' = \frac{A - \mu_A}{\sigma_A}$$

where

$$\mu_A = \frac{1}{n}\sum_{i=1}^{n} a_i$$

$$\sigma_A^2 = \frac{1}{n}\sum_{i=1}^{n}(a_i - \mu_A)^2 = \frac{1}{n}\left(\sum_{i=1}^{n} a_i^2 - 2\mu_A \sum_{i=1}^{n} a_i + \sum_{i=1}^{n} \mu_A^2\right)$$
$$= \frac{1}{n}\left(\sum_{i=1}^{n} a_i^2 - 2n\mu_A^2 + n\mu_A^2\right) = \frac{1}{n}\sum_{i=1}^{n} a_i^2 - \mu_A^2$$

z-normalisation transforms time series so that their mean is 0 and their standard deviation is 1, thus eliminating their differences in *offset* (or *baseline*, i.e. the mean value) and *amplitude* (the largest deviation from the mean value):

$$\mu_{A'} = \frac{1}{n}\sum_{1=n}^{n} a_i' = \frac{1}{n}\sum_{i=1}^{n}\frac{a_i - \mu_A}{\sigma_A} = \frac{1}{\sigma_A}\left(\frac{1}{n}\sum_{i=1}^{n} a_i - \frac{1}{n}\sum_{i=1}^{n}\mu_A\right)$$
$$= \frac{1}{\sigma_A}(\mu_A - \mu_A) = 0$$

$$\sigma_{A'}^2 = \frac{1}{n}\sum_{\text{i}=1}^{\text{n}}(a_i' - \mu_{A'})^2 = \frac{1}{n}\sum_{\text{i}=1}^{\text{n}}(a_i')^2$$
$$= \frac{1}{n}\sum_{\text{i}=1}^{\text{n}}\left(\frac{a_i - \mu_A}{\sigma_A}\right)^2 = \frac{1}{\sigma_\text{A}^2}\frac{1}{\text{n}}\sum_{i=1}^{n}(a_i^2 - 2a_i\mu_A + \mu_A^2)$$
$$= \frac{1}{\sigma_\text{A}^2}\left(\frac{1}{\text{n}}\sum_{\text{i}=1}^{\text{n}} a_i^2 - 2\mu_A\frac{1}{\text{n}}\sum_{i=1}^{n} a_i + \frac{1}{\text{n}}\sum_{i=1}^{n}\mu_A^2\right)$$
$$= \frac{1}{\sigma_\text{A}^2}\left(\frac{1}{\text{n}}\sum_{\text{i}=1}^{\text{n}} a_i^2 - 2\mu_A^2 + \mu_A^2\right) = \frac{1}{\sigma_\text{A}^2}\sigma_A^2 = 1$$

Figure 2.8 illustrates revealing the intrinsic similarity between two time series using z-normalisation. Figure 2.8a shows two seemingly very different time series, whereas in the case of series B, its mean value (solid line) and standard deviation area (dashed lines) are also indicated. Figure 2.8b shows the first step of z-normalisation: mean subtraction. As a result, both sequences are *shifted* so that their new mean is 0. However, their amplitudes are still very different. This is solved by *scaling* in the second step: dividing by the standard deviation. In this way, after applying z-normalisation, both series have the same mean value (0) and standard deviation (1) and their real similarity becomes clearly visible (Fig. 2.8c).

z-normalisation is a composition of two basic transformations of time series: *shifting* (or *offset translation*) and (amplitude) *scaling* [36]. Shifting means increasing the value of each time series point by an arbitrary real number q, and scaling means multiplying them by a real number p greater than 0. For example, the time series B in Fig. 2.8a was obtained by scaling A by 3, shifting it by 1.3 and adding some random noise: $B = 3A + 1.3 + \textit{noise}$.

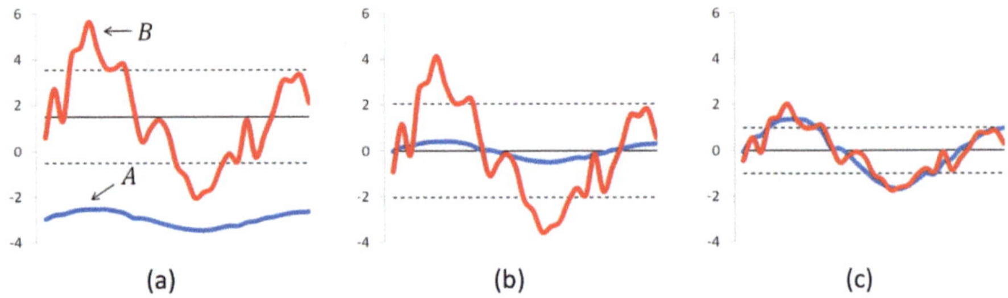

Fig. 2.8 Revealing the intrinsic similarity between time series using *z*-normalisation: two seemingly very different time series (**a**); the result of shifting their offsets (**b**) and then scaling their amplitudes (**c**)

If B is obtained by scaling and shifting the time series A (i.e. if $B = pA + q$), then $B' = A'$. Namely, based on the fact that

$$\mu_B = \frac{1}{n}\sum_{i=1}^{n} b_i = \frac{1}{n}\sum_{i=1}^{n}(pa_i + q) = p\frac{1}{n}\sum_{i=1}^{n} a_i + \frac{1}{n}\sum_{i=1}^{n} q = p\mu_A + q$$

$$\sigma_B^2 = \frac{1}{n}\sum_{i=1}^{n}(b_i - \mu_B)^2 = \frac{1}{n}\sum_{i=1}^{n}\left(pa_i + \mathrm{q} - p\mu_A - \mathrm{q}\right)^2 = \mathrm{p}^2\frac{1}{n}\sum_{i=1}^{n}(a_i - \mu_A)^2 = (\mathrm{p}\sigma_\mathrm{A})^2$$

it follows that

$$B' = \frac{B - \mu_B}{\sigma_B} = \frac{pA + q - p\mu_A - q}{p\sigma_A} = \frac{A - \mu_A}{\sigma_A} = A'$$

which implies that the *z*-normalised series shown in Fig. 2.8c would match exactly if B contained no noise.

It should be noted that the utility of *z*-normalisation depends on the domain. If the difference in offset and amplitude is an important characteristic of the analyzed time series, it should not be applied as it will nullify that feature. Furthermore, some other forms of normalisation can also be considered [37, 38], for example:

- *Min–max normalisation* (linearly) transforms the data points of a time series A from the range $[A_{min}, A_{max}]$ into a new range $[min, max]$ (often into [0, 1]):

$$A' = min + \frac{A - A_{min}}{A_{max} - A_{min}}(max - min)$$

- *Decimal scaling* moves the decimal point of the numbers in $A = (a_1, a_2, \ldots, a_n)$ so that they are all in the interval $(-1, 1)$:

$$A' = \frac{A}{10^d},$$

where d is the number of digits in the integer part of $max\{|a_i|\}$ (i.e. the smallest (non-negative) integer for which $max\{|a_i'|\} < 1$ or, equivalently, the smallest (non-negative) integer for which $max\{|a_i|\} < 10^d$).

2.2 The UCR Time Series Classification Archive

All the experiments presented in this book were performed relying on the UCR Time Series Classification Archive [39], the largest collection of labeled time-series datasets that is publicly available to researchers. Since its introduction in 2002, the number of included datasets has increased from 16 to 128, and it has become an important basis for benchmarking different methods related to time-series data mining [39].

The UCR datasets come form a variety of research domains, including: astronomy, sport, medicine, robotics, biology, seismology, food science, physics and others. Due to the diversity of data sources, the raw datasets are not uniform in all their properties: some of them contain missing values, some are z-normalised (while others are not), class enumeration is not consistent across all datasets, and some datasets contain time series of varying lengths. To ensure that all datasets are z-normalised, with no missing values, and that all the time series belonging to the same set have the same length, we applied Paparizzos' script [40].

The basic properties of the UCR datasets (prepared utilizing Paparizzo's script [40]) are summarized in Table 2.2. The number of classes ranges from 2 to 60 (column C), the length of the time series varies from 15 to 2844 (column L) and the size of the datasets is between 40 and 24,000 (the sum of the *Train* and *Test* columns). The experiments carried out for the purposes of this book were performed on datasets obtained by merging the appropriate UCR training and testing subsets. For a more compact presentation of the results, in some of the tables in the rest of this book, we will use the identification numbers listed in the *ID* column instead of the names of the datasets.

In the reminder of this section, in order to illustrate the diversity of the UCR archive, we will give a brief overview of some of the datasets. Additional information on the UCR datasets can be found on the following website: http://www.timeseriesclassification.com/dataset.php

Coffee (16). The time series in this dataset represent spectrographs of two variants of coffee beans: Arabica and Robusta. The data was collected using Fourier transform infrared spectrometer. It is considered that Arabica beans have a finer flavor, and the dataset was used to differentiate the two coffee species [41].

Crop (21). *Crop* (21) is a subset of a much larger dataset of satellite image time series (SITS) presented in [42]. SITS are time series obtained from series of satellite images showing the same geographic area at different times. Each time series describes the change of one pixel over time. The *Crop* (21) time series were extracted from 46 satellite images of an area near the city of Toulouse in France during 2006. Classes

Table 2.2 Basic properties of the UCR datasets: number of classes (C), number of time series in the train (Train) and the test (Test) set, length of the time series (L)

ID	Dataset	C	Train	Test	L	ID	Dataset	C	Train	Test	L
1	ACSF1	10	100	100	1,460	65	ItalyPowerDemand	2	67	1,029	24
2	Adiac	37	390	391	176	66	LargeKitchenAppliances	3	375	375	720
3	AllGestureWiimoteX	10	300	700	500	67	Lightning2	2	60	61	637
4	AllGestureWiimoteY	10	300	700	500	68	Lightning7	7	70	73	319
5	AllGestureWiimoteZ	10	300	700	500	69	Mallat	8	55	2,345	1,024
6	ArrowHead	3	36	175	251	70	Meat	3	60	60	448
7	Beef	5	30	30	470	71	MedicalImages	10	381	760	99
8	BeetleFly	2	20	20	512	72	MelbournePedestrian	10	1,200	2,450	24
9	BirdChicken	2	20	20	512	73	MiddlePhalanxOutlineAgeGroup	3	400	154	80
10	BME	3	30	150	128	74	MiddlePhalanxOutlineCorrect	2	600	291	80
11	Car	4	60	60	577	75	MiddlePhalanxTW	6	399	154	80
12	CBF	3	30	900	128	76	MixedShapesRegularTrain	5	500	2,425	1,024
13	Chinatown	2	20	345	24	77	MixedShapesSmallTrain	5	100	2,425	1,024
14	ChlorineConcentration	3	467	3,840	166	78	MoteStrain	2	20	1,252	84
15	CinCECGTorso	4	40	1,380	1,639	79	NonInvasiveFetalECGThorax1	42	1,800	1,965	750
16	Coffee	2	28	28	286	80	NonInvasiveFetalECGThorax2	42	1,800	1,965	750
17	Computers	2	250	250	720	81	OliveOil	4	30	30	570
18	CricketX	12	390	390	300	82	OSULeaf	6	200	242	427
19	CricketY	12	390	390	300	83	PhalangesOutlinesCorrect	2	1,800	858	80
20	CricketZ	12	390	390	300	84	Phoneme	39	214	1,896	1,024
21	Crop	24	7,200	16,800	46	85	PickupGestureWiimoteZ	10	50	50	361

(continued)

Table 2.2 (continued)

ID	Dataset	C	Train	Test	L	ID	Dataset	C	Train	Test	L
22	DiatomSizeReduction	4	16	306	345	86	PigAirwayPressure	52	104	208	2,000
23	DistalPhalanxOutlineAgeGroup	3	400	139	80	87	PigArtPressure	52	104	208	2,000
24	DistalPhalanxOutlineCorrect	2	600	276	80	88	PigCVP	52	104	208	2,000
25	DistalPhalanxTW	6	400	139	80	89	PLAID	11	537	537	1,345
26	DodgerLoopDay	7	78	80	288	90	Plane	7	105	105	144
27	DodgerLoopGame	2	20	138	288	91	PowerCons	2	180	180	144
28	DodgerLoopWeekend	2	20	138	288	92	ProximalPhalanxOutlineAgeGroup	3	400	205	80
29	Earthquakes	2	322	139	512	93	ProximalPhalanxOutlineCorrect	2	600	291	80
30	ECG200	2	100	100	96	94	ProximalPhalanxTW	6	400	205	80
31	ECG5000	5	500	4,500	140	95	RefrigerationDevices	3	375	375	720
32	ECGFiveDays	2	23	861	136	96	Rock	4	20	50	2,844
33	ElectricDevices	7	8,926	7,711	96	97	ScreenType	3	375	375	720
34	EOGHorizontalSignal	12	362	362	1,250	98	SemgHandGenderCh2	2	300	600	1,500
35	EOGVerticalSignal	12	362	362	1,250	99	SemgHandMovementCh2	6	450	450	1,500
36	EthanolLevel	4	504	500	1,751	100	SemgHandSubjectCh2	5	450	450	1,500
37	FaceAll	14	560	1,690	131	101	ShakeGestureWiimoteZ	10	50	50	385
38	FaceFour	4	24	88	350	102	ShapeletSim	2	20	180	500
39	FacesUCR	14	200	2,050	131	103	ShapesAll	60	600	600	512
40	FiftyWords	50	450	455	270	104	SmallKitchenAppliances	3	375	375	720
41	Fish	7	175	175	463	105	SmoothSubspace	3	150	150	15
42	FordA	2	3,601	1,320	500	106	SonyAIBORobotSurface1	2	20	601	70

(continued)

Table 2.2 (continued)

ID	Dataset	C	Train	Test	L	ID	Dataset	C	Train	Test	L
43	FordB	2	3,636	810	500	107	SonyAIBORobotSurface2	2	27	953	65
44	FreezerRegularTrain	2	150	2,850	301	108	StarLightCurves	3	1,000	8,236	1,024
45	FreezerSmallTrain	2	28	2,850	301	109	Strawberry	2	613	370	235
46	Fungi	18	18	186	201	110	SwedishLeaf	15	500	625	128
47	GestureMidAirD1	26	208	130	360	111	Symbols	6	25	995	398
48	GestureMidAirD2	26	208	130	360	112	SyntheticControl	6	300	300	60
49	GestureMidAirD3	26	208	130	360	113	ToeSegmentation1	2	40	228	277
50	GesturePebbleZ1	6	132	172	455	114	ToeSegmentation2	2	36	130	343
51	GesturePebbleZ2	6	146	158	455	115	Trace	4	100	100	275
52	GunPoint	2	50	150	150	116	TwoLeadECG	2	23	1,139	82
53	GunPointAgeSpan	2	135	316	150	117	TwoPatterns	4	1,000	4,000	128
54	GunPointMaleVersusFemale	2	135	316	150	118	UMD	3	36	144	150
55	GunPointOldVersusYoung	2	136	315	150	119	UWaveGestureLibraryAll	8	896	3,582	945
56	Ham	2	109	105	431	120	UWaveGestureLibraryX	8	896	3,582	315
57	HandOutlines	2	1,000	370	2,709	121	UWaveGestureLibraryY	8	896	3,582	315
58	Haptics	5	155	308	1,092	122	UWaveGestureLibraryZ	8	896	3,582	315
59	Herring	2	64	64	512	123	Wafer	2	1,000	6,164	152
60	HouseTwenty	2	40	119	2,000	124	Wine	2	57	54	234
61	InlineSkate	7	100	550	1,882	125	WordSynonyms	25	267	638	270
62	InsectEPGRegularTrain	3	62	249	601	126	Worms	5	181	77	900
63	InsectEPGSmallTrain	3	17	249	601	127	WormsTwoClass	2	181	77	900
64	InsectWingbeatSound	11	220	1,980	256	128	Yoga	2	300	3,000	426

indicate what a given pixel represents in the images (corn, wheat, dense building, wasteland, soybean, sunflower, eucalyptus, barley, poplars, lake, etc.).

ECG200 (30). *ECG200* (30) is one of the datasets that were introduced by Robert T. Olszewski through his PhD thesis [43]. It contains 200 ECG signals where each time series represents measurements recorded by one electrode during one heartbeat. There are two classes in this dataset: based on the analysis of domain experts, time series are identified as normal or as abnormal.

FaceFour (38). There are four classes in this dataset, each representing one person (one female and three males). The time series describe the contours of the heads of the four persons as they make different facial expressions (smiling, talking, laughing, frowning, etc.) [44]. The **FaceAll (37)** dataset was obtained in a similar way, with the number of different persons (and thus the number of classes) being 14 instead of 4. The **FaceUCR (39)** dataset represent a rotationally aligned version of the *FaceAll* (37) dataset. In addition, these two datasets were divided into training and test subsets in different ways.

FiftyWords (40). The *FiftyWords* (40) dataset [45] contains vertical projection profiles (normalised sums of pixel intensity values) of 50 common words (such as "the", "and", etc.) extracted from the *Word Spotting* dataset [46]. The *Word Spotting* dataset is composed of four-dimensional time series, each of which represents a different profile of 2381 word images from 10 consecutive pages of George Washington's handwritten manuscripts. The **WordSynonyms (125)** dataset is a variation of the *FiftyWords* (40) dataset remapped to 25 classes where the time series are horizontally flipped.

GunPoint (52). The *GunPoint* (52) dataset contains time series that describe the movement of the right hand in two scenarios. In the first scenario, a replica of a gun is taken out from a hip-mounted holster, aimed at a target for approximately one second and then returned into the holster. In the second scenario, the index finger is pointed at a target. In the initial position, the hand is placed next to the body in both cases. The data were obtained by recording the position of the right hand of one female and one male actor [44, 47]. **GunPointAgeSpan (53)**, **GunPointMaleVersusFemale (54)** and **GunPointOldVersusYoung (55)** are extensions of the original *GunPoint* (52) dataset with new time series recorded after 15 years with the help of the same actors [48]. Each of these datasets contains two classes. In the original version, the classes represent scenarios. In the new datasets, classes denote the actors (*GunPointAgeSpan* (53)), the years the data was collected (*GunPointOldVersusYoung* (55)) and the gender of the actors (*GunPointMaleVersusFemale* (54)).

ItalyPowerDemand (65). The time series in this dataset describe daily electrical power demand from Italy and are divided into two classes [49]. The first class contains data from the winter months (from October to Match) and the second one from the summer months (from April to September).

MelbournePedestrian (72). This dataset contains the numbers of pedestrians identified by the automated Pedestrian Counting System (PCS) of the City of Melbourne[3] during 2017 in ten selected city locations. The PCS was built to better understand pedestrian movement, which can facilitate urban planning and management, contribute to improving transport, examining the impact of various events on pedestrian activities and developing marketing strategies. The classes represent the 10 selected city locations, and the time series contain daily data collected every hour. **Chinatown (13)** is a subset of this dataset. It contains data collected from only one city location (Chinatown-Swanston St (North)). In addition, the series were classified in two classes, depending on whether their data were collected on weekends or weekdays.

Pig. Data contained in the **PigAirwayPressure (86)**, **PigArtPressure (87)** and **PigCVP (88)** datasets were collected by monitoring airway pressure (*PigAirwayPressure* (86)), arterial blood pressure (*PigArtPressure* (87)) and central venous pressure (*PigCVP* (88)) of 52 sedated healthy pigs before and after subjecting them to induced slow bleeding [50]. In the original dataset, two sequences of data correspond to each pig: one describes its condition before and the other one after the bleeding. In the UCR datasets, these sequences are divided into three consecutive parts of 2000 points each. The first part of each sequence represents one time series in the training set. The second and third parts represent two separate time series that are included in the test set. Thus, for each pig there are two time series in the training set and four in the test set. Classes denote the individual pigs.

PLAID (89). *PLAID* (Plug-Level Appliance Identification Dataset) [51] is a dataset of high-resolution electrical device measurements for load identification. It contains 11 classes of time series that represent current and voltage measurements from various types of household appliances such as washing machines, air conditioners, laptops, microwaves, hairdryers, fridges and others. Appliances of the same type are represented by several different brands and models.

SemgHand. The **SemgHandGenderCh2 (98)**, **SemgHandMovemenCh2 (99)**, **SemgHandSubjectCh2 (100)** datasets contain forearm sEMG (surface electromyography) signal data collected from five healthy individuals (three females and two males) of approximately the same age (between 20 and 22 years old). The subjects were asked to hold 6 different types of objects for 6 seconds in order to perform 6 typical hand grasp types (spherical, tip, palmar, lateral, cylindrical, hook). The experiments were repeated 30 times using a 2-channel EMG system. Time series were obtained by extracting the power spectrum from the raw data of channel 2 only. Three datasets were obtained that differ only in what time series classes represent: in *SemgHandGenderCh2* (98) classes correspond to gender, in *SemnHandMovemenCh2* (99) they are the different grasp types and in *SemgHandSubjectCh2* (100) they represent the subjects.

[3] http://www.pedestrian.melbourne.vic.gov.au/

SmoothSubspace (105). The purpose of creating this artificial dataset was to evaluate the performance of a new *k*-means type smooth subspace clustering algorithm called TSkmeans [52]. The 15-point time series are divided into three classes depending on the position of a smooth subsequence of 5 points in length. In the case of the first class, the smooth subsequence is at the beginning of the time series, in the case of the second one it is in the middle and in the case of the third one it is at the end. The remaining parts of the time series are randomly generated.

StarLightCurves (108). The three classes in this dataset represent different celestial objects (Cepheid, Eclipse Binary, RR Lyra) and the corresponding time series describe their light curves (light intensity as a function of time) [53].

TwoPatterns (117). *TwoPatterns* (117) is another artificial dataset. It was introduced by Piere Geurts in his PhD thesis [54]. Time series are built of two patterns (*u*—upward step, *d*—downward step) with some noise around them. Their class is determined by the permutation of the two patterns in them: *uu*, *ud*, *du*, *dd*.

UWaveGestureLibrary. In [55], for the purpose of evaluating a new method for gesture recognition based on one accelerometer, gesture samples were collected from eight users. Each user repeated 10 times each of the 8 selected gestures on seven different days within three weeks. The samples were recorded using a Nintendo Wii remote control which has a built-in three-axis accelerometer. The **UWaveGestureLibraryX (120)**, **UWaveGestureLibraryY (121)** and **UWaveGestureLibraryZ (122)** datasets contain accelerometer recordings on the respective axes. The **UWaveGestureLibraryAll (119)** dataset was obtained by concatenating time series from these datasets.

Yoga (128). This dataset contains time series generated from images that capture two actors (a man and a woman) changing various yoga poses in front of a green screen [45]. Both actors performed the same yoga movements. The images were converted to time series by measuring the distances between the points of the outline and the center. Time series belong to one of two classes, depending on the actor who performed the yoga movements.

2.3 Similarity Measures

Many time series mining methods (especially among those used for indexing, classification and clustering) rely heavily on time series similarity assessment using a (dis)similarity measure. Similarity and dissimilarity (or *distance*) measures represent intercorrelated concepts referred as *proximity measures* [6, 56]. They describe the degree of (dis)similarity between time series using numerical values, but in opposite ways. The more similar the time series are the higher the value of the similarity measures and the lower the value of the dissimilarity measures is.

Similarity and dissimilarity measures of (*n*-dimensional, univariate) time series can be formally defined as follows [57, 58]:

A *similarity measure* is a function $sim : R^n \times R^n \rightarrow R$ where for every $A, B \in R^n$ it holds that

$$1 = sim(A, A) \geq sim(A, B) = sim(B, A) \geq 0.$$

A *dissimilarity measure* is a function $dis : R^n \times R^n \rightarrow R$ where for every $A, B \in R^n$ it holds that

$$0 = dis(A, A) \leq dis(A, B) = dis(B, A) < \infty.$$

Similarity and dissimilarity measures can be transformed into each other using the following formulas:

$$dis(A, B) = 1 - sim(A, B),$$

$$sim(A, B) = \frac{1}{1 + dis(A, B)}.$$

A dissimilarity measure $d : R^n \times R^n \rightarrow R$ is called a *metric* if for every $A, B, C \in R^n$ it satisfies the following two conditions:

1. $d(A, B) = 0$ implies $A = B$,
2. $d(A, B) \leq d(A, C) + d(C, B)$.

The first condition (called *definiteness*) ensures that the distance between two time series A and B is 0 only if they are identical, i.e. when it is $A[i] = B[i]$ for every $1 \leq i \leq n$. When it comes to dissimilarity measures that do not satisfy this condition, the distance is allowed to be 0, even between time series that are not identical.

The second condition (called *triangle inequality*) makes it possible to speed up finding the time series that most closely resembles a given query series Q by estimating the lower bound of the distance between Q and the time series being checked. Let X denote the dataset being searched and let A, B, C_1 and C_2 be elements of X. Let the distances between the elements of X be known in advance. Furthermore, suppose that among the series already compared with Q, the most similar is A at a distance of 1. If it is determined that $d(Q, B) = 4$, and if it is known that $d(B, C_1) = 1$ (Fig. 2.9), the distance between Q and C_1 does not have to be calculated because based on the triangle inequality it follows that:

$$d(Q, C_1) \geq d(Q, B) - d(B, C_1) = 4 - 1 = 3 > 1 = d(Q, A).$$

Similarly, it is not necessary to calculate the distance between Q and C_2 to know that C_2 does not resemble Q more than A:

$$d(Q, C_2) \geq d(Q, B) - d(B, C_2) = 4 - 2 = 2 > 1 = d(Q, A).$$

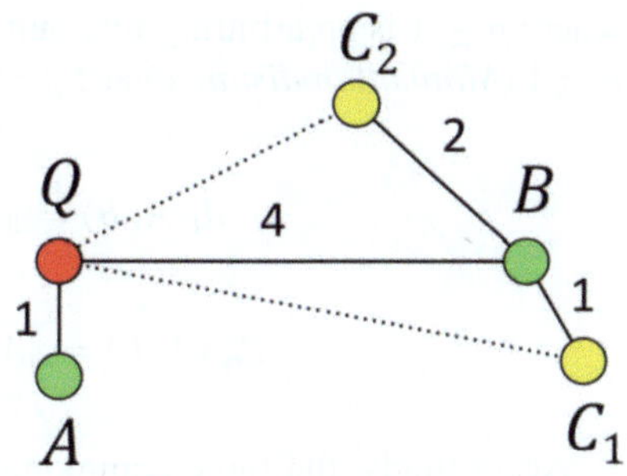

Fig. 2.9 Speeding up finding the most similar time series to Q utilizing the triangle inequality

Thus, if the distance measure d satisfies the triangle inequality, when searching for the time series that most resembles Q, the distance between Q and C need not be calculated if the lower bound of their distance is greater than or equal to the distance between Q and A, where A represents the time series that is currently the most similar to Q, i.e. if it is satisfied that $|d(Q, B) - d(B, C)| \geq d(Q, A)$.

Since the choice of an appropriate proximity measure represents a key issue in solving many tasks related to time series analysis and mining, a wide range of different methods for assessing the (dis)similarity between time series has been proposed [16, 22, 38, 59–64]. In the following subsections, we will provide an in-depth overview of those (di)similarity measures that are of prominent importance for the research which results are presented in this book. In doing so, in accordance with the widely adopted practice in the time series data mining community, we will use the terms similarity measure, dissimilarity measure and distance measure as synonyms.

2.3.1 Euclidean Distance (L_2)

One of the simplest and best-known distance measures within the domain of time series data mining is the *Euclidean distance* [9, 16, 62, 65–67]. It is calculated as the square root of the sum of the squared differences between the individual points of two same-length time series $A = (a_1, a_2, \ldots, a_n)$ and $B = (b_1, b_2, \ldots, b_n)$:

$$d_2(A, B) = \sqrt{\sum_{i=1}^{n} (a_i - b_i)^2}$$

The Euclidean distance belongs to the family of distance functions defined by the *Minkowski distance* (i.e. the L_p *norm* of the vector $A - B$):

$$d_p(A, B) = \|A - B\|_p = \sqrt[p]{\sum_{i=1}^{n} |a_i - b_i|^p}$$

where $p \geq 1$ is an arbitrary real number. In addition to $p = 2$ (Euclidean distance), $p = 1$ (*Manhattan distance*) and $p = \infty$ (*Chebyshev distance*) are also used:

$$d_1(A, B) = \|A - B\|_1 = \sum_{i=1}^{n} |a_i - b_i|$$

$$d_\infty(A, B) = \|A - B\|_\infty = \max_{i=1,\ldots,n} |a_i - b_i|$$

Accordingly, the three common distances based on L_p norms between the time series A and B shown in Fig. 2.10 are calculated as follows:

$$d_2(A, B) = \sqrt{(11 - 7)^2 + (12 - 6)^2 + \cdots + (10 - 4)^2} = \sqrt{491} \approx 22.16$$

$$d_1(A, B) = |11 - 7| + |12 - 6| + \cdots + |10 - 4| = 75$$

$$d_\infty(A, B) = max(|11 - 7|, |12 - 6|, \ldots, |10 - 4|) = 9$$

L_p norms (i.e. distances based on L_p norms) are easy to implement, fast to compute, have no additional parameters and are metrics which makes them suitable for indexing [19, 65, 68, 69]. However, they have some disadvantages that might negatively affect their applicability [25, 70]: (1) matching lengths are required for the time series and (2) they are sensitive to distortions in both the value (y) and the time (x) axis (in the sense that they give a pessimistic estimate of the distance between similar time series that are shifted, stretched or compressed along the axes).

An illustration of the sensitivity of L_p norms to distortions along the axes is given in Fig. 2.11. Series A_1 in Fig. 2.11a is similar to Q, but is shifted along the value axis. Figure 2.11b shows a copy of Q that is scaled along the value axis (A_2). The third example (Fig. 2.11c) presents a time-shifted variant (A_3) of the original sequence Q. In addition to shifting and scaling, the series A_1, A_2 and A_3 also contain a small amount of random noise. In contrast to A_1, A_2 and A_3 which show a noticeable similarity to Q, the series B presented in Fig. 2.11d visibly deviates from it. However, as can be

	1	2	3	4	5	6	7	8	9	10	11	12
A	11	12	13	18	22	18	15	14	12	9	8	10
B	7	6	8	12	13	12	9	6	4	3	3	4

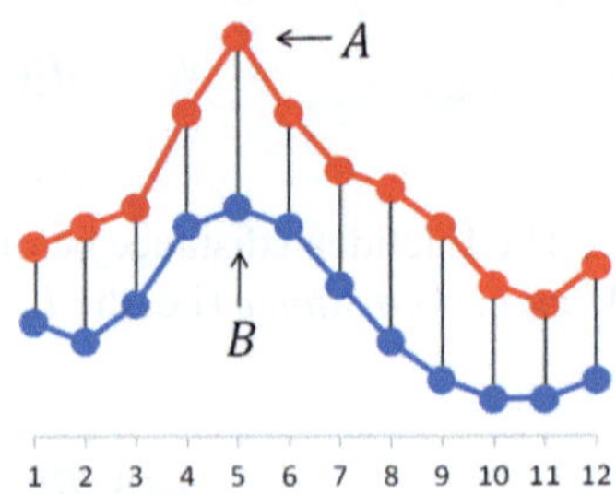

Fig. 2.10 Matching of time series points by L_p norms

seen from Table 2.3, L_p norms will recognize B as more similar to Q than A_1, A_2 or A_3. This means that, for example, the 1NN classifier would assign the class of B to Q and not the class to which A_1, A_2 or A_3 belongs.

Some of the shortcomings of L_p norms might be overcome by applying appropriate transformations to the time series before computing their distance. For example, the length of time series can be equalized by interpolation [47, 71–73] or by truncating the longer series [74]; value axis distortions such as shifting (Fig. 2.11a) and scaling (Fig. 2.11b) can be addressed by normalisation [75–77]. Though, the influence of distortions along the time axis (Fig. 2.11c) cannot be avoided by preprocessing transformations. This is clearly seen in Fig. 2.12 which depicts the z-normalised versions of Q, A_1, A_2, A_3 and B. The distances between the normalised series Q', A_1', A_2', A_3' and B' are given in Table 2.4. While the sequences A_1' and A_2' are almost identical to Q' (the barely noticeable deviations are due to noise in A_1 and A_2),

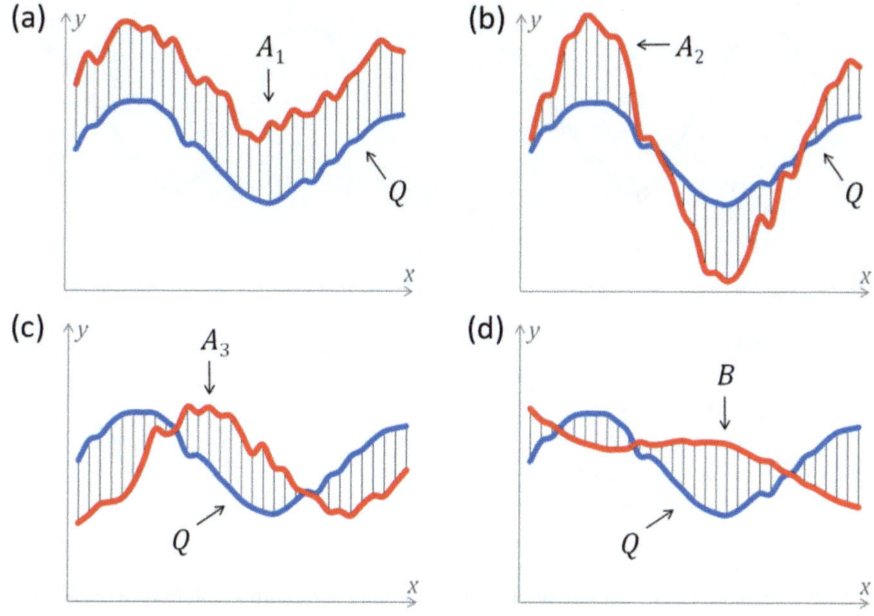

Fig. 2.11 Sensitivity of L_p norms to shifting (**a**) and scaling (**b**) along the value axis and to distortions along the time axis (**c**); an example of dissimilar series (**d**)

Table 2.3 Distances between the time series shown in Fig. 2.11

	Euclidean L_2	Manhattan L_1	Chebyshev L_∞
d(Q, A_1)	7.29	40.38	1.48
d(Q, A_2)	5.19	24.86	1.61
d(Q, A_3)	5.49	27.67	1.53
d(Q, B)	**4.51**	**21.67**	**1.47**

the dissimilarity level of A'_3 remained apparent. Thus, in the presence of distortions along the time axis, L_p norms should be replaced by more appropriate similarity measures, the best-known representatives of which will be presented in the following subsections.

In addition to sensitivity to distortions along the axes, L_p norms are also sensitive to noise and outliers, which is the consequence of the fact that every point of both series is included in the total distance between them (along with those representing noise or outliers) [16, 78–81]. For example, when comparing $Q = (4, 5, 6, 7, 8)$ to $A = (5, 6, 7, 8, 0)$ and $B = (4, 4, 3, 4, 4)$ using L_p norms (Fig. 2.13a), it turns out that B is more similar to Q than A (Table 2.5) which is obtained by shifting Q along the time axis, with its last point being an outlier. As Fig. 2.13b and Table 2.5 illustrate,

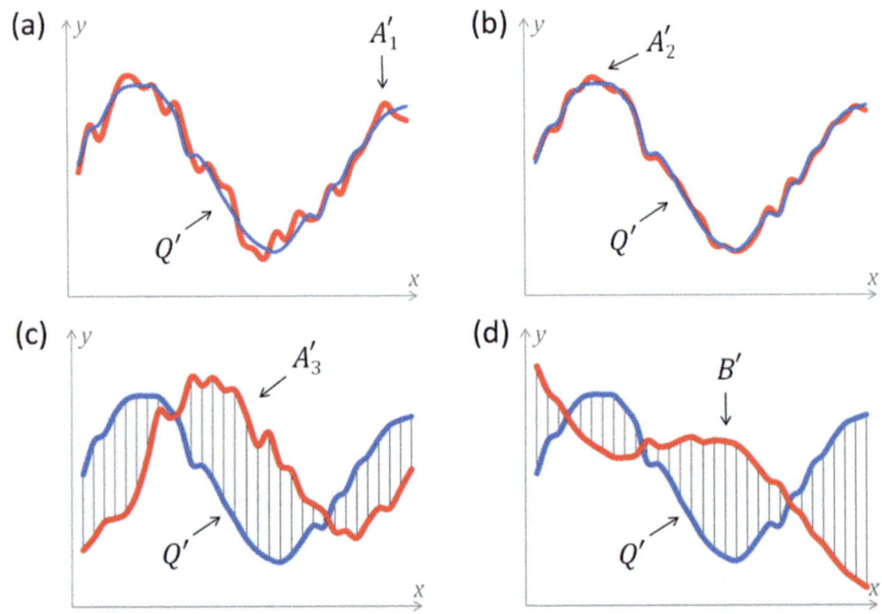

Fig. 2.12 The influence of z-normalisation on the sensibility of L_p norms to shifting (**a**) and scaling (**b**) along the value axis and to distortions along the time axis (**c**); an example of dissimilar series (**d**)

Table 2.4 Distances between the z-normalised time series depicted in Fig. 2.12

	Euclidean L_2	Manhattan L_1	Chebyshev L_∞
$d(Q', A'_1)$	1.07	5.14	0.36
$d(Q', A'_2)$	**0.43**	**2.06**	**0.15**
$d(Q', A'_3)$	8.34	41.93	2.37
$d(Q', B')$	8.48	40.04	3.16

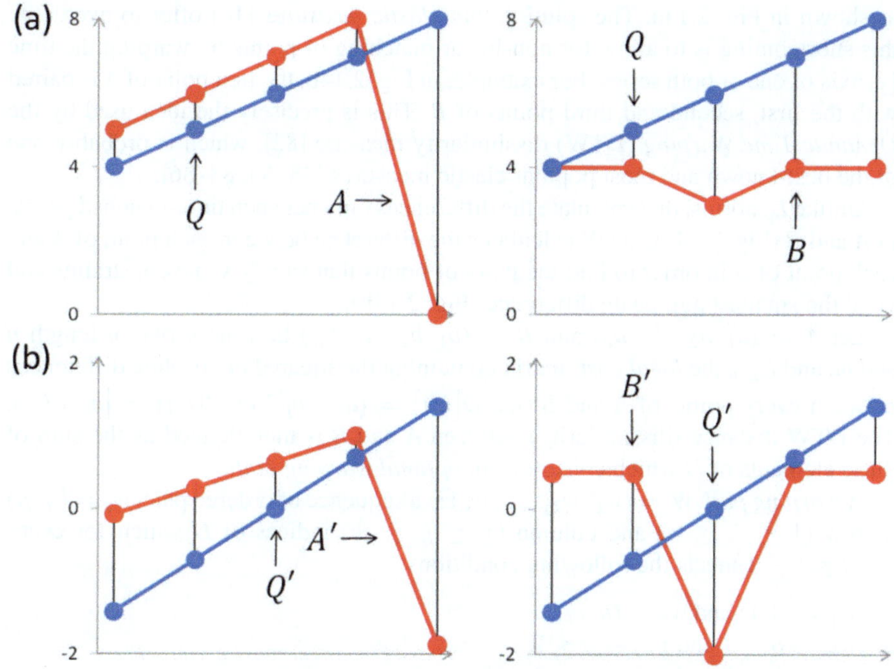

Fig. 2.13 Sensitivity of L_p norms to outliers: original time series (**a**); and their z-normalised versions (**b**)

Table 2.5 Distances between the time series shown in Fig. 2.13

	Euclidean L_2	Manhattan L_1	Chebyshev L_∞
$d(Q, A)$	8.25	12.00	8.00
$\mathbf{d(Q, B)}$	**5.92**	**11.00**	**4.00**
$d(Q', A')$	3.75	6.56	3.28
$\mathbf{d(Q', B')}$	**3.16**	**6.24**	**2.00**

in the general case, applying z-normalisation will not resolve L_p norms sensitivity to noise and outliers.

2.3.2 *Dynamic Time Warping (DTW)*

The adverse effects of distortions along the value (y) axis on L_p norms can be neutralized by applying normalisation, however it does not eliminate their sensitivity to shiftings and scalings along the time (x) axis, which is a consequence of the linear matching of time series points: the ith point of A is matched with the ith point of B, as shown in Fig. 2.14a. The solution that *elastic measures* [16] offer to overcome this shortcoming is to allow for non-linear matching of points by warping the time (x) axis of one or both series. For example, in Fig. 2.14b, the first point of A is paired with the first, second and third points of B. This is precisely the idea used by the *Dynamic Time Warping* (DTW) dissimilarity measure [82], which is probably one of the best-known and most popular elastic measures [38, 60, 83–86].

Unlike L_p norms, that calculate the differences only between time-matched points of A and B (Fig. 2.14a), DTW calculates the difference between each point of A and each point of B in order to find the pairs of points that satisfy some constraints and yield the smallest aggregate difference (Fig. 2.14b).

Let $A = (a_1, a_2, \ldots, a_n)$ and $B = (b_1, b_2, \ldots, b_m)$ be time series of length n and m, and $D_{n,m}$ the *local cost* matrix containing the squared or absolute differences between every points of A and B (i.e. $D[i,j] = (a_i - b_j)^2$ or $D[i,j] = |a_i - b_j|$). The DTW distance (dissimilarity) between A and B is then defined as the sum of those elements of D which constitute an *optimal warping path.*

A *warping path* $W = (w_1, w_2, \ldots, w_k)$ is a sequence of ordered pairs $w_t = (i_t, j_t)$ of row ($1 \le i_t \le n$) and column ($1 \le j_t \le m$) indices of D which for every $1 \le t \le k - 1$ meets the following conditions:

1. $w_1 = (1,1)$ and $w_k = (n, m)$,
2. $i_{t+1} - i_t \le 1$ and $j_{t+1} - j_t \le 1$,
3. $i_{t+1} - i_t \ge 0$ and $j_{t+1} - j_t \ge 0$.

The first condition is called the *boundary condition* and it ensures that the warping starts by matching a_1 (the first point of A) with b_1 (the first point of B) and ends by matching a_n (the last point of A) with b_m (the last point of B). The second condition (called the *continuity condition*) guarantees that the matching does not skip any points of A or B: the elements of the warping path must represent adjacent cells in D. The third condition (called the *monotonity condition*) prohibits the warping path from

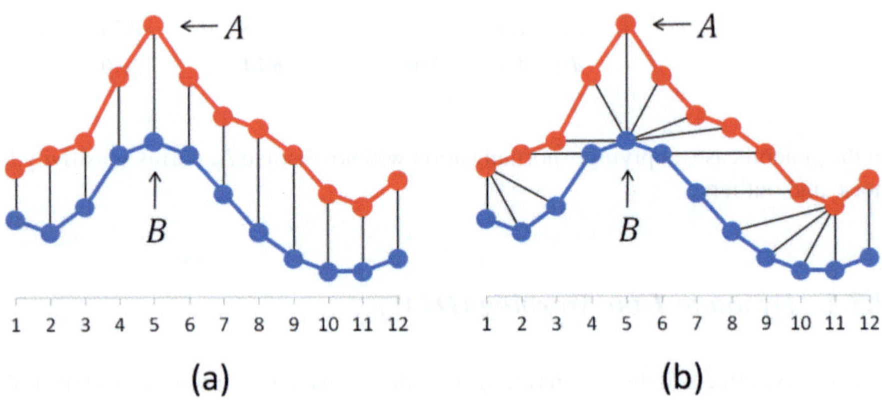

Fig. 2.14 Matching time series points: linear (**a**) and elastic (**b**)

going back in time (meaning that W can advance only upwards or/and to the right in D).

As the longest warping path goes first along one of the two time series and then along the other one, and since the length of the shortest path cannot be less than the length of the longer series (otherwise some of its points would not be included), it follows that $max(n, m) \leq k \leq n + m - 1$.

The total number of different warping paths leading from the point (1,1) to the point (n, m) of the local cost matrix D can be calculated based on the following considerations. Each warping path contains diagonal, horizontal and vertical steps between adjacent elements of D. As the number of diagonal steps d ranges between 0 and $min(n, m) - 1$, the number of vertical steps is $n - 1 - d$ and the number of horizontal steps is $m - 1 - d$. Since the total number of steps within a given warping path is equal to the sum of the number of diagonal, vertical and horizontal steps (i.e. $n + m - 2$), the total number of different warping paths is given by the following formula:

$$\sum_{d=0}^{min(n,m)-1} P_{d,n-1-d,m-1-d}^{n+m-2} = \sum_{d=0}^{min(n,m)-1} \frac{(n+m-2)!}{d!(n-1-d)!(m-1-d)!}$$

Each warping path $W = (w_1, w_2, \ldots, w_k)$ represents one particular way of matching the points of time series A and B, and the *warping cost* is calculated as the sum of the local costs determined by the elements of the cost matrix D that the given warping path includes:

$$cost(W) = \sum_{t=1}^{k} cost(w_t) = \sum_{t=1}^{k} D\left[i_t, j_t\right]$$

A warping path is said to be *optimal* if it has the lowest cost among all possible warping paths. The DTW distance between the time series A and B is defined as the cost of their optimal warping path:

$$d_{DTW}(A, B) = \min_{W} cost(W)$$

Figure 2.15 shows two examples of warping paths within the local cost matrix D of time series A and B introduced in Fig. 2.10. The result of the diagonal warping path W_a in Fig. 2.15a is the linear matching of the points of A and B shown in Fig. 2.14a, and since the local cost was calculated as the square of the difference between the time series points, the warping cost is equal to the square of the Euclidean (L_2) distance between A and B:

$$\begin{aligned} cost(W_a) &= (11-7)^2 + (12-6)^2 + \cdots + (10-4)^2 \\ &= 16 + 36 + \cdots + 36 = 491 = d_2^2(A, B) \end{aligned}$$

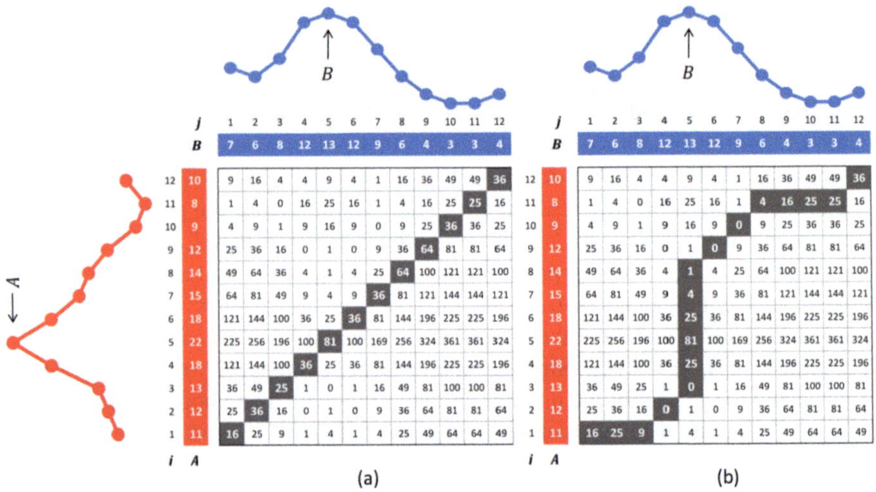

Fig. 2.15 Examples of warping paths: $\boldsymbol{W_a}$ (a), $\boldsymbol{W_b}$ (b)

If the local cost was calculated as the absolute difference between the points, the cost of the diagonal warping path would be equal to the square of the Manhattan (L_1) distance between A and B.

The warping path W_b in Fig. 2.15b is optimal, which means that its cost is the smallest among the costs of all possible warping paths between A and B, therefore $cost(W_b)$ also represents their DTW distance:

$$d_{DTW}(A, B) = cost(W_b) = (11-7)^2 + (11-6)^2 + (11-8)^2 + \cdots + (8-3)^2 + (8-3)^2 + (10-4)^2 = 16 + 25 + 9 + \cdots + 25 + 25 + 36 = 292$$

The non-linear alignment of the points of A and B corresponding to the optimal warping path W_b is given in Fig. 2.14b.

The Euclidean (or the Manhattan) distance (more precisely, its squared value) represent the cost of one possible warping path, which may not be the optimal one. DTW finds the matching between the points of the time series that will minimize the warping cost, and the distance between the series is the cost of that optimal warping path.

The DTW distance between $A = (a_1, a_2, \ldots, a_n)$ and $B = (b_1, b_2, \ldots, b_m)$ can be computed using *dynamic programming* as $d_{DTW}(A, B) = DTW[n, m]$, where DTW represents the *warping matrix* defined by the following recursive formula:

$$DTW[i,j] = \begin{cases} 0 & i = j = 0 \\ \infty & (i = 0 \text{ and } j > 0) \\ & \text{or } (i > 0 \text{ and } j = 0) \\ d(a_i, b_j) & \\ \quad + min \begin{cases} DTW[i, j-1] \\ DTW[i-1, j-1] \\ DTW[i-1, j] \end{cases} & i, j > 0 \end{cases}$$

where $0 \leq i \leq n$, $0 \leq j \leq m$ and $d(a_i, b_j)$ denotes the local cost expressed as the squared or the absolute difference between a_i and b_j (i.e. $D[i,j]$). In this book, we applied the squared difference.

Computing the DTW distance using the warping matrix *DTW* eliminates the need to generate the local cost matrix *D* and to find the optimal warping path among all possible warping paths. The warping matrix corresponding to the time series *A* and *B* introduced in Fig. 2.10 is shown in Fig. 2.16. The distance between *A* and *B* is obtained from the upper right cell of the *DTW* matrix:

$$d_{DTW}(A, B) = DTW[n, m] = DTW[12.12] = 292$$

Since each cell of the warping matrix contains the accumulative cost of the local optimal warping path leading from cell (1,1) to the given cell, the optimal warping path can be traced back from cell (*n*, *m*) to cell (1,1) by choosing the neighboring cell with the smallest value (to the left, diagonally and below) at each step (Fig. 2.16). Figure 2.17 depicts the matchings between the points of the time series *A* and *B* induced by their optimal warping path. As can be seen from this example, while in

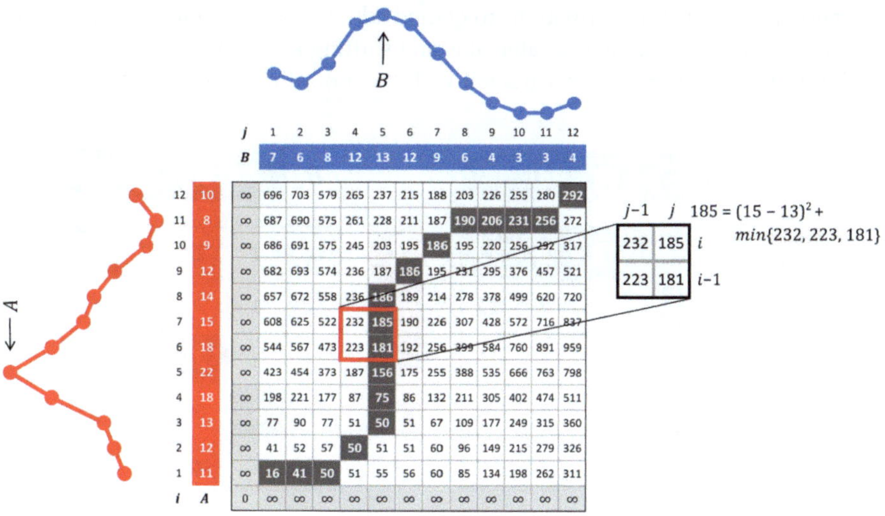

Fig. 2.16 Computing DTW using dynamic programming

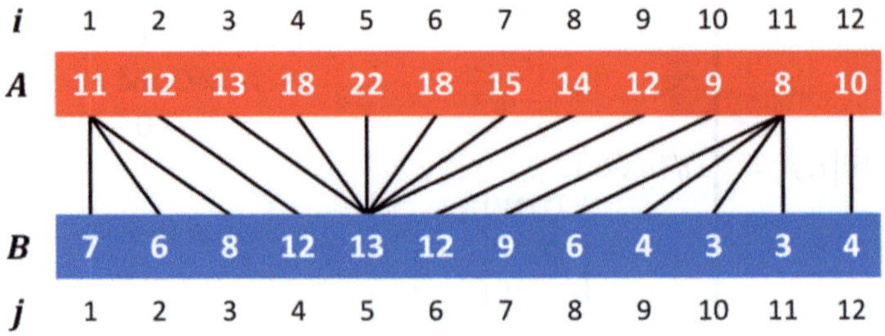

Fig. 2.17 Non-linear matching of the points of the time series A and B induced by their optimal warping path

the case of DTW the correspondence between the points can be *one-to-one* (between $A[2]$ and $B[4]$, $A[9]$ and $B[6]$, $A[10]$ and $B[9]$ and between $A[12]$ and $B[12]$), *one-to-many* (between $A[1]$ and $B[1{:}3]$ and between $A[11]$ and $B[8{:}11]$) and *many-to-one* (between $A[3{:}8]$ and $B[5]$), in the case of L_p norms it can only be *one-to-one* (Fig. 2.10). Using dynamic programming, the sum of squared differences of the matched points is obtained in cell $DTW[n, m]$.

It should be noted that the optimal warping path need not to be unique, there may be multiple optimal alignments of time series points with the same (minimum) cost. Figure 2.18 shows two possible optimal warping paths for the time series $A = (0, 1, 1, 1)$ and $B = (1, 2, 3, 4)$.

Owing to elastic aligning of points, unlike L_p norms, DTW is successful in recognizing similar sequences that are shifted and/or scaled along the time axis. This can be seen in Table 2.6 which presents the DTW distances between the time series Q and three other series (A_1, A_2, A_3 and B) introduced in the previous subsection (their DTW alignments are depicted in Fig. 2.19). Wherein, A_1, A_2 and A_3 were obtained by applying various transformations to Q (and adding some random noise): shifting (A_1) and scaling (A_2) along the value axis and shifting along the time axis (A_3). The auxiliary time series B was constructed with the aim of making its difference with

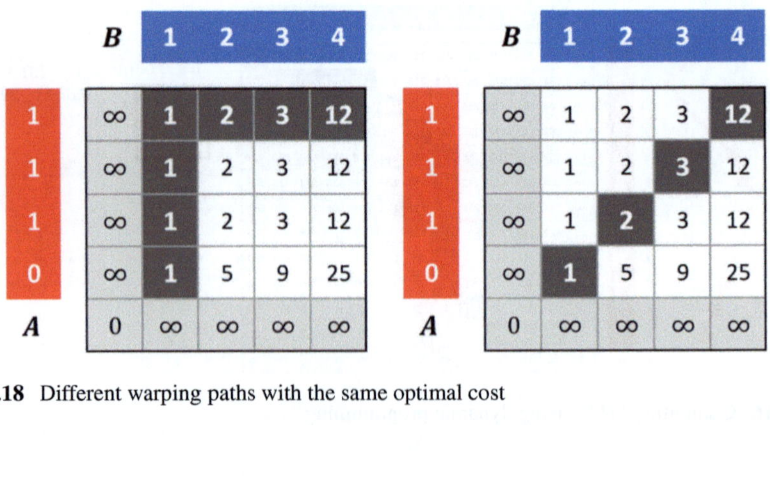

Fig. 2.18 Different warping paths with the same optimal cost

Table 2.6 DTW distances between Q and three other time series before (Fig. 2.19) and after z-normalisation (Fig. 2.20)

	Before	After
A_1	32.73	0.95
A_2	22.21	**0.18**
A_3	**6.69**	9.50
B	10.27	43.89

respect to Q visually evident. According to DTW, the series that most closely resembles Q is A_3 (in contrast to the L_2, L_1 and L_∞ norms according to which it is B—see Table 2.3). These examples also demonstrate that DTW, just like L_p norms, is sensitive to shifting and scaling along the value axis which can be avoided by normalising the series [35, 87, 88]. As it can be seen from Table 2.6, after normalisation, B' has the highest dissimilarity value (the alignments of the normalised sequences are shown in Fig. 2.20).

Since each point of one time series must be matched with at least one point of the other one (and vice versa), DTW is sensitive to noise and outliers [80, 81, 89, 90], similarly to L_p norms. As an example, the (dis)similarity of the time series $Q = (4, 5, 6, 7, 8)$, $A = (5, 6, 7, 8, 0)$ and $B = (4, 4, 3, 4, 4)$ shown in Fig. 2.21a can be considered. The last point of A represents an outlier, and the others were obtained by simply shifting the points of Q along the value axis. Despite the fact that

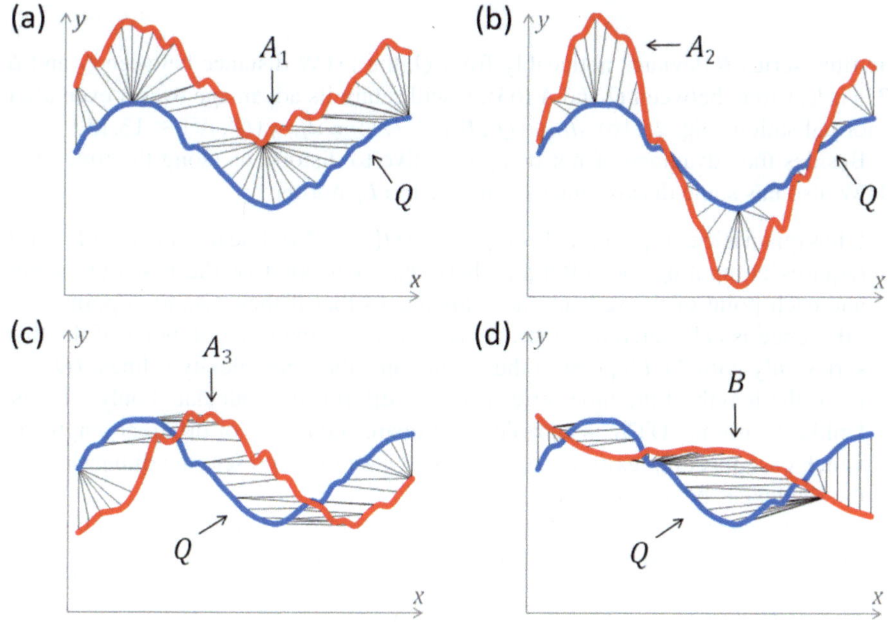

Fig. 2.19 Effects of various distortions on DTW: shifting (**a**) and scaling (**b**) along the value axis and distortions along the time axis (**c**); an example of dissimilar series (**d**)

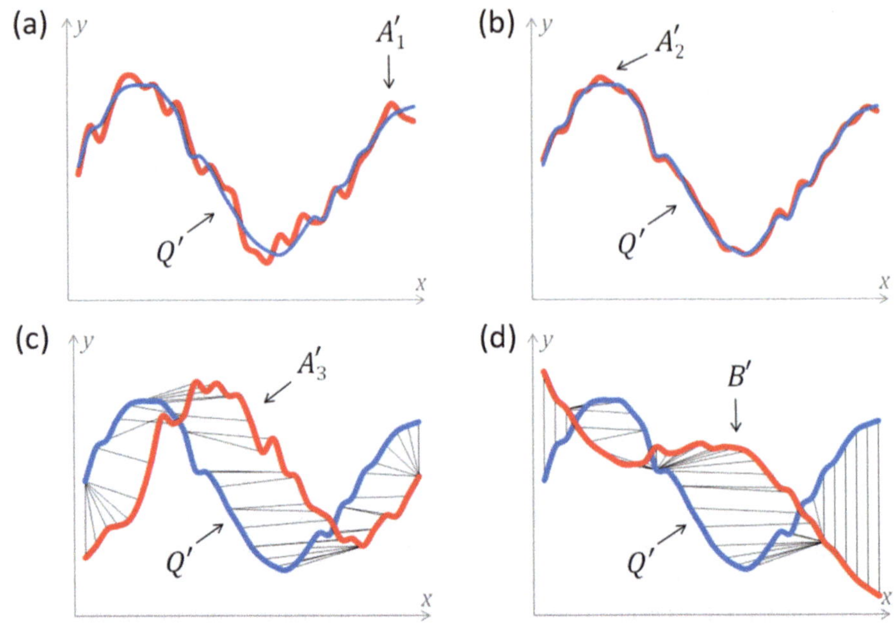

Fig. 2.20 The influence of z-normalisation on DTW in the presence of shifting (**a**) and scaling (**b**) along the value axis and distortions along the time axis (**c**); an example of dissimilar series (**d**)

the time series B deviates noticeably from Q, the DTW distance between Q and B (31) is less than between Q and A (65). B will retain its advantage over A even after z-normalisation (Fig. 2.21b): $d_{DTW}\left(Q', B'\right) \approx 7.92 < d_{DTW}\left(Q', A'\right) \approx 13.15$.

Besides the advantage of not being sensitive to distortions along the time axis, DTW also has some disadvantages compared to L_p norms:

- It has quadratic computational complexity ($O\left(n^2\right)$). Nonlinear matching of points requires computing the difference between each point of the first time series and each point of the second one. This means that if they contain n points, the difference is calculated n^2 times. As L_p norms compare the ith point of the first series only with the ith point of the second one, their complexity is linear ($O(n)$), i.e. if the length of the time series is n, the difference is calculated only n times. Unlike L_p norms, DTW is capable to compare sequences of different lengths in which case its complexity is $O(nm)$, where n and m represent the dimensions of the sequences being compared.
- It is not a metric as it does not satisfies the *triangle inequality*. This can be seen, for example, by comparing the following three time series:

$$A = (0, 0, 0, 0),$$
$$B = (2, 2, 2, 3),$$
$$C = (2, 1, 1, 1).$$

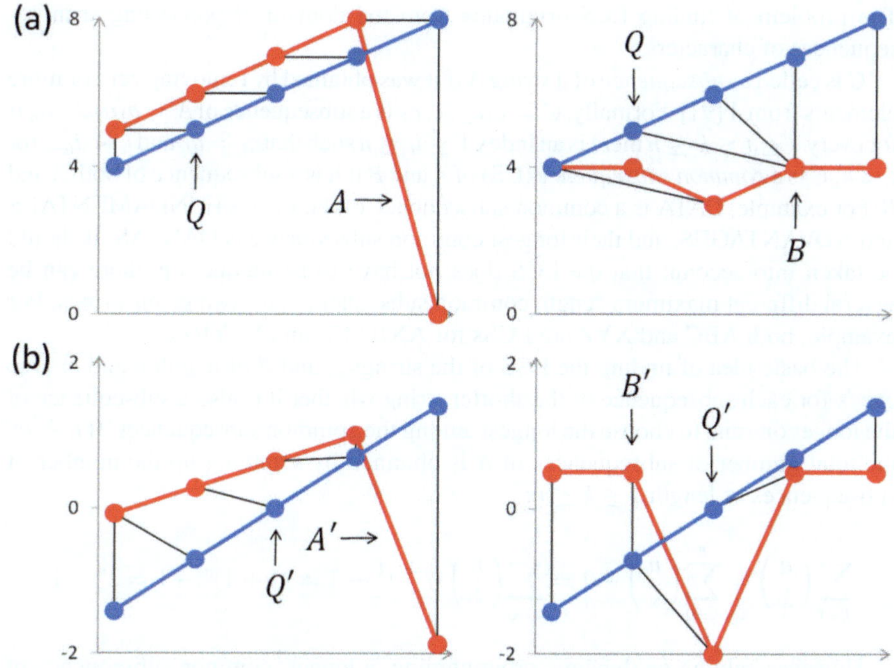

Fig. 2.21 Sensitivity of DTW to outliers: original time series (**a**); and their z-normalised versions (**b**)

The L_p norms satisfy the triangle inequality:

$$4.58 \approx \sqrt{21} = d_2(A, B) \leq d_2(A, C) + d_2(C, B) = \sqrt{7} + \sqrt{6} \approx 5.10,$$

$$9 = d_1(A, B) \leq d_1(A, C) + d_1(C, B) = 5 + 4 = 9,$$

$$3 = d_\infty(A, B) \leq d_\infty(A, C) + d_\infty(C, B) = 2 + 2 = 4,$$

but DTW does not:

$$21 = d_{DTW}(A, B) > d_{DTW}(A, C) + d_{DTW}(C, B) = 7 + 6 = 13.$$

2.3.3 *Longest Common Subsequence (LCS)*

The LCS dissimilarity measure is based on the idea of expressing the distance between time series dependent on the length of their longest common subsequence.

The problem of finding LCS originates from the domain of processing strings - sequences of characters.

C is called a *subsequence* of a *string* A if it was obtained by removing zero or more elements from A [91]. Formally, $C = c_1c_2 \dots c_k$ is a subsequence of $A = a_1a_2 \dots a_n$ if for every $1 \leq t \leq k \leq n$ there is an index $1 \leq i_t \leq n$ such that $c_t = a_{i_t}$ and $i_t < i_{t+1}$ for $t < k$. C is a *common subsequence* (CS) of A and B if it is a subsequence of both A and B. For example, DATA is a common subsequence of the strings FUNDAMENTALS and ADVANTAGES, and their longest common subsequence is DANTAS. It should be taken into account that the LCS does not have to be unique, i.e. there can be several different maximum-length common subsequences for two given strings. For example, both ABC and XYZ are LCSs for AXBYCZ and XAYBZC.

The basic idea of finding the LCS of the strings A and B of length n and m is to check for each subsequence of the shorter string whether it is also a subsequence of the longer one and to choose the longest among the common subsequences. If $n < m$, the total number of subsequences of A is obtained by summing up the number of subsequences of length $1 \leq k \leq n$:

$$\sum_{k=1}^{n}\binom{n}{k} = \sum_{k=0}^{n}\binom{n}{k} - 1 = \sum_{k=0}^{n}\binom{n}{k}1^{n-k}1^{k} - 1 = (1+1)^n - 1 = 2^n - 1$$

However, relying on dynamic programming, a longest common subsequence of two strings $A = a_1a_2 \dots a_n$ and $B = b_1b_2 \dots b_m$ can be obtained as $LCS(A, B) = LCS(n, m)$ using the following recursive formula [91–93]:

$$LCS(n,m) = \begin{cases} \emptyset & n = 0 \text{ or } m = 0 \\ LCS(n-1, m-1) + a_n & n, m > 0 \text{ and } a_n = b_m \\ longest\{LCS(n, m-1), LCS(n-1, m)\} & n, m > 0 \text{ and } a_n \neq b_m \end{cases}$$

where $\emptyset$ represents the empty string, and $+$ is the concatenation operator.

Namely, let $C = c_1c_2 \dots c_k$ be an LCS of $A[1{:}i]$ and $B\left[i : j\right]$, then:

- If $a_i = b_j$, then $c_k = a_i$ and $C[1{:}k-1]$ is an LCS of $A[1{:}i-1]$ and $B\left[i : j-1\right]$. If we assume that $c_k \neq a_i$, it follows that $C + a_i$ is a common subsequence of $A[1{:}i]$ and $B\left[i : j\right]$ of length $k+1$ which is in contradiction with the initial assumption that C is one of their longest common subsequences. As $C[1{:}k-1]$ is a common subsequence of $A[1{:}i-1]$ and $B\left[i : j-1\right]$, if we suppose that it is not one of their longest ones, it follows that there exists a common subsequence X of $A[1{:}i-1]$ and $B\left[i : j-1\right]$ such that $|X| > |C| = k$. Then, $X + a_i$ is a CS of $A[1{:}i]$ and $B\left[i : j\right]$, which is, again, in contradiction with the initial assumption that C is one of their LCSs ($|X + a_i| > k + 1 > |C| = k$).
- If $a_i \neq b_j$, there are two possibilities:
  - If $c_k \neq a_i$, then C is an LCS of $A[1{:}i-1]$ and $B\left[i : j\right]$. Since C is an LCS of $A[1{:}i]$ and $B\left[i : j\right]$ and $c_k \neq a_i$, it means that C is a CS of $A[1{:}i-1]$ and $B\left[i : j\right]$. If we assume that C is not one of their longest CSs, it follows that there

is a longer common subsequence W of them (i.e. $|W| > |C| = k$). However, since W is also CS of $A[1{:}i]$ and $B[i : j]$, this contradicts the initial assumption that C is one of their longest CSs.

- If $c_k \neq b_j$, then C is an LCS of $A[1{:}i]$ and $B[i : j - 1]$. Since C is an LCS of $A[1{:}i]$ and $B[i : j]$ and $c_k \neq b_j$, it means that C is a CS of $A[1{:}i]$ and $B[i : j - 1]$. If we assume that C is not one of their longest CSs, it follows that there is a longer common subsequence W of them (i.e. $|W| > |C| = k$). However, since W is also CS of $A[1{:}i]$ and $B[i : j]$, this contradicts the initial assumption that C is one of their longest CSs.

Based on the recursive formula for finding an LCS of two strings $A = a_1a_2 \ldots a_n$ and $B = b_1b_2 \ldots b_m$, an analogous formula for calculating its length can be given by means of an n-by-m matrix (Fig. 2.22) using dynamic programming:

$$LCS[i,j] = \begin{cases} 0 & i = 0 \text{ or } j = 0 \\ LCS[i-1,j-1]+1 & i,j > 0 \text{ and } a_i = b_j \\ max\{LCS[i,j-1], LCS[i-1,j]\} & i,j > 0 \text{ and } a_i \neq b_j \end{cases}$$

where $0 \leq i \leq n$, $0 \leq j \leq m$ and $LCS[i,j]$ is the length of a longest common subsequence of $A[1{:}i]$ and $B[1{:}j]$. Hence, $|LCS(A, B)| = LCS[n, m]$.

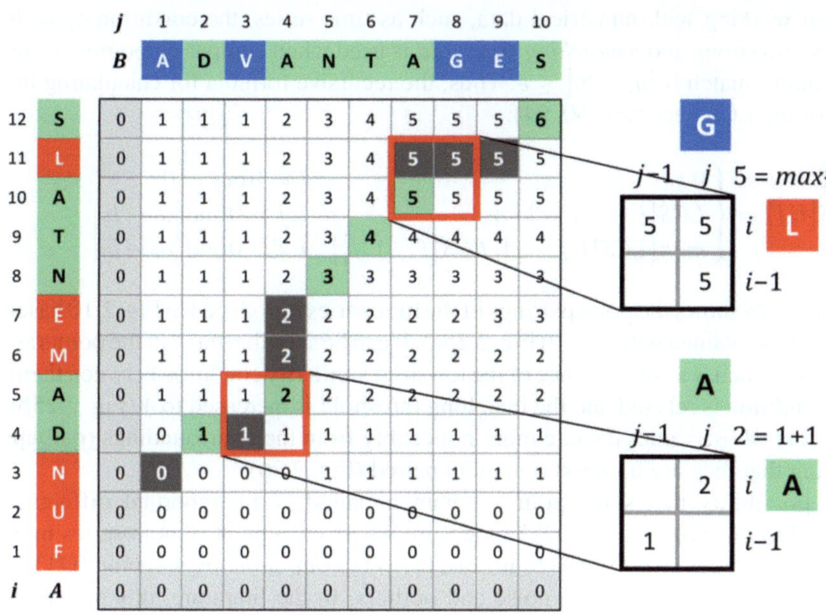

Fig. 2.22 Finding the length of the LCS of two strings using dynamic programming

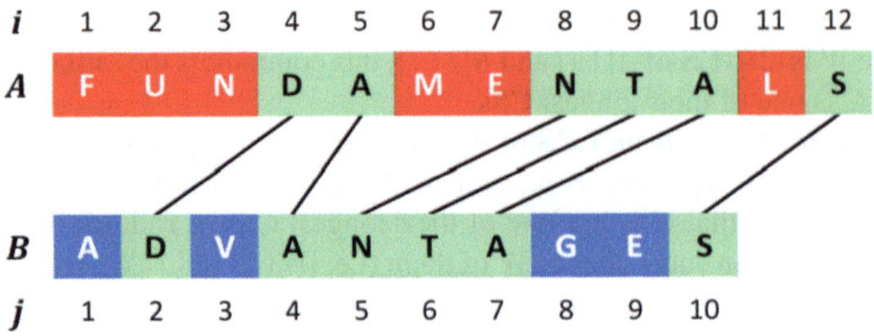

Fig. 2.23 The longest common subsequence (DANTAS) of the strings FUNDAMENTALS and ADVANTAGES

The matchings between the elements of the strings A i B and their LCS (Fig. 2.23) can be reconstructed from their LCS matrix starting from cell (n, m) and $LCS(A, B) = \varnothing$ as follows (Fig. 2.22):

- if $A[n] = B[m] = x$, x is added to LCS as a prefix (i.e. $LCS(A, B) = x + LCS(A, B)$) and the reconstruction continues through cell $(n-1, m-1)$,
- if $A[n] \neq B[m]$, depending on whether $LCS[n, m-1] \geq LCS[n-1, m]$, cell $(n, m-1)$ or $(n-1, m)$ is selected,
- the reconstruction stops when the value of the selected cell becomes 0 (zero).

When working with numerical data, such as time series, the condition $a_i = b_j$ might be too strong and a *matching threshold* is used when comparing points: a_i and b_j are said to match if $|a_i - b_j| \leq \varepsilon$. Thus, the recursive formula for calculating the length of the LCS becomes [60, 94]:

$$LCS[i,j] = \begin{cases} 0 & i = 0 \text{ or } j = 0 \\ LCS[i-1,j-1] + 1 & i,j > 0 \text{ and } |a_i - b_j| \leq \varepsilon \\ max\{LCS[i,j-1], LCS[i-1,j]\} & i,j > 0 \text{ and } |a_i - b_j| > \varepsilon \end{cases}$$

Figure 2.24 shows the point pairings of the time series introduced in Fig. 2.10 based on their LCS obtained with $\varepsilon = 0$ (Fig. 2.25a). It can be seen that most of the points are not paired, which is a consequence of the too strict similarity condition between them. If this condition is relaxed and the matching threshold is increased to 4 (Fig. 2.25b), only points whose values deviate more noticeably from their surroundings (perhaps indicating that they are outliers) remain unpaired (Fig. 2.26).

The possibility that some points remain unpaired is an important difference compared to L_p norms and the DTW measure where each point of both series must be matched through *one-to-one* (L_p norms, DTW) or *one-to-many* mapping (DTW), which makes them sensitive to noise and outliers. In the literature, this feature of LCS is referred to as *one-to-none* mapping of points [22]. Based on these examples, it can be also noticed that LCS generates *one-to-one* matches between similar points (which can be time-shifted, for example, between $A[2]$ and $B[4]$ in Fig. 2.24).

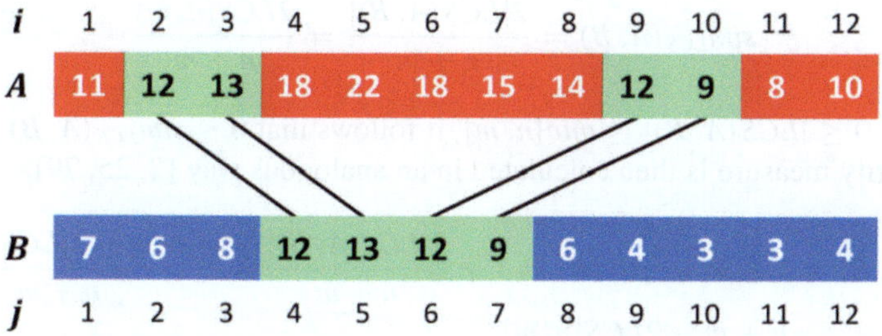

Fig. 2.24 The longest common subsequence of the time series A and B obtained with $\varepsilon = 0$

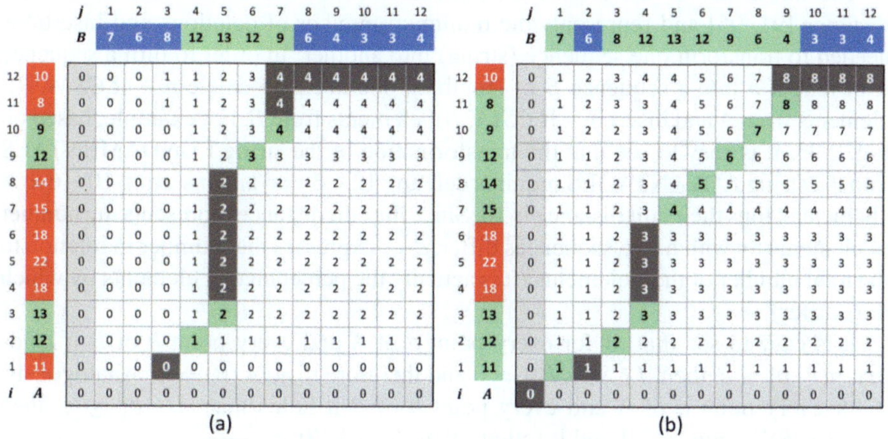

Fig. 2.25 Computing LCS using dynamic programming with $\varepsilon = 0$ (**a**) and $\varepsilon = 4$ (**b**)

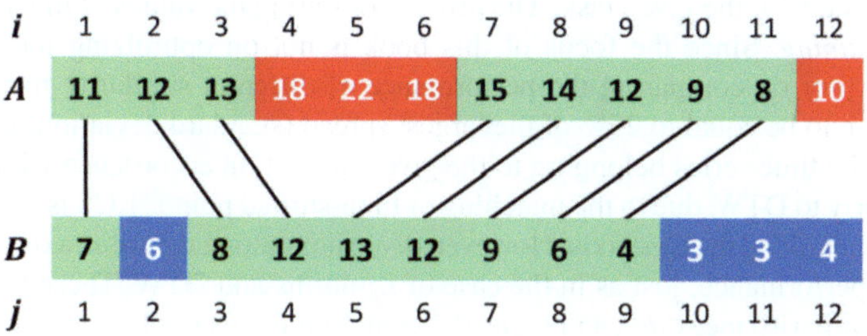

Fig. 2.26 The longest common subsequence of the time series A and B obtained with $\varepsilon = 4$

The similarity measure between two time series $A = (a_1, a_2, \ldots, a_n)$ and $B = (b_1, b_2, \ldots, b_n)$ as a function of the length of their longest common subsequence can be expressed by the following formula:

$$sim_{LCS}(A, B) = \frac{2|LCS(A, B)|}{n + m} = \frac{2LCS[n, m]}{n + m}$$

where, as $0 \leq |LCS(A, B)| \leq min\{n, m\}$, it follows that $0 \leq sim_{LCS}(A, B) \leq 1$. The dissimilarity measure is then calculated in an analogous way [2, 25, 79]:

$$d_{LCS}(A, B) = 1 - sim_{LCS}(A, B) = 1 - \frac{2|LCS(A, B)|}{n + m} = \frac{n + m - 2|LCS(A, B)|}{n + m}$$
$$= \frac{n + m - 2LCS[n, m]}{n + m}$$

The numerator of the dissimilarity measure formula is a variation of the *edit distance* [91, 95] and represents the minimum number of deletions and insertions needed to transform one sequence (string) into another: in order to turn a sequence A of length n into a sequence B of length m, one needs to delete $n - |LCS(A, B)|$ elements from A and insert $m - |LCS(A, B)|$ elements from B. For example, based on Fig. 2.23, it is easy to see that the transformation of the string FUNDAMENTALS into the string ADVANTAGES requires deleting $12 - 6 = 6$ and inserting $10 - 6 = 4$ characters, thus their edit distance is 10. Since the edit distance represents the number of elements by which two sequences differ, d_{LCS} expresses that number in relation to the total number of their elements (it is actually the percentage of elements by which they differ).

It is easy to see that, if for every point a of A and every point b of B it holds that $|a - b| > \varepsilon$, then $|LCS(A, B)| = 0$ and therefore $d_{LCS}(A, B) = 1$. Analogously, if for every point a of A and every point b of B it holds that $|a - b| \leq \varepsilon$, then $|LCS(A, B)| = min\{n, m\}$ and it follows that $d_{LCS}(A, B) = \frac{|n-m|}{n+m}$.

The optimal value of the matching threshold parameter depends on the data that the time series represents and is usually chosen from some interval so that it produces the best results for the given task. The process of setting the values of the parameters is called *tuning*. Since the focus of this book is not on optimizing the matching threshold (nor on comparing the performance of different similarity measures), it was chosen to be equal to 25% of the largest spread (standard deviation) of the data points in the time series belonging to the given dataset, in accordance with [81].

Similarly to DTW, due to the matching of time-shifted points, LCS is not sensitive to distortions along the time axis. However, distortions along the value axis negatively affect its performance, just as in the case of L_p norms and DTW. These traits of the LCS dissimilarity measure can be clearly seen with the example time series shown in Fig. 2.27. They were used for the same purpose in the previous subsections: with some random noise added, A_1 (A_2) was obtained by shifting (scaling) Q along the value axis, A_3 is the result of shifting it along the time axis, and B was constructed so that it deviates visibly from Q.

Based on the LCS distances shown in Table 2.7, it can be seen that, before applying z-normalisation, B looks more similar to Q than A_1 and A_2, but A_3 is the most similar. This is also evident if the length of their longest common subsequences is compared (the time series contains 31 points). The negative effects of shifting and scaling along

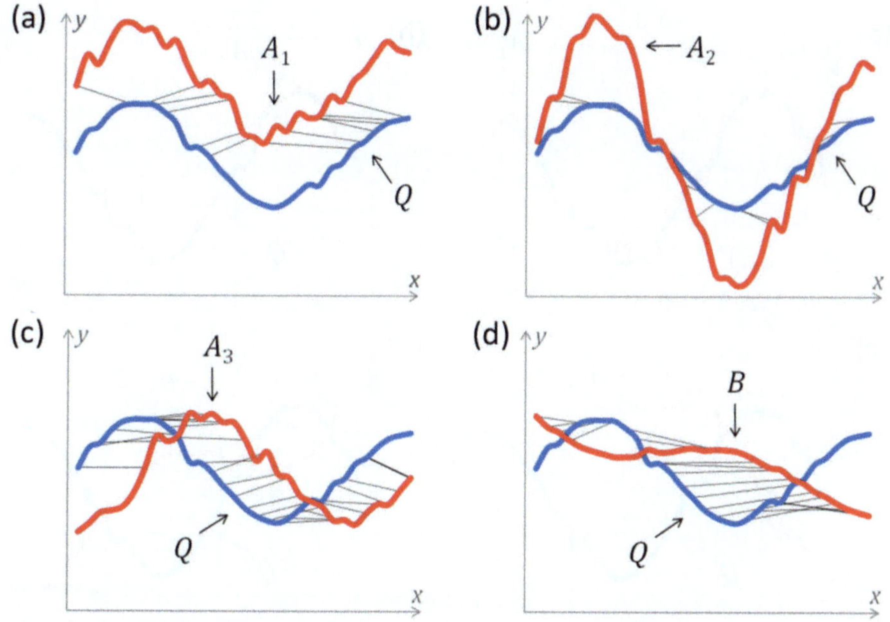

Fig. 2.27 Effects of various distortions on LCS: shifting (**a**) and scaling (**b**) along the value axis and distortions along the time axis (**c**); an example of dissimilar series (**d**)

the value axis are neutralized by normalisation and the order of the series (from the most to the least similar to Q) changes to: A'_2, A'_1, A'_3, B'. The alignments of the points of the normalised series are presented in Fig. 2.28. When calculating the distances, the value of the parameter ε was set to one-fourth of the highest standard deviation of the sequences [81]: ≈ 0.38 (before normalisation) and 0.25 (after normalisation).

The robustness of the LCS measure to noise and outliers [96–98] is a consequence of the ε-limited matching of points. The time series $Q = (4, 5, 6, 7, 8)$, $A = (5, 6, 7, 8, 0)$ and $B = (4, 4, 3, 4, 4)$ in Fig. 2.29a illustrate this phenomenon well. By setting ε to a quarter of their maximum standard deviation as recommended in [81], while the first four points of A are paired, the last one is too far from the

Table 2.7 LCS distances between Q and four other time series and the length of the corresponding LCSs before (Fig. 2.27) and after z-normalisation (Fig. 2.28)

	Before ($\varepsilon \approx 0.38$)		After ($\varepsilon = 0.25$)	
	d_{LCS}	$\lvert LCS \rvert$	d_{LCS}	$\lvert LCS \rvert$
A_1	0.61	12	0.23	24
A_2	0.55	14	**0.00**	**31**
A_3	**0.19**	**25**	0.42	18
B	0.45	17	0.58	13

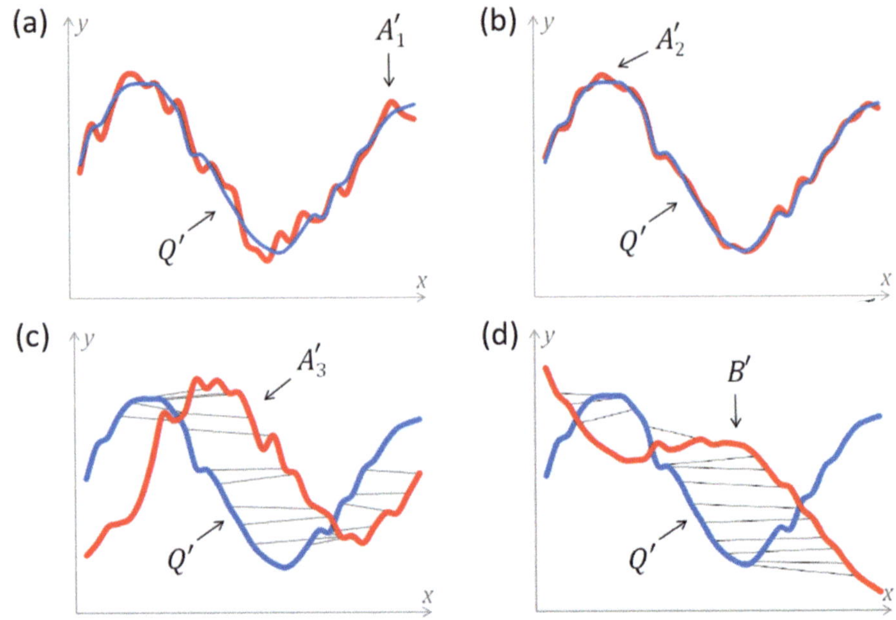

Fig. 2.28 The influence of z-normalisation on LCS in the presence of shifting (**a**) and scaling (**b**) along the value axis and distortions along the time axis (**c**); an example of dissimilar series (**d**)

points of Q to be matched with any of them. When it comes to B, each of its points (except the third one) has the same value as the first point of Q, however, since LCS does not support *one-to-many*, but only *one-to-one* pairings of similar points, only its first point will be matched with the first point of Q. After z-normalisation, the first point of B' is matched with the fourth point of Q' (the others remain unmatched), but as two points of A' are paired with the points of Q', A' is still more similar to Q' than B'. The distances between the time series are given in Table 2.8.

However, it should be taken into account that the results are highly dependent on the value of the matching threshold. For example, if the value of ε is set to 4, while all 5 points of B will be paired with the corresponding points of Q, the last point of A remains unpaired. In this way, B becomes more similar to Q than A.

In relation to L_p norms, LCS shows the same deficiencies as DTW:

- It's computational complexity is quadratic: calculating the distance between two series of length n and m requires $n \times m$ comparisons of their points.
- It does not satisfy the triangle inequality, and thus it is not a metric. For example, taking $\varepsilon = 1$, for $A = (0, 0, 0, 0)$, $B = (2, 2, 2, 3)$ and $C = (2, 1, 1, 1)$ we get:

$$1 = d_{LCS}(A, B) > d_{LCS}(A, C) + d_{LCS}(C, B) = 0.25 + 0.25 = 0.5.$$

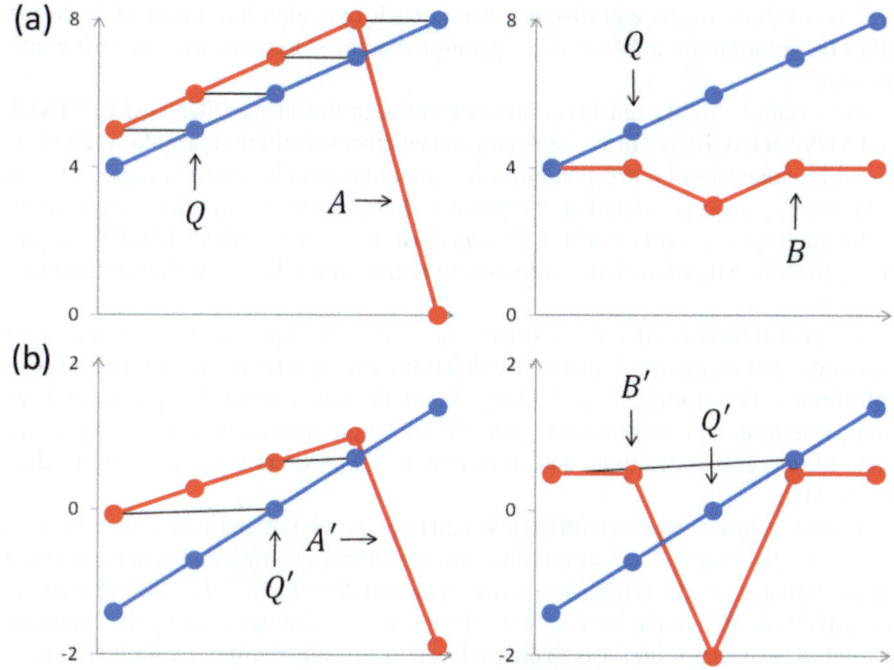

Fig. 2.29 Robustness of LCS to outliers: original time series (**a**); and their z-normalised versions (**b**)

Table 2.8 LCS distances between Q and the series A and B in Fig. 2.29 and the length of the corresponding LCSs before and after z-normalisation

	Before ($\varepsilon \approx 0.70$)		After ($\varepsilon = 0.25$)	
	d_{LCS}	$\|LCS\|$	d_{LCS}	$\|LCS\|$
A	**0.2**	**4**	**0.6**	**2**
B	0.8	1	0.8	1

2.3.4 Edit Distance with Real Penalty (ERP)

The ERP dissimilarity measure [99] was created with the aim of combining the advantage of L_p norms of being metrics (but they are sensitive to distortions along the time axis) and the DTW and LCS measures of being robust to distortions along the time axis due to non-linear alignment of the points (but they are not metrics because they do not satisfy the triangle inequality).

Similar to LCS, ERP is based on the *edit distance* of strings (sequences of characters) [91], which represents the minimum cost of edit operations (insertion, deletion and substitution) required to convert one string to another. The *Levenshtein distance*

[95] is a variant of the edit distance where each operation has a cost of 1, thus it gives the minimum number of edit operations required to transform one string into another.

For example, the Levenshtein distance between the strings FUNDAMENTALS and ADVANTAGES is 8 or 10, depending on whether substitutions are allowed or not. Possible sequences of corresponding edit operations are illustrated in Figs. 2.30 and 2.31. Here, ↓ denotes a deletion, ↑ represents an insertion, ↕ stands for a substitution and matchings are marked with |. To transform the string ADVANTAGES into the string FUNDAMENTALS, it is sufficient to replace the deletions with insertions and vice versa.

It can also be noted that the optimal sequence of edit operations does not have to be unique and that pairs of successive deletions and insertions can be replaced by a substitution. For example, Fig. 2.32 depicts an alternative optimal sequence of deletions, insertions and substitutions where the substitutions were obtained by replacing pairs of successive deletions and insertions of the optimal sequence presented in Fig. 2.31.

Formally, in accordance with [91], an edit operation is defined as an ordered pair of characters $(a, b) \neq (\varnothing, \varnothing)$ where Ø represents an empty string that is not contained in the alphabet of the strings $A = a_1a_2 \ldots a_n$ and $B = b_1b_2 \ldots b_m$. Edit operations are usually written in the form $a \rightarrow b$. Hence, $a \rightarrow \emptyset$ denotes deleting the character a from A, inserting b into A is expressed as $\emptyset \rightarrow b$, substituting a with b is written as $a \rightarrow b$ where $a, b \neq \emptyset$ and $a \neq b$ and edit operations of the form $a \rightarrow a$ are *matchings*. In the literature, substitutions are often called as *mismatchings*; insertions and deletions are denoted as *indels*, and the inserted and deleted elements are referred to as *gaps* [100].

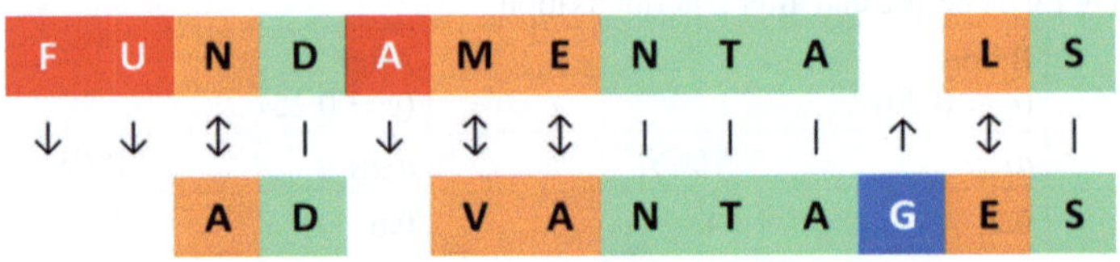

Fig. 2.30 An optimal sequence of deletion, insertion and substitution operations that transforms the string FUNDAMENTALS into the string ADVANTAGES and vice versa

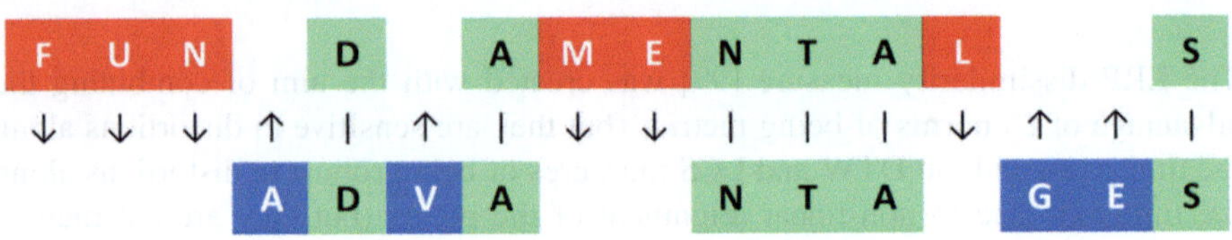

Fig. 2.31 An optimal sequence of deletion and insertion operations that transforms the string FUNDAMENTALS into the string ADVANTAGES and vice versa

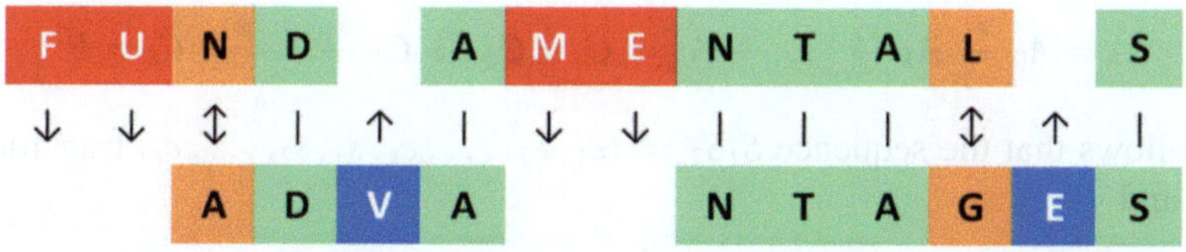

Fig. 2.32 An alternative optimal sequence of edit operations that transforms the string FUNDAMENTALS into the string ADVANTAGES and vice versa

B is a result of applying the edit operation $e = a \to b$ to A, which is denoted as $A \xrightarrow{e} B$, if there are strings X and Y such that $A = XaY$ and $B = XbY$. If $e = a \to b$ transforms A into B, then $\overline{e} = b \to a$ transforms B into A:

$$A = XaY \underset{b \to a}{\overset{a \to b}{\leftrightarrow}} XbY = B$$

B is obtained from A via the sequence of edit operations (or *edit sequence*) $S = (e_1, e_2, \ldots, e_k)$ (denoted as $A \xrightarrow{S} B$) if there is a sequence of strings $A = A_0, A_1, \ldots, A_k = B$ such that $A_{i-1} \xrightarrow{e_i} A_i$ for $1 \le i \le k$. If $S = (e_1, e_2, \ldots, e_k)$ transforms A into B, then the inverse edit sequence $\overline{S} = (\overline{e_k}, \ldots, \overline{e_2}, \overline{e_1})$ transforms B into A through the sequence of strings $B = A_k, \ldots, A_1, A_0 = A$ where $A_i \xrightarrow{\overline{e_i}} A_{i-1}$ for $k \le i \le 1$:

$$A = A_0 \underset{\overline{e_1}}{\overset{e_1}{\leftrightarrow}} A_1 \underset{\overline{e_2}}{\overset{e_2}{\leftrightarrow}} A_2 \underset{\overline{e_3}}{\overset{e_3}{\leftrightarrow}} \ldots \underset{\overline{e_k}}{\overset{e_k}{\leftrightarrow}} A_k = B$$

For example, transforming the string FUNDAMENTALS into the string ADVANTAGES as depicted in Fig. 2.30 involves applying the following sequence of edit operations: F $\to$ Ø, U $\to$ Ø, N $\to$ A, D $\to$ D, A $\to$ Ø, M $\to$ V, E $\to$ A, N $\to$ N, T $\to$ T, A $\to$ A, Ø $\to$ G, L $\to$ E and S $\to$ S. The inverse sequence transforms the string ADVANTAGES into the string FUNDAMENTALS: S $\to$ S, E $\to$ L, G $\to$ Ø, A $\to$ A, T $\to$ T, N $\to$ N, A $\to$ E, V $\to$ M, Ø $\to$ A, D $\to$ D, A $\to$ N, Ø $\to$ U and Ø $\to$ F.

Let *cost* be an arbitrary non-negative function which determines the cost (weight, score) of each edit operation. The cost of an edit sequence $S = (e_1, e_2, \ldots, e_k)$ is defined as the sum of the costs of the involved operations:

$$cost(S) = \sum_{i=1}^{k} cost(e_i) \ge 0$$

Let $S_1 = (e_1, e_2, \ldots, e_k)$ and $S_2 = (s_1, s_2, \ldots, e_t)$ be edit sequences such that $A \xrightarrow{S_1} C$ and $C \xrightarrow{S_2} B$. Then there is a sequence of strings $A = A_0, A_1, \ldots, A_k = C = C_0, C_1, \ldots, C_t = B$ for which

$$A = A_0 \xrightarrow{e_1} A_1 \xrightarrow{e_2} \cdots \xrightarrow{e_k} A_k = C = C_0 \xrightarrow{s_1} C_1 \xrightarrow{s_2} \cdots \xrightarrow{s_t} C_t = B$$

It follows that the sequence $S_1S_2 = (e_1, e_2, \ldots, e_k, s_1, s_2, \ldots, e_t)$ transforms A into B and

$$cost(S_1S_2) = \sum_{i=1}^{k} cost(e_i) + \sum_{i=1}^{t} cost(s_i) = cost(S_1) + cost(S_2)$$

The edit distance between two strings $A = a_1a_2 \ldots a_n$ and $B = b_1b_2 \ldots b_m$ defined as the minimum cost of all the edit sequences S that transform A into B:

$$d_{ED}(A, B) = \begin{cases} 0 & A = B = \emptyset \\ \min\limits_{S} cost(S) \geq 0 & A \neq \emptyset \text{ or } B \neq \emptyset \end{cases}$$

has several important properties (assuming that the cost function is non-negative):

1. **Triangle inequality.** The edit distance satisfies the triangle inequality regardless the underlying cost function.

   Let A, B and C be arbitrary strings, and $d_{ED}(A, B) = cost(S)$, $d_{ED}(A, C) = cost(S_1)$ and $d_{ED}(A, B) = cost(S_2)$. Then S_1S_2 also transforms A into B. If we assume that d_{ED} does not satisfy the triangle inequality for A, B and C, it follows that

   $$\begin{aligned} cost(S) = d_{ED}(A, B) > d_{ED}(A, C) + d_{ED}(C, B) = cost(S_1) + cost(S_2) \\ = cost(S_1S_2) \end{aligned}$$

   which means that the edit sequence S_1S_2 has a lower cost than S, therefore it cannot be $d_{ED}(A, B) = cost(S)$, which is a contradiction.
2. **Symmetry.** If the cost function is symmetric, the edit distance is also symmetric.

   If the cost function is symmetric, i.e. if $cost(e = a \to b) = cost(\overline{e} = b \to a)$ for every edit operation e, then $cost(S) = cost(\overline{S})$ for every edit sequence S. It follows that the edit distance is symmetric as well. Namely, if S is a minimum-cost edit sequence that transforms A into B, then $\overline{S}$ must be a minimum-cost edit sequence that transforms B into A:

   $$d_{ED}(A, B) = cost(S) = cost(\overline{S}) = d_{ED}(B, A)$$

3. **Identity.** If $cost(a \to b) > 0$ for every $a \neq b$ and $cost(a \to a) = 0$ for every a, then $d_{ED}(A, B) = 0$ if and only if $A = B$, for every two strings A and B.

   Let $cost(a \to b) > 0$ for every $a \neq b$ and $cost(a \to a) = 0$ for every a, and let A and B be two arbitrary strings.

   If $0 = d_{ED}(A, B) = cost(S)$ then $A = B = \emptyset$ or $cost(e) = 0$ for every edit operation e of S because the $cost$ function is non-negative. Furthermore, based on the assumption made in this point, it follows that every operation of S must be a matching (i.e. they are of the form $a \to a$), because the cost of non-matching operations is greater than 0 (zero). This implies that $A = B$ (otherwise S should contain at least one non-matching operation as well).

Let $A = B$. If $A = B = \varnothing$, then $d_{ED}(A, B) = 0$ by definition. If $A = B \neq \varnothing$, then $A = B = a_1 a_2 \ldots a_n$ for some $n > 0$. The edit sequence $S = (a_1 \to a_1, a_2 \to a_2, \ldots, a_n \to a_n)$ containing only matchings transforms A into B and its cost is minimal: $0 = cost(S) = d_{ED}(A, B)$.

4. **Metric**. Based on the properties 1–3 it follows that if the cost function is symmetric, matchings have zero cost, and non-matching operations have positive costs, the edit distance is a metric (which implies that if the cost function is a metric, then the edit distance is also a metric).

The cost function of the Levenshtein distance assigns 1 to each edit operation that modifies the source string:

$$cost_{LEV3}(a \to b) = \begin{cases} 0 & a = b \\ 1 & a \neq b \end{cases}$$

If only deletions and insertions are allowed, the following cost function is used instead:

$$cost_{LEV2}(a \to b) = \begin{cases} 0 & a = b \\ 2 & a \neq b \text{ and } a, b \neq \varnothing \\ 1 & a = \varnothing \text{ or } b = \varnothing \end{cases}$$

Since both of these cost functions fulfills the requirements of properties 2 and 3, the Levenshtein distances d_{LEV3} and d_{LEV2} are metrics.

It is easy to see that, as d_{LEV2} gives the minimum number of characters to delete from strings A and B in order to reduce them to their longest common subsequence (see Fig. 2.31), it follows that:

$$|LCS(A, B)| = \frac{|A| + |B| - d_{LEV2}(A, B)}{2}$$

The edit distance between two strings $A = a_1 a_2 \ldots a_n$ and $B = b_1 b_2 \ldots b_m$ can be calculated as $d_{ED}(A, B) = E[n, m]$ based on the following recursive formula [91]:

$$E[i, j] = \begin{cases} 0 & i = j = 0 \\ E[i, j-1] + cost(\varnothing \to b_j) & i = 0 \text{ and } j > 0 \\ E[i-1, j] + cost(a_i \to \varnothing) & i > 0 \text{ and } j = 0 \\ min \begin{cases} E[i-1, j-1] + cost(a_i \to b_j) \\ E[i, j-1] + cost(\varnothing \to b_j) \\ E[i-1, j] + cost(a_i \to \varnothing) \end{cases} & i, j > 0 \end{cases}$$

where $0 \leq i \leq n$ and $0 \leq j \leq m$.

The edit operations shown in Figs. 2.30 and 2.31 were obtained by tracing the optimal editing paths starting from the cell (n, m) of the editing matrices presented in Fig. 2.33. Along the editing paths, the red cells denote deletions from A, the blue ones are insertions into A, the cells painted in orange are mismatchings, and the green ones represent matchings.

ERP applies the idea of edit distance in the domain of time series by defining the cost of edit operations as the distance between the points of the time series [99]:

$$cost_{ERP}(a \to b) = \begin{cases} |a - b| & a, b \neq \varnothing \\ |b - g| & a = \varnothing \\ |a - g| & b = \varnothing \end{cases}$$

where g is a constant value. In this way, the cost of replacing point a with point b is equal to their absolute difference, and the cost of deleting or inserting a point is determined by its distance from the reference point g. Although other L_p norms can also be used to calculate the distance between points and g can have any value, in this book we use only L_1 and $g = 0$ as in [99].

By replacing the general *cost* function with $cost_{ERP}$ in the formula of E, the ERP distance between the time series A and B is given by $d_{ERP}(A, B) = ERP[n, m]$ [99]:

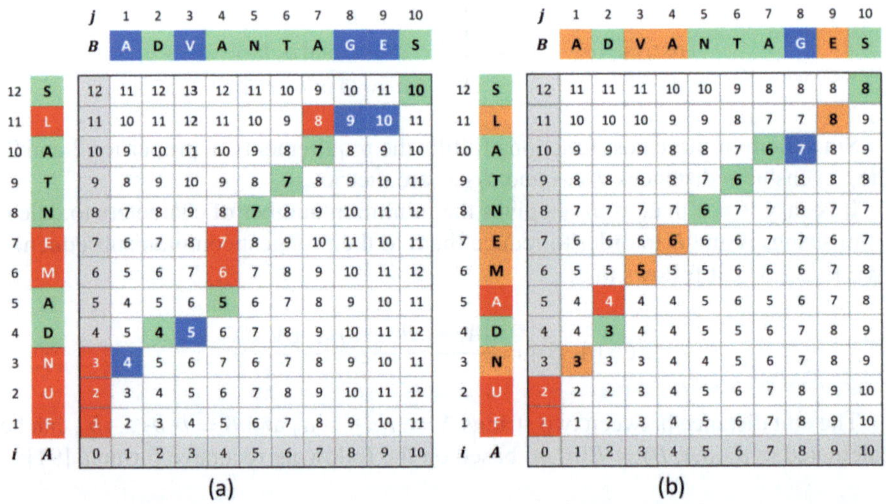

Fig. 2.33 Transforming the string FUNDAMENTALS into the string ADVANTAGES using deletions and insertions (**a**) and deletions, insertions and substitutions (**b**)

$$ERP[i,j] = \begin{cases} 0 & i = j = 0 \\ ERP[i,j-1] + |b_j - g| & i = 0 \text{ and } j > 0 \\ ERP[i-1,j] + |a_i - g| & i > 0 \text{ and } j = 0 \\ min\begin{cases} ERP[i-1,j-1] + |a_i - b_j| \\ ERP[i,j-1] + |b_j - g| \\ ERP[i-1,j] + |a_i - g| \end{cases} & i,j > 0 \end{cases}$$

where $n = |A|$, $m = |B|$ and $0 \leq i \leq n$, $0 \leq j \leq m$.

Since $cost_{ERP}$ satisfies the requirements of properties 2 and 3, d_{ERP} represents a metric. Therefore, unlike DTW and LCS (and EDR), ERP satisfies the triangle inequality.

An example of time series point matching applying the ERP dissimilarity measure is given in Fig. 2.34. To transform the series A into the series B, the points connected by a solid line should be replaced (Fig. 2.34a). The points of A shown in the form of circles ($i = 4, 5, 6, 7$) should be deleted (or, equivalently, they should be inserted into B *after* the point $B[3]$ with which they are connected with dashed lines). And finally, the points of the series B indicated by circles ($j = 9, 10, 11, 12$) should be inserted into A *after* the point $A[12]$ with which they are connected with dashed lines (or, equivalently, they should be deleted from B).

The points to be deleted or inserted are matched with the point 0 (zero) which acts as the reference point g, as illustrated in Fig. 2.34b. The matchings were obtained by tracing back the editing path from cell (12,12) to cell (0,0) of the editing matrix presented in Fig. 2.35. The distance between A and B is 75, which can be read from $E[12, 12]$ and is equal to the sum of the absolute values of the differences between the matched points (taking care to match the gaps with g):

$$\begin{aligned} d_{ERP}(A, B) = {} & |20 - 12| + |14 - 11| + |9 - 13| + |6 - g| + |4 - g| \\ & + |3 - g| + |5 - g| + |13 - 17| + |15 - 18| + |16 - 17| \\ & + |14 - 14| + |11 - 11| + |9 - g| + |8 - g| + |8 - g| + |9 - g| \end{aligned}$$

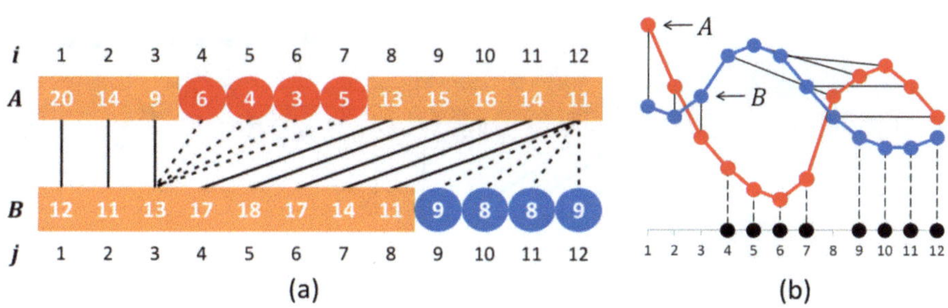

Fig. 2.34 An optimal alignment of the points of time series by the ERP distance measure with $g = 0$

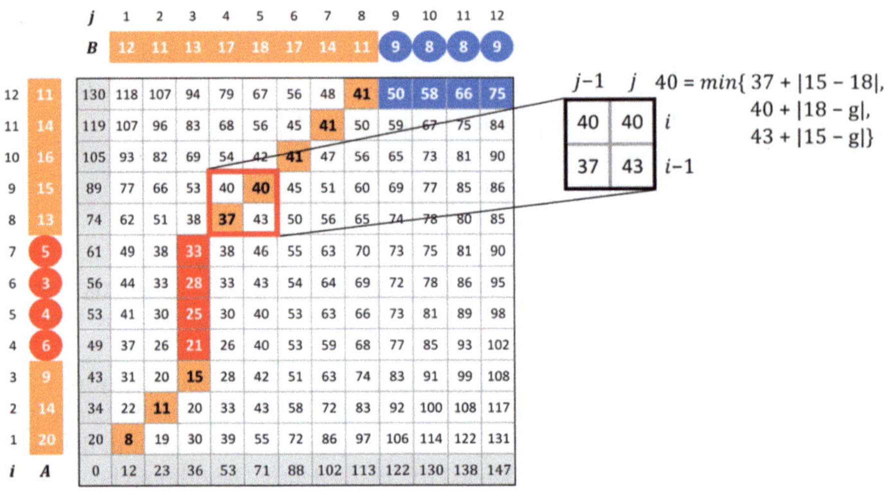

Fig. 2.35 Computing ERP using dynamic programming with $g = 0$

Figure 2.36 shows the optimal matching of the time series introduced in Fig. 2.34, with the value of the reference point g set to 21. Note that $A[1]$ is not connected to any point of B: to transform A into B, it needs to be deleted from A or inserted *before* $B[1]$ (Fig. 2.37). In this case, the distance is 69.

These two examples show the dependence of the optimal alignment of the points and the distance from the reference point. Furthermore, it can also be seen that ERP supports only *one-to-one* mappings of the points which can be shifted in time. The shifts are the result of matching the gaps (i.e. the points to be deleted or inserted) with the reference point g (recall that in the case of the LCS measure they remain unpaired).

Without gaps, a linear alignment of points is obtained as it can be seen in Fig. 2.38 on the example of the time series introduced in Fig. 2.10. In this case, since the reference point is too far, the editing path includes only the elements of the diagonal

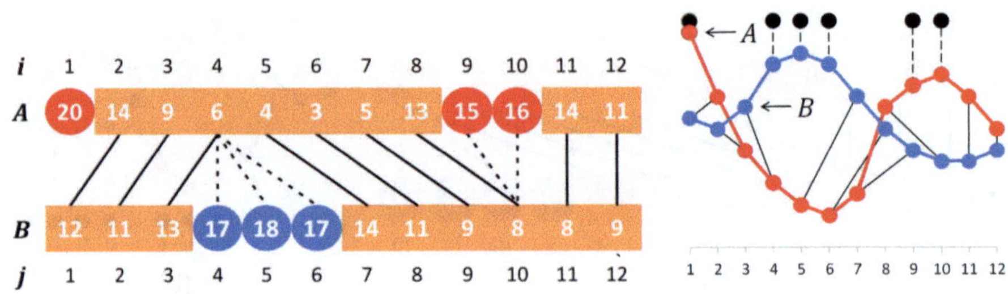

Fig. 2.36 An optimal alignment of the points of time series by the ERP distance measure with $g = 21$

i	*A*													
	j		1	2	3	4	5	6	7	8	9	10	11	12
	B		12	11	13	17	18	17	14	11	9	8	8	9
12	11	122	113	103	95	91	88	84	77	67	67	70	71	69
11	14	112	103	93	85	81	78	74	67	65	67	68	67	67
10	16	105	96	86	78	74	71	67	62	62	64	61	62	62
9	15	100	91	81	73	69	66	62	57	57	59	56	57	57
8	13	94	85	75	67	63	60	56	53	53	53	50	51	53
7	5	86	77	67	59	55	52	52	51	49	45	46	49	55
6	3	70	61	51	43	39	40	42	43	41	43	46	51	59
5	4	52	43	33	25	25	28	32	33	37	41	46	53	65
4	6	35	26	16	12	16	19	23	30	36	42	49	62	74
3	9	20	11	5	13	17	20	24	31	39	47	60	73	85
2	14	8	3	11	19	23	26	30	37	47	59	72	85	97
1	20	1	8	18	26	30	33	37	44	54	66	79	92	104
i	*A*	0	9	19	27	31	34	38	45	55	67	80	93	105

Fig. 2.37 Computing ERP using dynamic programming with $g = 21$

of the editing matrix and the ERP is reduced to the L_1 norm, similarly as the DTW is reduced to the square of the Euclidean distance.

Figure 2.39 shows the effect of different distortions on the ERP ($g = 0$) dissimilarity measure by comparing the time series Q with the series A_1, A_2, A_3 and B where the first three were obtained by different transformations of Q (A_1 by shifting along the value axis, A_2 by scaling along the value axis, and A_3 by shifting along the time axis) and B is composed so that it is visibly different from Q. Moreover, A_1, A_2, A_3 and B also contain a small amount of random noise. For clarity, the charts

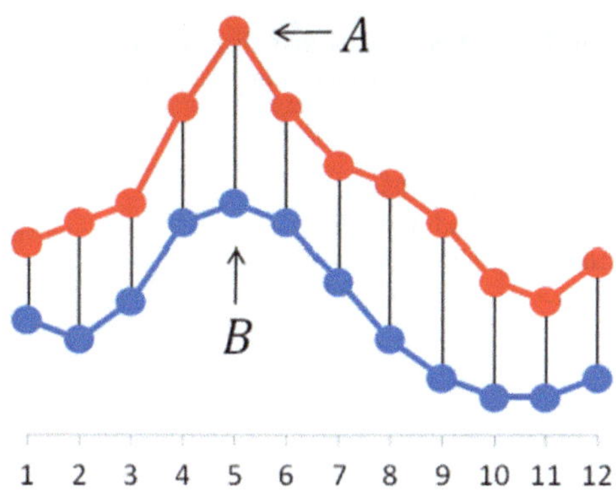

Fig. 2.38 Without gaps the ERP distance is reduced to the L_1 norm

in Fig. 2.39 do not show the matching of the gaps with the reference point g. The distances obtained using two different values of the reference point g (0 and 10) are shown in Table 2.9.

When $g = 0$, A_3 is the most similar to Q, however A_1 and A_2 are at a significantly greater distance from it than B, suggesting that due to the matching of time-shifted points, the ERP might be robust to time-axis distortions but it is sensitive to value-axis distortions. This is also supported by the results obtained for $g = 10$. Namely, when the reference point g is far enough, the ERP is reduced to L_1, which supports only linear matching of points and is therefore sensitive to distortions along the time axis (the results shown in column $g = 10$ of Table 2.9 are identical to the Manhattan distance (L_1) results in Table 2.3). Consequently, series B becomes the most similar in terms of ERP as well.

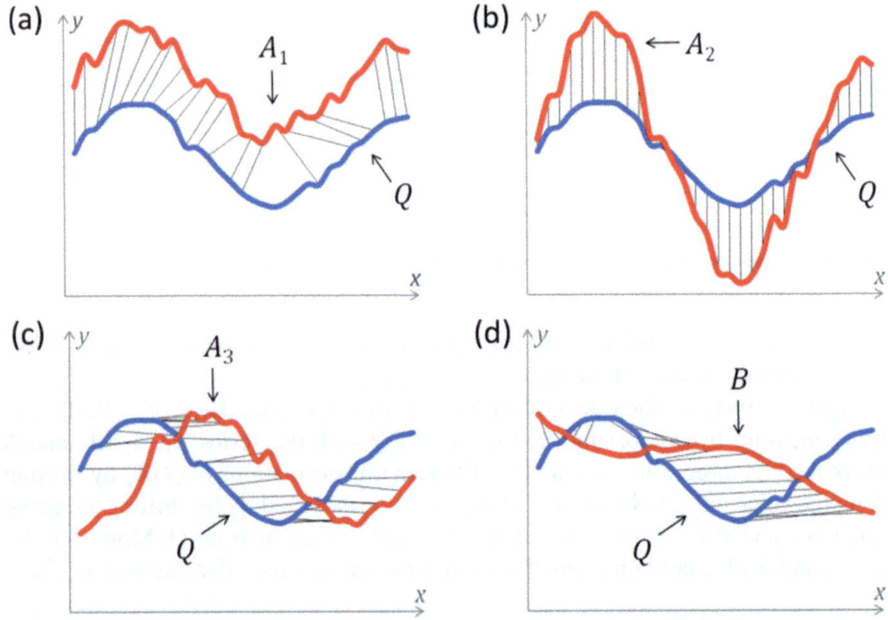

Fig. 2.39 Effects of various distortions on ERP ($g = 0$): shifting (**a**) and scaling (**b**) along the value axis and distortions along the time axis (**c**); an example of dissimilar series (**d**)

Table 2.9 ERP distances between Q and four other time series shown in Fig. 2.39

	$g = 0$	$g = 10$
A_1	40.38	40.38
A_2	24.86	24.86
A_3	**9.30**	27.67
B	13.66	**21.67**

These two examples demonstrate the sensitivity of ERP to the choice of reference point as well. It should be noted that, while in the case of A_2 ERP is reduced to L_1 even for $g = 0$, when it comes to A_1, gaps are present and the actual value of the distance is somewhat smaller than the displayed value which is rounded to two decimal places.

The normalised versions of the time series shown in Fig. 2.39 are presented in Fig. 2.40, in which the matching of the gaps with the reference point g has been omitted for greater clarity, in the same way as in Fig. 2.39. The distances between the time series are given in the Table 2.10 for two values of g (0 and 10). These results indicate that z-normalisation has a favorable effect on ERP when it comes to distortions along the value axis, but they also strengthen the view that it might be quite dependent on the value of g, especially in terms of resistance to distortions along the time axis.

LCS is robust to outliers by skipping non-matching (indel) points. As the ERP measure aligns such points with a reference point g, their (absolute) distance from g is taken into account when calculating the dissimilarity between time series. In addition, when the reference point is too far from the time series, ERP is reduced to L_1. As a result, ERP can be sensitive to noise and outliers [81, 96, 101], which can be seen in the example time series ($Q = (4, 5, 6, 7, 8)$, $A = (5, 6, 7, 8, 0)$, $B = (4, 4, 3, 4, 4)$) shown in Fig. 2.41, where $g = 0$. In Fig. 2.41a, the first point of Q and the last

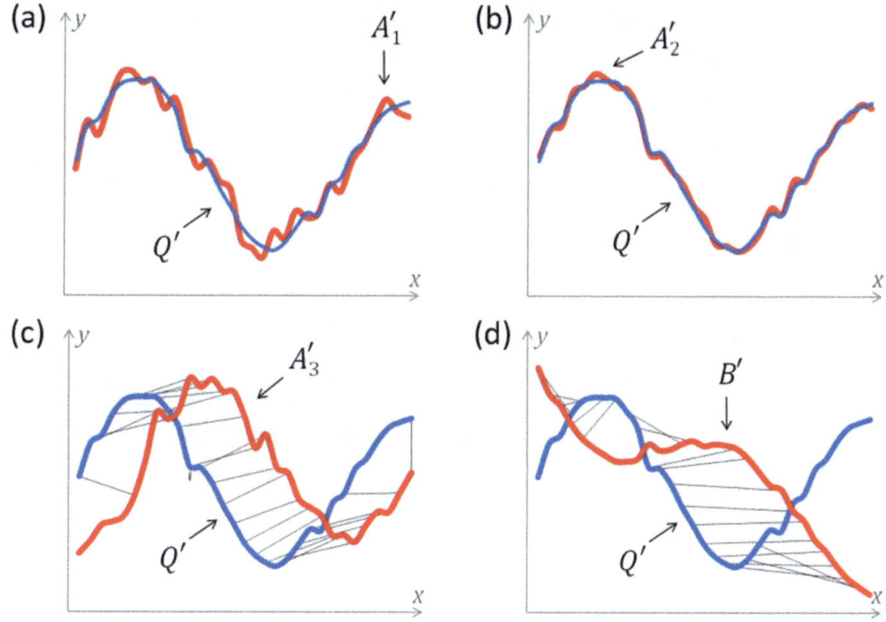

Fig. 2.40 The influence of z-normalisation on ERP ($g = 0$) in the presence of shifting (**a**) and scaling (**b**) along the value axis and distortions along the time axis (**c**); an example of dissimilar series (**d**)

Table 2.10 ERP distances between Q and four other time series after z-normalisation (Fig. 2.40)

	$g = 0$	$g = 10$
A_1	4.94	5.14
A_2	**2.06**	**2.06**
A_3	18.44	41.93
B	18.25	40.04

point of A are matched with g and the alignment between Q and B is linear (without gaps). Since $A[5] = 0 = g$, point $A[5]$ is effectively skipped, similarly as in the case of LCS in Fig. 2.29a. Consequently, $d_{ERP}(Q, A) = 4 < 11 = d_{ERP}(Q, B)$. However, if g is too far away (e.g. if $g = 10$), the pairing of Q and A also reduces to linear (without gaps) which gives $d_{ERP}(Q, A) = 12 > 11 = d_{ERP}(Q, B)$. After normalization (with $g = 0$), gaps are further away from g when comparing Q and A than when calculating the distance between Q and B, resulting in B being more similar to Q than A: $d_{ERP}(Q, A) \approx 4.64 > 3.41 \approx d_{ERP}(Q, B)$. If g is far enough (e.g. $g = 10$), ERP reduces to the Manhattan distance: $d_{ERP}(Q, A) \approx 6.56 > 6.24 \approx d_{ERP}(Q, B)$ (see Table 2.5).

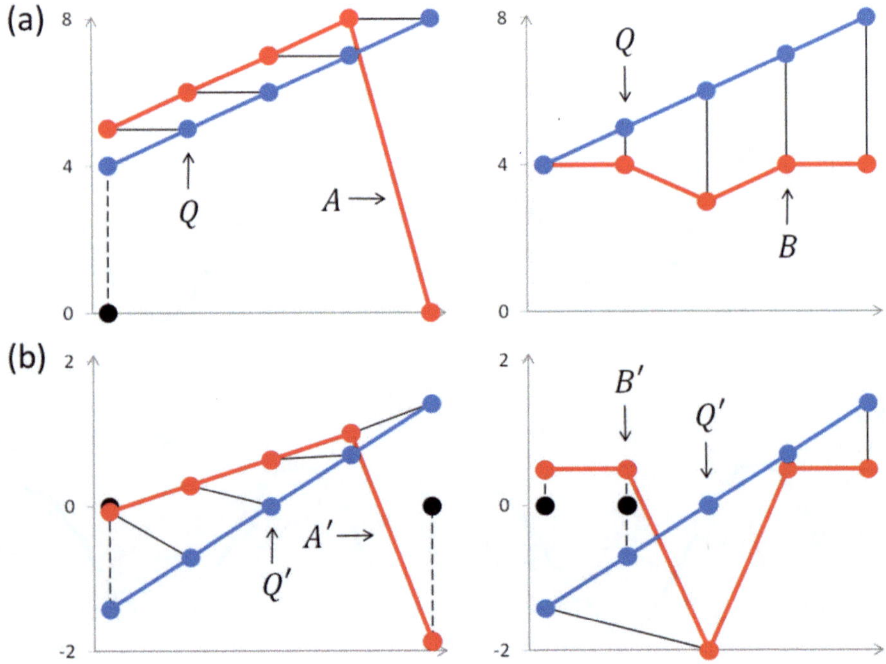

Fig. 2.41 The influence of the reference point on the sensitivity of ERP to outliers: before (**a**) and after (**b**) z-normalisation

2.3.5 Edit Distance on Real Sequence (EDR)

EDR [81] represents an adaptation of the Levenshtein distance d_{LEV3} with three edit operations (deletion, addition, substitution - see Edit Distance with Real Penalty (ERP)) to sequences of real numbers utilising the following cost function [102]:

$$cost_{EDR}(a \to b) = \begin{cases} 0 & |a - b| \leq \varepsilon \text{ and } a, b \neq \varnothing \\ 1 & \text{otherwise} \end{cases}$$

where $\varepsilon \geq 0$ is the matching threshold parameter introduced with LCS.

Let $A = (a_1, a_2, \ldots, a_n)$ and $B = (b_1, b_2, \ldots, b_m)$ be time series of length n and m. The EDR distance between A and B is equal to the minimal number of deletions, insertions and substitutions required to transform A into B (and vice versa) which, relying on dynamic programming, is calculated as $d_{EDR}(A, B) = EDR[n, m]$ using the following recursive formula:

$$EDR[i,j] = \begin{cases} 0 & i = j = 0 \\ EDR[i, j-1] + 1 & i = 0 \text{ and } j > 0 \\ EDR[i-1, j] + 1 & i > 0 \text{ and } j = 0 \\ min \begin{cases} EDR[i-1, j-1] + cost_{EDR}(a_i \to b_j) \\ EDR[i, j-1] + 1 \\ EDR[i-1, j] + 1 \end{cases} & i, j > 0 \end{cases}$$

Figure 2.42 shows the alignment of the points of the time series A and B obtained by applying the EDR distance measure with $\varepsilon = 0$. The minimal sequence of edit operations that converts one of the series into the other one consists of 12 substitutions, which gives $d_{EDR}(A, B) = 12$. However, if the edit operations are limited to only deletions and insertions (no substitutions), the EDR becomes an adaptation of the Levenshtein distance d_{LEV2} on which the LCS measure is indirectly based. This can be easily seen by comparing Fig. 2.24 with Fig. 2.43. The only difference between them is that in the case of Fig. 2.24 the emphasis is on the matching points and in the case of Fig. 2.43 the focus is on the points to be deleted or inserted (they are depicted as red and blue circles, and matchings are represented by dashed lines). However, the maximum number of matchings ($|LCS(A, B)|$) and the minimum number of insertions and deletions ($d_{EDR2}(A, B)$) required to transform A into B are interrelated values:

$$16 = d_{EDR2}(A, B) = n + m - 2|LCS(A, B)| = 12 + 12 - 2 \times 4,$$

where $cost_{EDR2}$ is an adaptation of $cost_{LEV2}$ to time series:

$$cost_{EDR2}(a \to b) = \begin{cases} 0 & |a - b| \leq \varepsilon \text{ and } a, b \neq \varnothing \\ 2 & |a - b| > \varepsilon \text{ and } a, b \neq \varnothing \\ 1 & \text{otherwise} \end{cases}$$

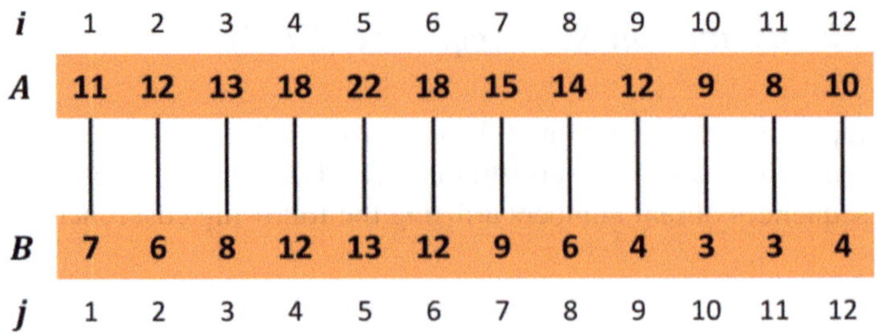

Fig. 2.42 An optimal alignment of the points of time series by the EDR distance measure with $\varepsilon = 0$

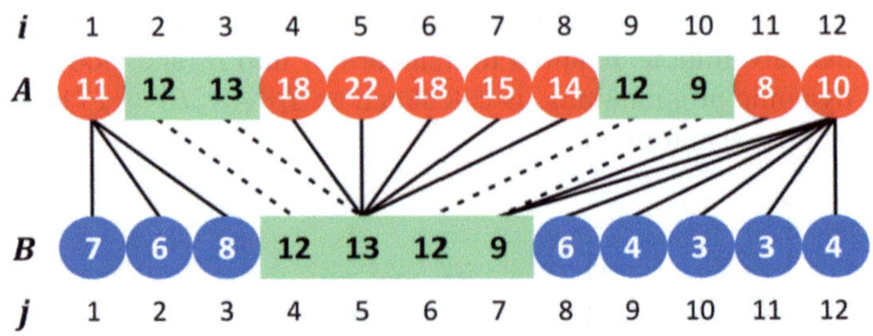

Fig. 2.43 If only deletions and insertions are allowed (without substitutions), the optimal number of edit operations can also be obtained by finding the LCS

Based on the above considerations, it is clear that LCS and EDR are related distance measures. Both techniques assess the similarity of two series by calculating the minimum number of edit operations required to convert one series into the other. In doing so, LCS allows only deletions and insertions and calculates indirectly by looking for the longest common subsequence, while EDR calculates directly and allows substitutions as well.

It should also be noted that, like in the case of the other elastic measures discussed in the book, the optimal alignment of the points need not to be unique for EDR either. For example, instead of 12 substitutions as shown in Fig. 2.42, A and B can also be optimally aligned as illustrated in Fig. 2.44, using two deletions, two insertions and eight substitutions. This example also demonstrates that EDR supports the same types of point-mappings as DTW: *one-to-one* (for example, between $A[2]$ and $B[4]$), *one-to-many* (between $A[1]$ and $B[1:3]$) and *many-to-one* (between $A[10:12]$ and $B[12]$).

As in the case of LCS, the EDR distance strongly depends on the value of the parameter ε that determines the matching threshold between time series points. For example, if we increase its value from 0 to 4, the number of edit operations required to transform the time series A into B decreases from 12 (Figs. 2.42 and 2.44) to 7 (Fig. 2.45): $B[2]$ should be inserted between $A[1]$ and $A[2]$ (or, equivalently, it should be deleted), the points from $A[4]$ to $A[6]$ should be deleted (or inserted between

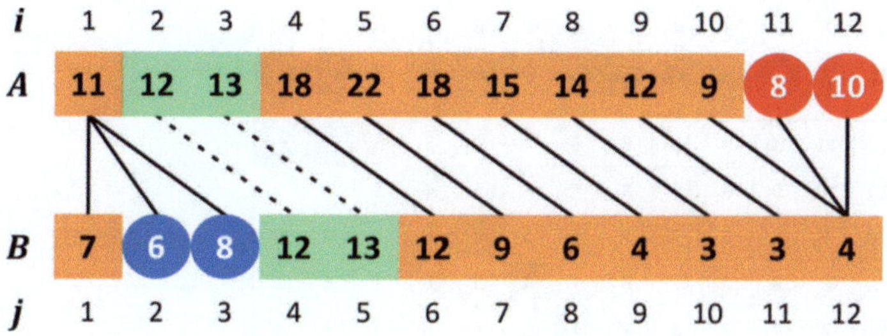

Fig. 2.44 An alternative optimal alignment of the points of time series by the EDR distance measure with $\varepsilon = 0$

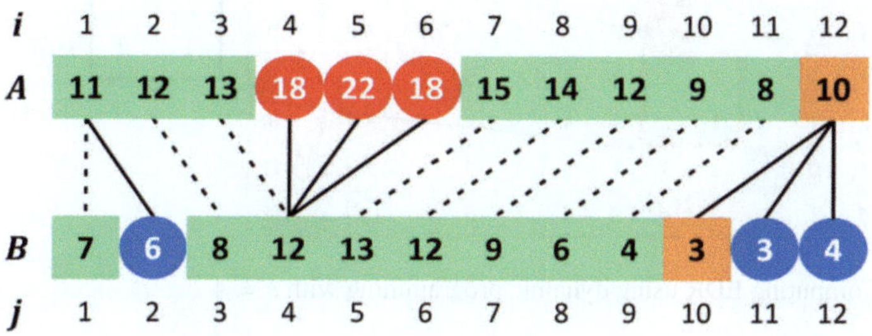

Fig. 2.45 An optimal alignment of the points of time series by the EDR distance measure with $\varepsilon = 4$

$B[4]$ and $B[5]$), $A[12]$ should be replaced with $B[10]$ (or vice versa) and $B[11]$ and $B[12]$ should be inserted after $A[12]$ (or they should be deleted). The editing matrix corresponding to $\varepsilon = 4$ is shown in Fig. 2.46. The optimal sequence of edit operations can be traced back from $EDR[n, m]$ to $EDR[0, 0]$.

Similar to the LCS measure, the best value of the matching threshold parameter ε depends on the data that the time series represents and it is usually chosen from some interval so that it produces the best results for the given task. It is easy to notice that, if for every point a of A and every point b of B it holds that $|a - b| > \varepsilon$, then $d_{EDR}(A, B) = max\{n, m\}$ (namely, $min\{n, m\}$ points should be substituted and the remaining ones should be deleted/inserted). Furthermore, if for every point a of A and every point b of B it holds that $|a - b| \leq \varepsilon$, then $d_{EDR}(A, B) = |n - m|$ (excess points should be deleted from the longer series, or they should be inserted into the shorter one).

Since the focus of this book is not on fine tuning the matching threshold, it was chosen to be equal to 25% of the maximum of data points deviations from their mean value in time series belonging to the given dataset, as recommended in [81].

In order to transform A into B, some points of B (A) may need to be inserted into A (B) (or deleted from B (A)). These points are paired with the point after which they should be added. For example, in Fig. 2.45, $A[4]$, $A[5]$ and $A[6]$ are all connected to $B[4]$ as they should be added between $B[4]$ and $B[5]$. If the optimal editing path leaves the point $(0, 0)$ of the EDR matrix through points $(1, 0)$ or $(0, 1)$ (instead of

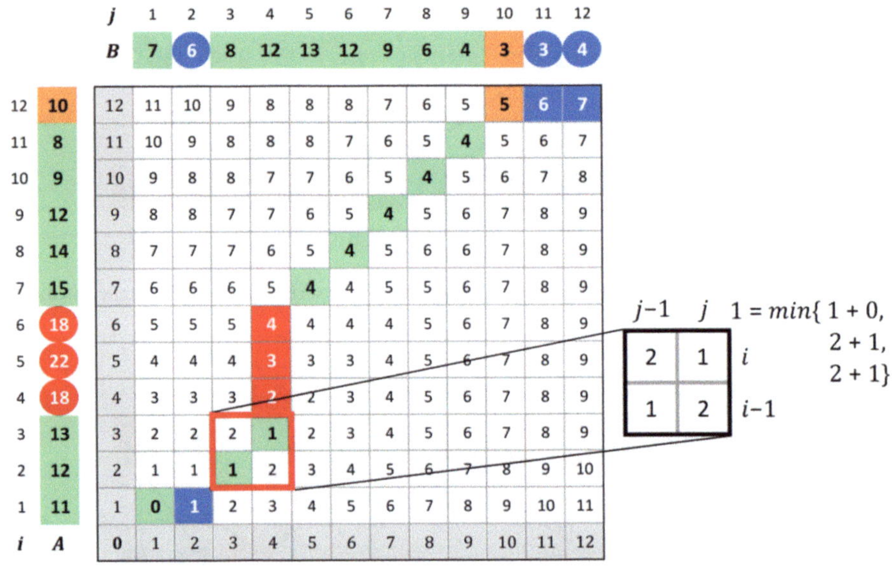

Fig. 2.46 Computing EDR using dynamic programming with $\varepsilon = 4$

(1, 1) as in Fig. 2.45), some points of A (or B) will not be paired with any of the points of B (or A). For the sake of visualization, such points are connected to the first point of the other series (*in front of* which they should be inserted) in Figs. 2.47, 2.48 and 2.49. Furthermore, since in the case of EDR we are interested in deletions, insertions and substitutions, in these figures the matching points of the optimal editing path are not connected (i.e. the lines between the points a and b such that $|a - b| \leq \varepsilon$ are omitted).

In a similar manner to the other distance measures discussed in this book, to illustrate the dependence of EDR on shifting and scaling along the value axis and its robustness to distortions along the time axis, Fig. 2.47 depicts the optimal alignments of the points between the time series Q and four series derived from it by shifting (A_1) and scaling (A_2) along the value axis, by shifting (A_3) along the time axis and with the aim of providing an example of a series (B) that deviates visibly from Q. Some random noise was also added to each of the four auxiliary series. Based on the distances shown in Table 2.11, it is easy to see that without normalization, A_3 is the most similar to Q, but the intrinsic similarity of A_1 and A_2 is not obvious when compared to B. After z-normalisation, the points become close enough for EDR to detect the clear similarity of A_1 and A_2 to Q and the apparent dissimilarity of B. The alignments of the points of the normalized series are presented in Fig. 2.48. As recommended in [81], the matching threshold value is set to one-fourth of the highest standard deviation of the time series.

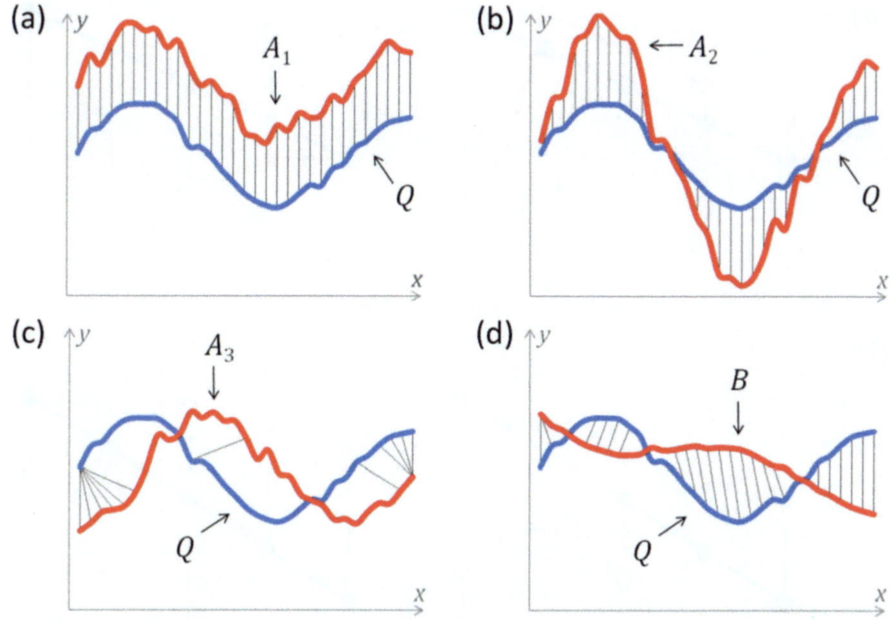

Fig. 2.47 Effects of various distortions on EDR: shifting (**a**) and scaling (**b**) along the value axis and distortions along the time axis (**c**); an example of dissimilar series (**d**)

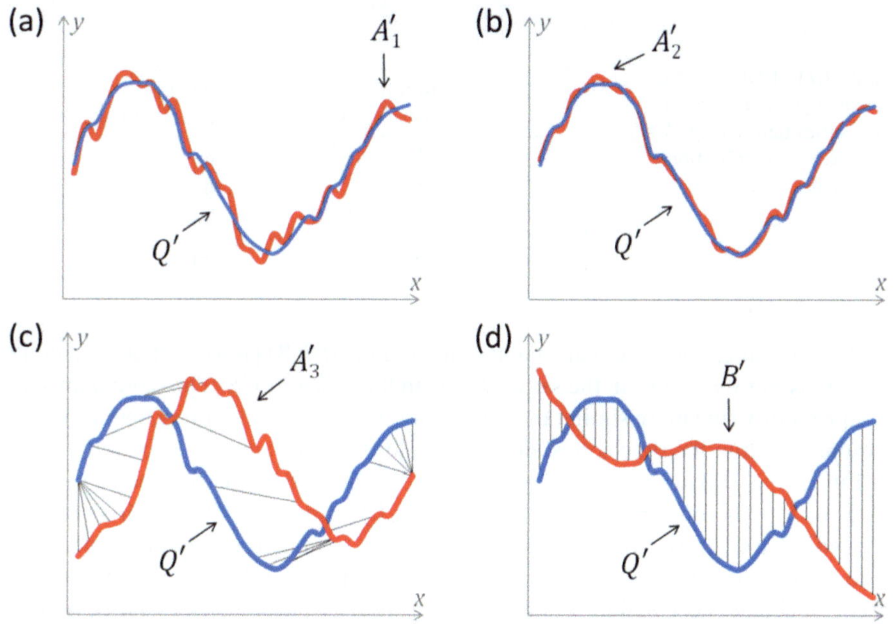

Fig. 2.48 The influence of z-normalisation on EDR in the presence of shifting (**a**) and scaling (**b**) along the value axis and distortions along the time axis (**c**); an example of dissimilar series (**d**)

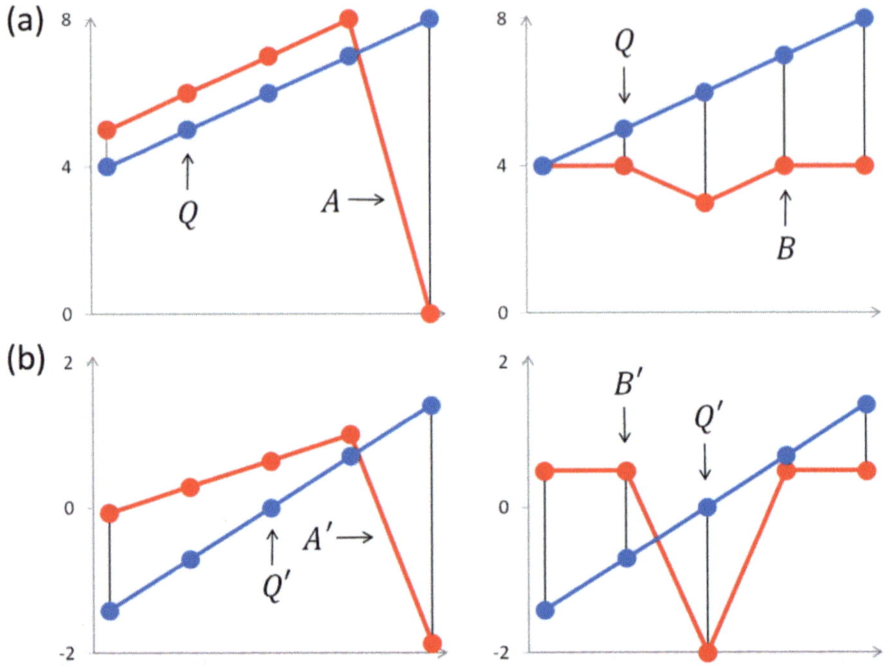

Fig. 2.49 Sensitivity of EDR to outliers: original time series (**a**); and their z-normalised versions (**b**)

Table 2.11 EDR distances between Q and four other time series before (Fig. 2.47) and after z-normalisation (Fig. 2.48)

	Before ($\varepsilon \approx 0.38$)	After ($\varepsilon = 0.25$)
A_1	31	8
A_2	22	**0**
A_3	**12**	22
B	22	29

The robustness of EDR against noise and outliers [60, 81] is a result of ε-matching the points of time series, in the same way as in the case of LCS. The more points are in ε-proximity (the more similar the time series are), the fewer editing operations are required to align the time series and the smaller their distance is, as it is illustrated in Fig. 2.49a. Taking $\varepsilon \approx 0.70$ as recommended in [81], the first point of $Q = (4, 5, 6, 7, 8)$ is too far from $A = (5, 6, 7, 8, 0)$ to be matched and it should be deleted from Q. In addition, the last point of A is too far from any of the points of Q to be matched and therefore it should be inserted into Q. Thus, the EDR distance between Q and A is equal to 2. When comparing $B = (4, 4, 3, 4, 4)$ with Q, only its first point is close enough to be matched (with the first point of Q), the rest of its points need to be substituted with the corresponding points of Q (or vice versa) to make the two series match. It follows that $4 = d_{EDR}(Q, B) > d_{EDR}(Q, A) = 2$, which is consistent with the visual perception that A is more similar to Q than B is. The same result is obtained after applying z-normalisation (Fig. 2.49b), with $\varepsilon = 0.25$.

It should be emphasized that, analogously to LCS, an inappropriate choice of the matching threshold can reduce the effectiveness of EDR in validly identifying the similarity between time series. For example, if we weaken the matching criterion and increase the value of ε to 4, each point of B will be matched with a corresponding point of Q, but the last point of A will remain unmatched. As a result, A will appear more dissimilar to Q than B: $0 = d_{EDR}(Q, B) < d_{EDR}(Q, A) = 1$.

Since EDR and LCS are conceptually related measures, EDR shows the same disadvantages compared to L_p norms:

- The computational complexity of EDR is quadratic ($O(nm)$, where n and m denotes the length of the time series being compared).
- EDR is not a metric as it does not satisfies the triangle inequality: taking $\varepsilon = 1$, for $A = (0, 0, 0, 0)$, $B = (2, 2, 2, 3)$ and $C = (2, 1, 1, 1)$ we get:

$$4 = d_{EDR}(A, B) > d_{EDR}(A, C) + d_{EDR}(C, B) = 1 + 1 = 2.$$

2.3.6 A Comparative Overview of the Properties of Distance Measures

For an easier comparison of the distance measures discussed in this book, this subsection is devoted to a concise presentation of their definitions and main properties (Table 2.12). In the formulas below, A and B denote time series of length n and m, i.e. $A = (a_1, a_2, \ldots, a_n)$ and $B = (b_1, b_2, \ldots, b_m)$ (in the case of L_p norms, it must be $n = m$).

L_p norms

$$d_p(A, B) = \|A - B\|_p = \sqrt[p]{\sum_{i=1}^{n} |a_i - b_i|^p}$$

where $p \geq 1$ is an arbitrary real number

Table 2.12 An overview of the main properties of the distance measures discussed in this book

	L_p	DTW	LCS	ERP	EDR
Time series must have equal lengths	✓	X	X	X	X
Alignment of points	Linear	Non-linear[a]	Non-linear[a]	Non-linear[a]	Non-linear[a]
Mapping of points	One-to-one	One-to-one, one-to-many, many-to-one	One-to-one, one-to-none	One-to-one	One-to-one, one-to-many, many-to-one
Complexity	Linear	Quadratic	Quadratic	Quadratic	Quadratic
Satisfies the triangle inequality	✓	X	X	✓	X
Robust to shifting and scaling along the value (y) axis	X[b]	X[b]	X[b]	X[b]	X[b]
Robust to shifting and scaling along the time (x) axis	X	✓	✓	✓[c]	✓
Robust to noise and outliers	X	X	✓[d]	X[c]	✓[d]

[a] Linear alignment of points is also possible: when the optimal warping (editing) path corresponds to the diagonal of the warping (editing) matrix
[b] Can be addressed by normalization
[c] Depending on the value of the reference point g
[d] Depending on the value of the matching threshold ε

DTW—Dynamic Time Warping

$$d_{DTW}(A, B) = DTW[n, m]$$

$$DTW[i,j] = \begin{cases} 0 & i = j = 0 \\ \infty & (i = 0 \text{ and } j > 0) \text{ or } (i > 0 \text{ and } j = 0) \\ d(a_i, b_j) + min\begin{cases} DTW[i, j-1] \\ DTW[i-1, j-1] \\ DTW[i-1, j] \end{cases} & i, j > 0 \end{cases}$$

LCS—Longest Common Subsequence

$$d_{LCS}(A, B) = \frac{n + m - 2LCS[n, m]}{n + m}$$

$$LCS[i, j] = \begin{cases} 0 & i = 0 \text{ or } j = 0 \\ LCS[i-1, j-1] + 1 & i, j > 0 \text{ and } |a_i - b_j| \leq \varepsilon \\ max\{LCS[i, j-1], LCS[i-1, j]\} & i, j > 0 \text{ and } |a_i - b_j| > \varepsilon \end{cases}$$

where $\varepsilon \geq 0$ is the matching threshold

ERP—Edit distance with Real Penalty

$$d_{ERP}(A, B) = ERP[n, m]$$

$$ERP[i, j] = \begin{cases} 0 & i = j = 0 \\ ERP[i, j-1] + |b_j - g| & i = 0 \text{ and } j > 0 \\ ERP[i-1, j] + |a_i - g| & i > 0 \text{ and } j = 0 \\ min\begin{cases} ERP[i-1, j-1] + |a_i - b_j| \\ ERP[i, j-1] + |b_j - g| \\ ERP[i-1, j] + |a_i - g| \end{cases} & i, j > 0 \end{cases}$$

where g is a constant value

EDR—Edit Distance on Real sequence

$$d_{EDR}(A, B) = EDR[n, m]$$

$$EDR[i, j] = \begin{cases} 0 & i = j = 0 \\ EDR[i, j-1] + 1 & i = 0 \text{ and } j > 0 \\ EDR[i-1, j] + 1 & i > 0 \text{ and } j = 0 \\ min\begin{cases} EDR[i-1, j-1] + cost_{EDR}(a_i \to b_j) \\ EDR[i, j-1] + 1 \\ EDR[i-1, j] + 1 \end{cases} & i, j > 0 \end{cases}$$

where

$$cost_{EDR}(a \to b) = \begin{cases} 0 & |a - b| \leq \varepsilon \text{ and } a, b \neq \varnothing \\ 1 & \text{otherwise} \end{cases}$$

and $\varepsilon \geq 0$ is the matching threshold

2.4 Global Constraints

The main advantage of the presented elastic measures compared to L_p norms is their robustness to shifting and scaling along the time (x) axis due to nonlinear matching of the time series points. However, this is achieved by comparing each point of one of the series with each point of the other one, which slows down the computation significantly. For example, since the length of the time series in the *FiftyWords* (40) dataset is 270, to calculate the distance between its two arbitrary representatives, L_p norms use 270 comparisons, but the elastic measures require $270 \times 270 = 72{,}900$.

The degree to which distance computation slows down when using elastic measures can be clearly seen in Fig. 2.50, which depicts the seconds (rounded to one decimal place) required to generate a (symmetric) *distance matrix* for the *FiftyWords* (40) dataset using different measures, on a computer with an AMD Ryzen 7 5700G processor and 32 GB of memory (relying on a single-threaded implementation built using our *Framework for Analysis and Prediction* [103–105]).

Let $S = \{S_1, S_2 \ldots, S_t\}$ be a dataset containing t time series. The *distance matrix* $DM_{t,t}$ of S is defined by

$$DM\left[i, j\right] = d\left(S_i, S_j\right)$$

where $d\left(S_i, S_j\right)$ represents the distance between S_i and S_j for $1 \leq i, j \leq t$.

Since every distance measure discussed in this book is symmetric, it is sufficient to obtain only the distances between S_i and S_j for $j \leq i$ and $1 \leq i \leq t$ when generating distance matrices. As the *FiftyWords* (40) dataset contains a total of 905 elements, it follows that the distance must be calculated between $905 \times 906/2 = 409965$ pairs of time series. Figure 2.50 indicates that the elastic measures take approximately 200 (LCS) to 650 (ERP) times more time to complete these calculations than L_2, clearly highlighting the need to improve their speed. When interpreting these results, however, it should also be taken into account that the formulas of the elastic measures are more complex than the formulas of the L_p norms, which also contributes to the speed difference.

Let $A = (a_1, a_2, \ldots, a_n)$ and $B = (b_1, b_2, \ldots, b_n)$ be two time series of the same length n. A direct approach to improving the speed of calculating the distance between A and B when using elastic measures is to limit their elasticity, i.e. the number of

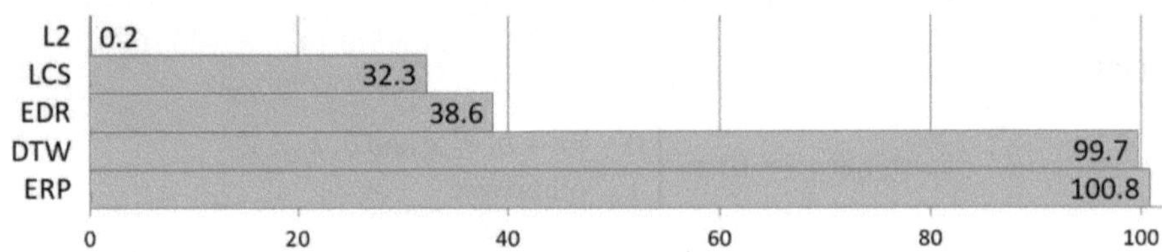

Fig. 2.50 Time (in seconds, rounded to one decimal place) required to generate a (symmetric) distance matrix for the *FiftyWords* (40) dataset using different distance measures

points of B with which a given point of A is compared. While the *unconstrained* elastic measures compare every point of A with every point of B, their *constrained* versions only consider a specific subset of points of B: $A[i]$ is compared with the points in a bounded (and not necessarily symmetric) neighborhood of $B[i]$. This is enforced by restricting the warping paths to a (symmetric) region (called the *warping window* [82]) around the diagonal of the warping matrix as illustrated in Fig. 2.51, which presents the two most commonly used *global constraints* [79, 106–110]: the *Sakoe-Chiba band* [111] and the *Itakura parallelogram* [112].

The time difference between the compared points in the case of the Sakoe-Chiba band cannot be greater than $0 \le w < n$ (Fig. 2.51a). Since the warping window is bounded by the lines $j = i - w$ and $j = i + w$, only those points $A[i]$ and $B[j]$ that satisfy the condition $|i - j| \le w$ are compared. Accordingly, the formula for calculating the elements of the warping matrix should be adjusted. For example, when it comes to DTW, the definition consistent with the Sakoe-Chiba band of width w reads as follows:

$$DTW[i,j] = \begin{cases} 0 & i = j = 0 \\ \infty & (i = 0 \text{ and } j > 0) \text{ or } (i > 0 \text{ and } j = 0) \\ \infty & i,j > 0 \; and \; |i - j| > w \\ d(a_i, b_j) & \\ \quad + min \begin{cases} DTW[i,j-1] \\ DTW[i-1,j-1] \\ DTW[i-1,j] \end{cases} & i,j > 0 \; and \; |i - j| \le w \end{cases}$$

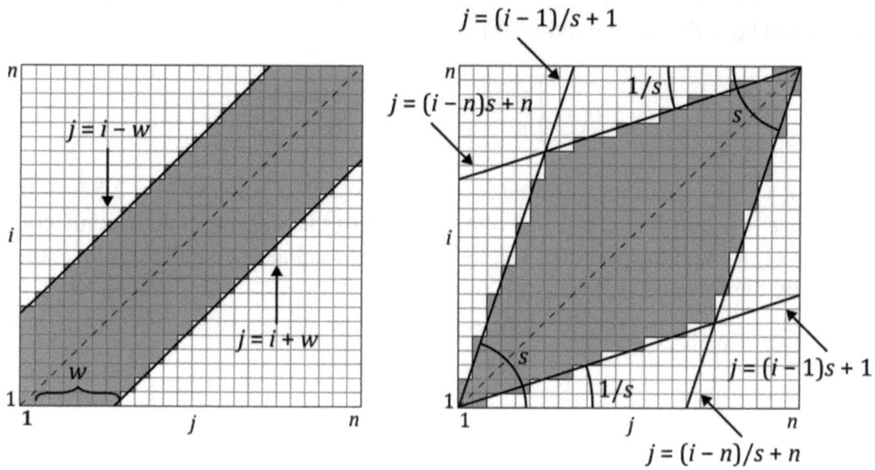

Fig. 2.51 The most commonly used global constraints: the Sakoe-Chiba band (**a**) and the Itakura parallelogram (**b**)

Unlike the Sakoe-Chiba band whose width is constant, the Itakura parallelogram uses a variable warping-window width that depends on the position of the points in the time series (Fig. 2.51b). Points at the beginning and at the end of the series have less elasticity than those in the middle. The rate of widening and narrowing of the allowed area is determined by the slopes $s \geq 1$ and $1/s$. Thus, point $A[i]$ is compared only with points $B[j]$ that satisfy the condition $min(i) \leq j \leq max(i)$ where

$$min(i) = max\left(\frac{i-1}{s} + 1,\ (i-n)s + n\right)$$

$$max(i) = min\left((i-1)s + 1,\ \frac{i-n}{s} + n\right)$$

Therefore, the recursive formula for calculating DTW can be written as:

$$DTW[i,j] = \begin{cases} 0 & i = j = 0 \\ \infty & i + j > 0 \text{ and } (j < min(i) \text{ or } j > max(i)) \\ d(a_i, b_j) + min\begin{cases} DTW[i,j-1] \\ DTW[i-1,j-1] \\ DTW[i-1,j] \end{cases} & i,j > 0 \text{ and } min(i) \leq j \leq max(i) \end{cases}$$

In this book, in order to make the performance of the Sakoe-Chiba and Itakura constraints easier to compare, we will define the slope s of the Itakura parallelogram so that the warping windows generated by the Itakura parallelogram and the Sakoe-Chiba band have the same width d (Fig. 2.52):

$$d = \begin{cases} w & w + n \text{ is even} \\ w + 1 & w + n \text{ is odd} \end{cases}$$

Let (x, y) represent the coordinates of the intersection point X (Fig. 2.52) of the lines $j = i - w$ and $j = n + 1 - i$ (the diagonal of the warping matrix connecting the point $(1, n)$ with the point $(n, 1)$), the slope s of the Itakura parallelogram is then defined by the formula

$$s = \frac{y-1}{x-1}$$

where

$$x = \frac{n+1-w}{2}$$

$$y = n + 1 - x$$

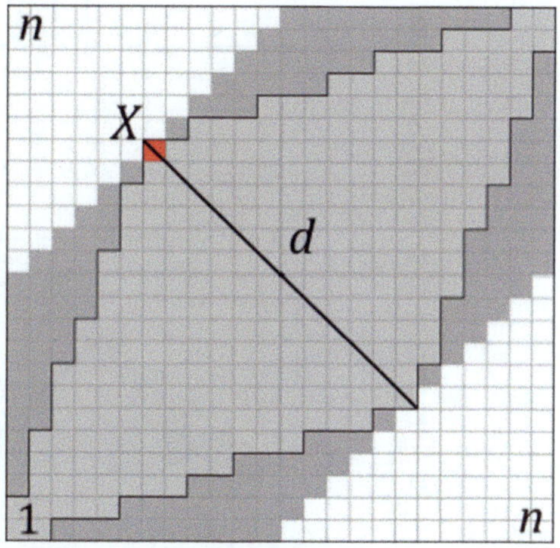

Fig. 2.52 Delimiting the warping window by the Sakoe-Chiba band and the Itakura parallelogram of the same width

For the sake of generalization, instead of the absolute width w, the size of the Sakoe-Chiba band is usually defined relatively, as a percentage of the length of the time series. The relative width is commonly denoted by r, i.e. $w = min(n - 1, n \times r)$.

Figure 2.53 shows the warping windows obtained by applying the Sakoe-Chiba band of relative width $r = 25\%$ and the corresponding Itakura parallelogram. When searching for the optimal warping path, only those warping paths that do not leave these windows are considered. Thus, pairs of points outside these areas need not to be compared.

Within the warping windows, the points of A can be matched with any number of adjacent points of B. For example, without applying global constraints, point $A[12]$ can be matched with any sequence of adjacent points from the entire length of B. Utilizing the Sakoe-Chiba band of width $r = 25\%$ restricts the matchings to the subseries $B[6{:}18]$, and applying the corresponding Itakura parallelogram narrows the admissible region to $B[8{:}16]$ (Fig. 2.54). In both cases, the allowed region represents a symmetric neighborhood of the point $B[12]$. The Sakoe-Chiba band tolerates a temporal deviation of width w, and the Itakura parallelogram is somewhat stricter. When it comes to the Sakoe-Chiba band, the width of the allowed shift from the observed point is limited only by the length of the time series, but in the case of the Itakura parallelogram it also depends on the slopes s and $1/s$, which is clearly seen in Figs. 2.53 and 2.54 on the examples of the points $A[4]$ and $A[21]$.

Figure 2.55 illustrates the warping windows obtained by applying Sakoe-Chiba bands and Itakura parallelograms of different sizes. Recall that the time series of the *FiftyWords* (40) dataset consists of 270 points. This means that the warping matrix contains 270×270 cells (not counting the auxiliary row and column used for initialization). Without applying global constraints, the warping window spans the entire matrix and to obtain the distance between two time series A and B, the value of all the $270 \times 270 = 72900$ cells must be evaluated. Applying global constraints reduces the number of warping matrix cells covered by the warping window, as

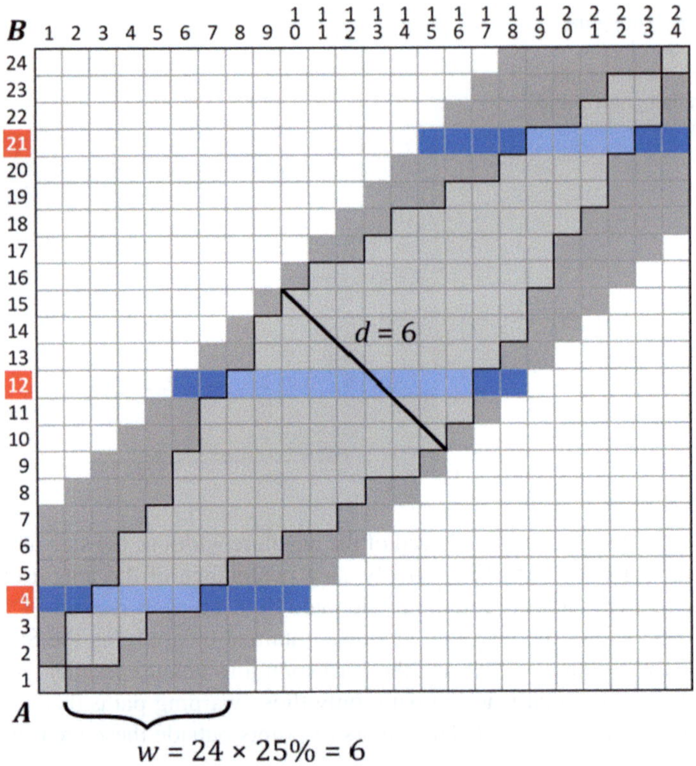

Fig. 2.53 Warping windows generated by the Sakoe-Chiba band and the Itakura parallelogram of the same width

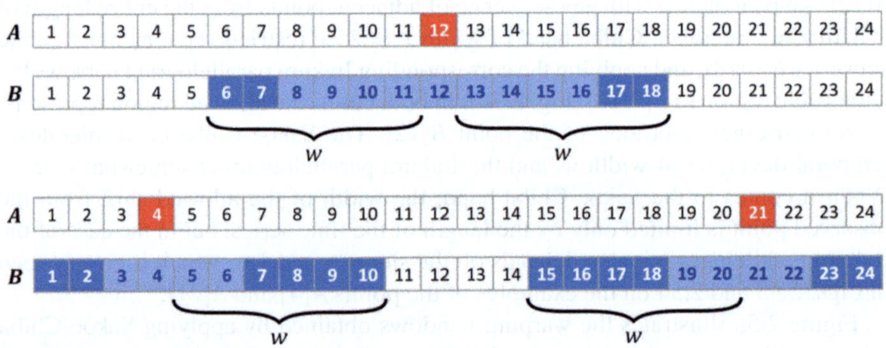

Fig. 2.54 Permissible matching areas of selected points (colored in red) of time series *A*. The Sakoe-Chiba band includes both dark blue and light blue points of *B*, and the Itakura parallelogram extends only to the light blue ones

shown in Fig. 2.56. For example, with $r = 50\%$, when using the Sakoe-Chiba band, the warping window covers approximately 75% of the warping matrix and about 50% when applying the Itakura parallelogram. With $r = 0\%$, the warping window is reduced to the diagonal and elasticity is completely disabled, i.e. $A[i]$ is compared exclusively with $B[i]$.

However, it should be taken into account that the implementation of global constraints complicates the recursive formulas of elastic measures, which negatively affects the speed of calculation. This can be clearly seen in the graphs of Fig. 2.57 which show the times required to generate symmetric distance matrices for the *FiftyWords* (40) dataset using different values of r (75%, 50%, 25%, 20% and 10%). The obtained results are expressed in percentages in relation to the time it took to generate a symmetric distance matrix relying on the corresponding unconstrained measure. It is evident that for $r > 50\%$ the acceleration is minimal (or non-existent) both when using the Sakoe-Chiba band and when using the Itakura parallelogram.

In addition to speeding up the calculation, by applying global constraints, pathological warpings (matching a short subsequence of one of the time series with a long subsequence of the other one) can also be avoided [106, 107, 113]. Examples of optimal warping paths containing such pathological matchings are given in Fig. 2.58. The presented warping matrices were obtained by computing the distance between two selected time series (belonging to different classes) of the *ItalyPowerDemand* (65) dataset (Fig. 2.59) relying on the unconstrained elastic measures. In the case of

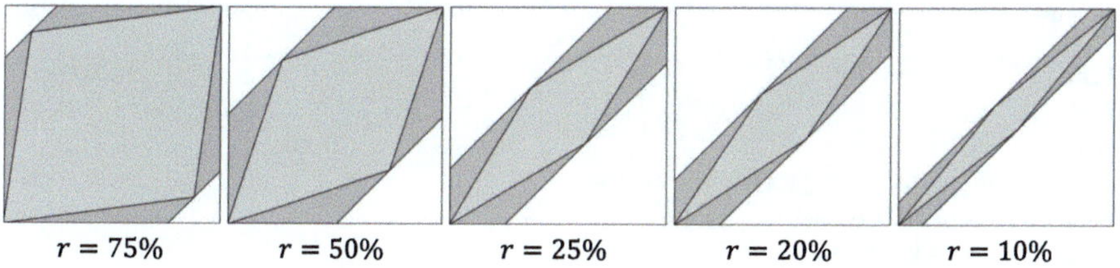

Fig. 2.55 Warping windows obtained by applying Sakoe-Chiba bands and Itakura parallelograms of different sizes

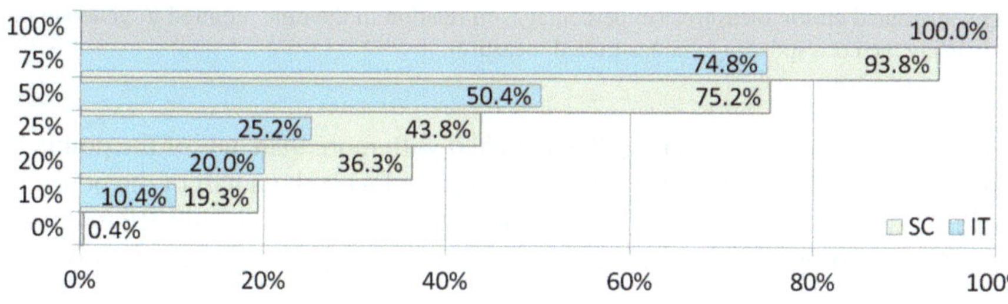

Fig. 2.56 The number of comparisons of points of two arbitrary time series of the *FiftyWords* (40) dataset when calculating their distance using elastic measures constrained by the Sakoe-Chiba band and the Itakura parallelogram of different sizes (in percentages compared to the number of comparisons when using unconstrained elastic measures)

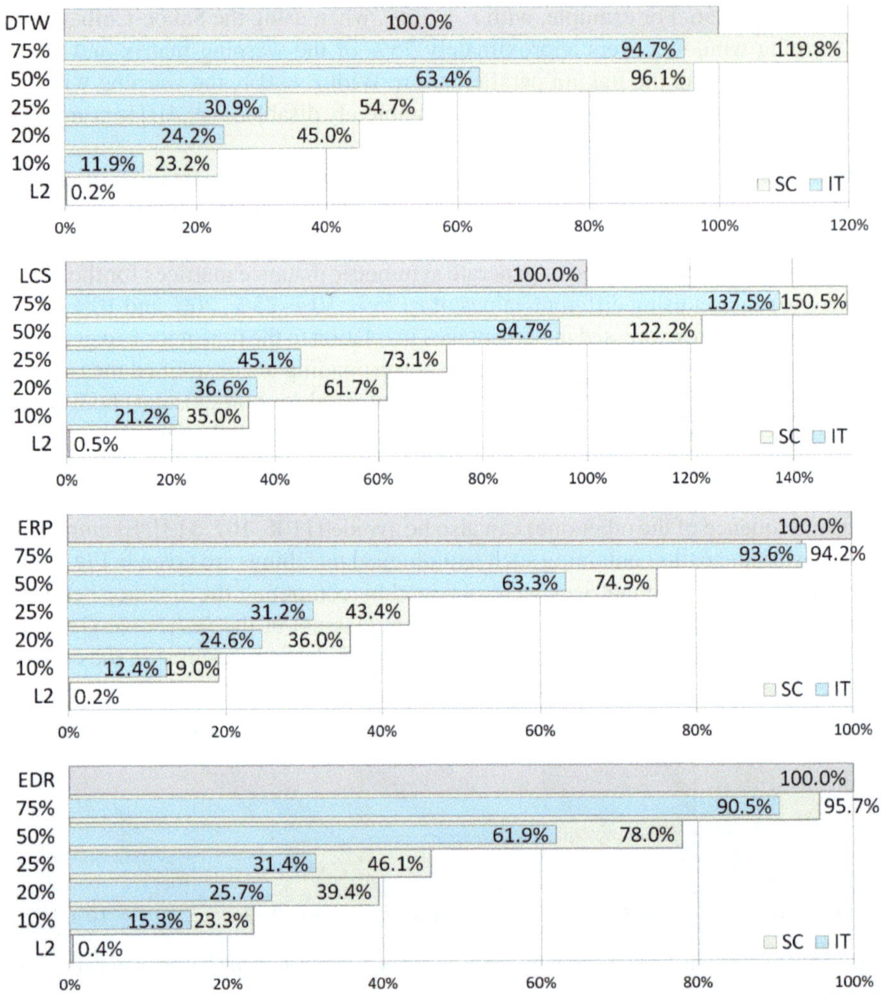

Fig. 2.57 Time required to generate a (symmetric) distance matrix for the *FiftyWords* (40) dataset using constrained elastic measures (in percentages in relation to the time required to generate the matrix using the corresponding unconstrained measure)

DTW, it can be easily seen that $A[1]$ is matched with more than half of the points of B (specifically, with $B[1{:}16]$) and that $B[24]$ is paired with about half of the points of A (with $A[12{:}24]$). Similar matchings are present in both the LCS and ERP matrices. The only exception is EDR where the optimal warping path closely follows the diagonal.

Applying global constraints narrows the warping window around the diagonal, bringing the optimal warping path closer to linear matching of points, as can be seen in Figs. 2.60, 2.61, 2.62 and 2.63, which show the warping matrices obtained by applying constrained elastic measures to the time series depicted in Fig. 2.59. In

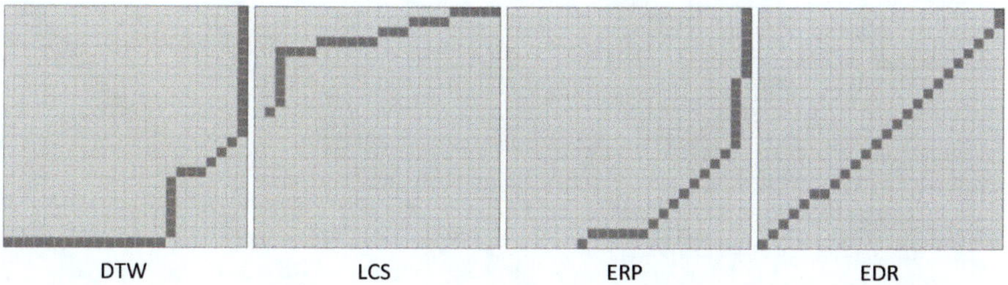

Fig. 2.58 Optimal warping paths of the unconstrained elastic measures when calculating the distance between the time series of the *ItalyPowerDemand* (65) dataset shown in Fig. 2.59

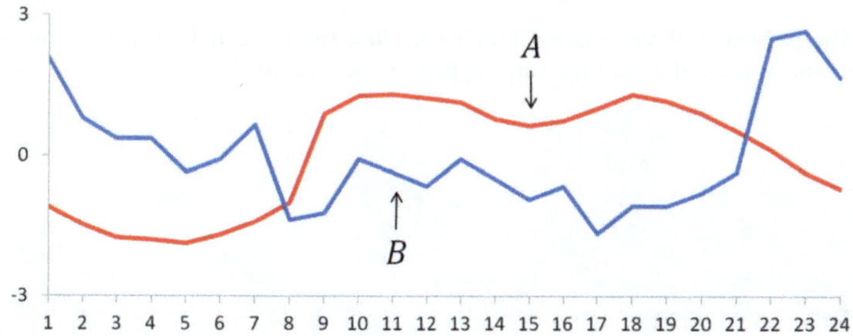

Fig. 2.59 Two time series of the *ItalyPowerDemand* (65) dataset belonging to different classes

these figures, the first row shows the matrices corresponding to the Sakoe-Chiba band and the second row the matrices generated by applying the Itakura parallelogram of the same size.

As indicated in Table 2.13, constraining the warping window also affects the estimation of the distance between time series. Since narrowing the warping window

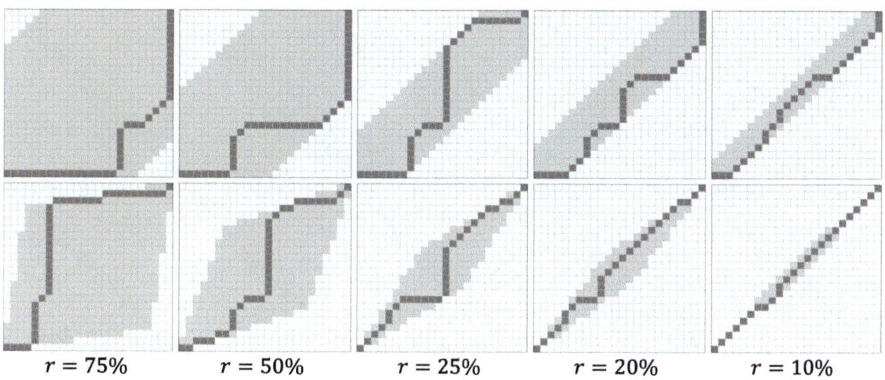

Fig. 2.60 The influence of the Sakoe-Chiba band (first row) and Itakura parallelogram (second row) of different sizes on the optimal warping path in the case of DTW

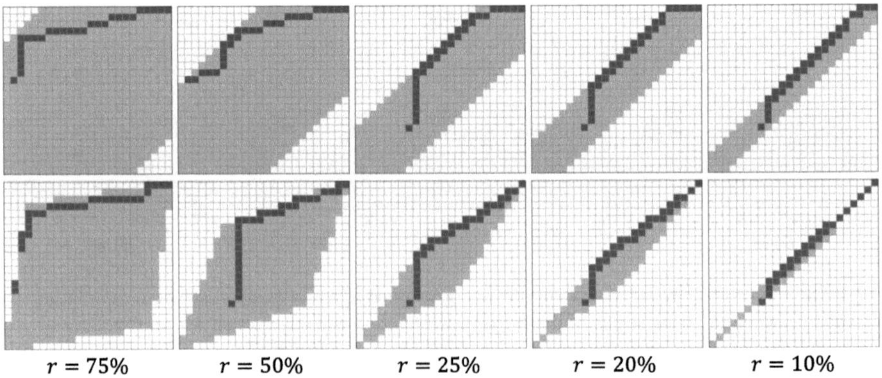

Fig. 2.61 The influence of the Sakoe-Chiba band (first row) and Itakura parallelogram (second row) of different sizes on the optimal warping path in the case of LCS

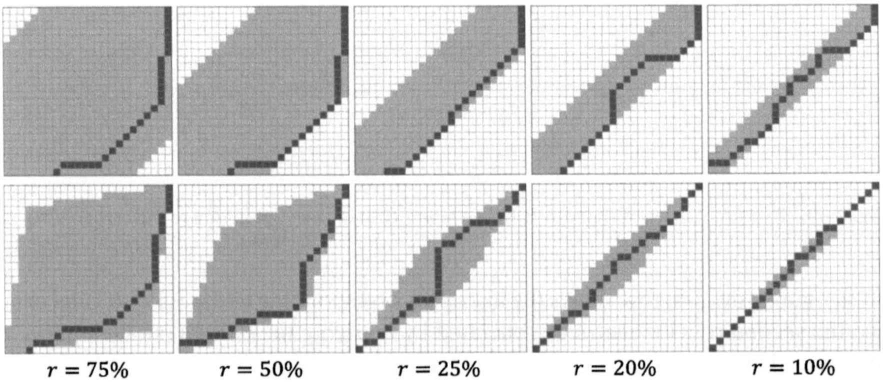

Fig. 2.62 The influence of the Sakoe-Chiba band (first row) and Itakura parallelogram (second row) of different sizes on the optimal warping path in the case of ERP

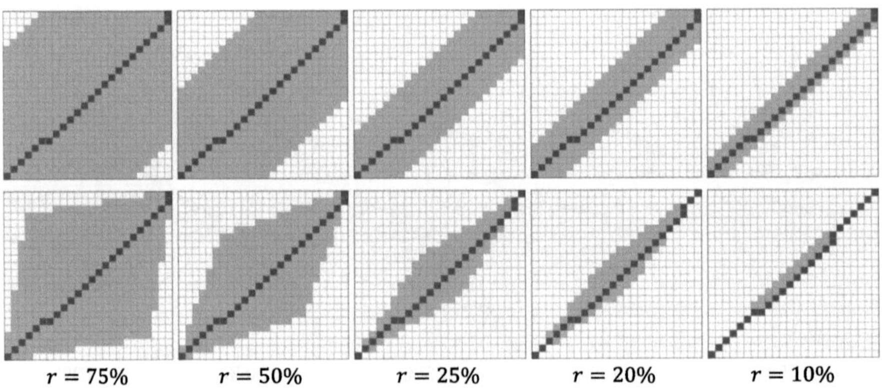

Fig. 2.63 The influence of the Sakoe-Chiba band (first row) and Itakura parallelogram (second row) of different sizes on the optimal warping path in the case of EDR

does not allow finding the optimal warping path (in terms of minimizing its total cost), the distance becomes larger and larger. However, in relation to classification accuracy (and especially when using DTW in combination with the Sakoe-Chiba band), the impact of global constraints turns out to be beneficial [16, 109, 110, 114–120], which we will investigate in great detail in this book.

When implementing global constraints (especially the Itakura parallelogram), special attention should be paid to the recursive definition of LCS. The following issues should be addressed:

1. When $|a_i - b_j| \leq \varepsilon$, $LCS[i,j]$ is defined as $LCS[i-1,j-1]+1$, but the cell $[i-1,j-1]$ may not be included in the warping window.

   Consider the time series $A = (1,0,1,1,0,0,0,1)$ and $B = (0,1,0,0,1,1,0,1)$. Their Itakura-constrained ($r = 50\%$) warping matrix obtained with $\epsilon = 0$ is depicted in Fig. 2.64(1). The problem in question occurs in the case of cells marked in orange ([2, 4], [3, 6], [32] and [6, 3]). Based on the given matrix, it turns out that the longest common subsequence of A and B is 10001 instead of 010001 (Fig. 2.65(1)).
2. When $|a_i - b_j| > \varepsilon$, $LCS[i,j]$ is defined as $max\{LCS[i,j-1], LCS[i-1,j]\}$, however neither cell $[i,j-1]$ nor cell $[i-1,j]$ needs to be part of the warping window.

   This issue leads to a segmented calculation of LCS, individually for pairs of subsequences defined by warping window areas between which there is no information transfer. In Fig. 2.64(2) showing the warping matrix of the time series $C = (0,0,1,1,0,0,1,1)$ and $D = (0,1,1,1,0,1,0,1)$ (utilizing Itakura parallelogram of width $r = 25\%$ and $\varepsilon = 0$), there are three such independent segments (the problematic cells are marked in orange):

   - The first segment compares $C[1] = 0$ and $D[1] = 0$ and $LCS[1,1]$ finds that the length of their LCS is 1.
   - The second segment compares $C[2{:}6]$ and $D[2{:}6]$ and $LCS[6,6]$ finds that the length of their LCS is 3 (their LCS is 110). However, as $C[2] \neq D[2]$, due

Table 2.13 The distance between the time series shown in Fig. 2.59 estimated by different elastic measures constrained applying Sakoe-Chiba bands (SC) and Itakura parallelograms (IT) of different sizes

r		100%	75%	50%	25%	20%	10%	0%
DTW	SC	33.56	33.56	33.70	47.85	51.44	59.10	65.99
	IT	33.56	43.56	54.86	61.80	62.94	63.50	65.99
LCS	SC	0.75	0.75	0.75	0.88	0.88	0.88	1.00
	IT	0.75	0.79	0.88	0.92	0.92	0.92	1.00
ERP	SC	19.72	19.72	19.72	26.88	27.89	32.55	37.70
	IT	19.72	22.39	26.64	33.10	34.76	37.70	37.70
EDR	SC	23.00	23.00	23.00	23.00	23.00	23.00	24.00
	IT	23.00	23.00	23.00	23.00	23.00	24.00	24.00

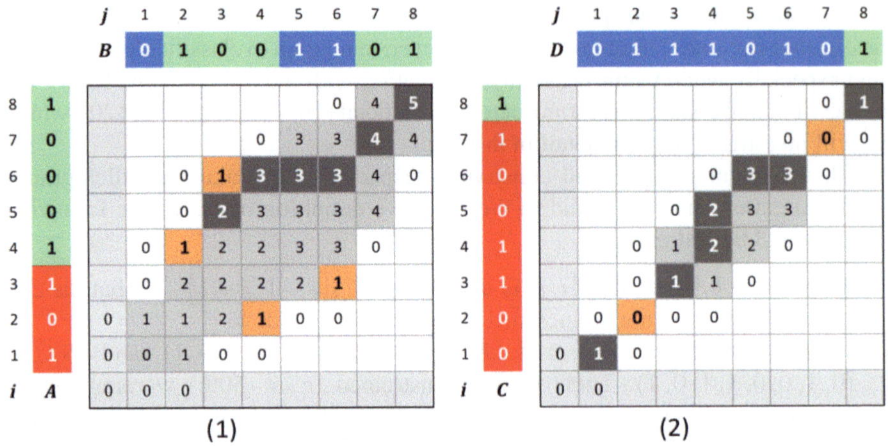

Fig. 2.64 Issues with the definition of the LCS measure when applying global constraints

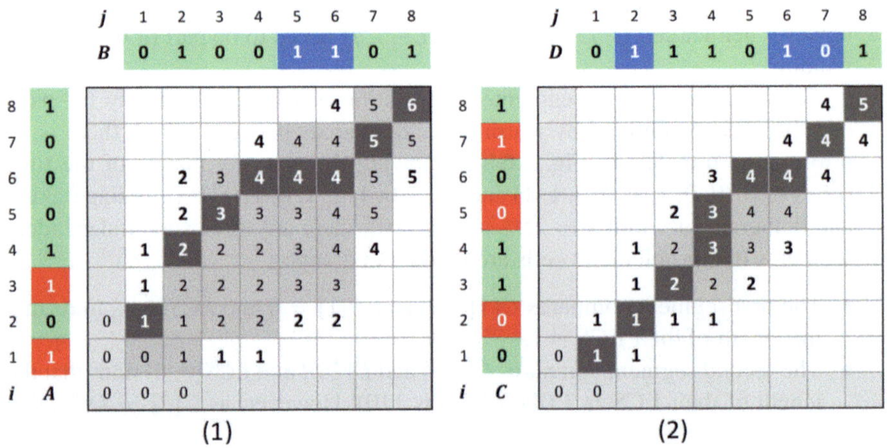

Fig. 2.65 Overcoming the issues of the LCS measure when applying global constraints with adequate initialization of the warping matrix

to the issue in question, the information that the length of the LCS in the first segment is 1 is lost.

- The third segment compares $C[7{:}8]$ and $D[7{:}8]$ and the length of their LCS is $LCS[8,8] = 1$. Again, because of $C[7] \neq D[7]$, the information about the length of the LCS of the previous segment(s) is lost. As a consequence, the LCS of C and D is wrongly equated with the surplus LCS of the last segment and $|LCS(C,D)| = LCS[8,8] = 1$ (instead of 01101, as can be seen in Fig. 2.65(2)).

Both of these issues can be remedied by careful initialization of the warping matrix taking care of adequate information transfer across its columns and rows, as illustrated in Fig. 2.65.

2.5 Distance Matrices

All the experiments within this book were computationally quite demanding and time consuming. Therefore, with the aim to speed them up, we computed the distances between time series according to all discussed distance measures and (in the case of elastic measures) in combination with the considered warping-window widths of both types of global constraints and stored the calculated values as distance matrices. This section presents the tools used to generate the matrices and to perform the experiments whose results are presented in Chaps. 4–6.

In order to classify an unclassified time series, distance-based classifiers must know the (dis)similarity between the given series and each example found in the training set. To optimize efficiency, pairwise similarity metrics between time series within the datasets can be pre-calculated and stored as distance matrices where each element represents the similarity between two specific series.

Table 2.14 shows the distance matrix calculated employing Dynamic Time Warping on the first 10 elements from the *BeetleFly* (8) dataset. Since DTW is a symmetric distance measure (i.e. $d(A, B) = d(B, A)$ for every A and B), only a lower triangular matrix needs to be computed. Each matrix element (i, j) contains the distance between ith and jth time series from the dataset. For the purposes of the experiments, based on the thus obtained distance matrices, we also generated the corresponding neighbor matrices where the jth element of the ith row represents the index of the jth nearest neighbor of the ith time series. Table 2.15 shows the first 10 rows and columns of the neighbor matrix generated from the distance matrix presented in Table 2.14.

For the purposes of the experiments, the results of which are presented in this book, slightly more than 3.8 TB matrices (distance and neighbor) were generated. As part of that process, the similarity between the time series was calculated 326,301,967,538 times.

To facilitate the generation of distance matrices (and based on them also the corresponding matrices of neighbors), a dedicated Java application has been developed using the services of our FAP (Framework for Analysis and Prediction) library [103]. This application offers the flexibility to interrupt and resume the matrix generation process, as well as the option to migrate serialized Java objects to another computer for continued computation. To facilitate matrix generation in a distributed environment (composed of, for example, several computer centers and faculty classrooms), we have also developed an extensible agent-based system called ADiS (Agent-based Distributed computing System) [121]. These functionalities will be described in more detail in the following subsections.

Table 2.14 Part of the distance matrix of the *BeetleFly* (8) dataset generated by applying the unconstrained DTW (the numbers shown are rounded to three decimal places)

	1	2	3	4	5	6	7	8	9	10
1	0									
2	77.890	0								
3	58.029	105.884	0							
4	114.909	42.643	132.272	0						
5	106.540	60.656	134.723	85.100	0					
6	110.290	18.089	139.862	29.402	26.051	0				
7	97.308	85.720	101.785	87.794	76.444	66.323	0			
8	96.745	28.128	150.848	9.594	84.960	27.093	97.424	0		
9	90.863	89.683	105.371	85.023	109.975	96.906	32.557	86.431	0	
10	100.527	103.488	127.715	105.815	109.624	94.204	69.053	114.796	69.732	0

Table 2.15 The first 10 rows and columns of the neighbor matrix obtained from the distance matrix shown in Table 2.14

	1	2	3	4	5	6	7	8	9	10
1	3	38	28	2	33	23	30	37	32	9
2	6	8	26	4	30	28	5	23	17	33
3	13	35	15	16	1	32	38	39	14	40
4	8	25	6	30	26	2	17	29	12	38
5	6	28	23	2	33	7	34	36	17	39
6	2	5	8	4	30	26	28	25	23	17
7	9	27	36	32	29	33	39	25	13	18
8	4	25	6	30	2	26	12	23	17	28
9	27	7	24	29	32	33	11	16	14	31
10	21	22	24	29	31	37	34	39	28	33

2.5.1 Framework for Analysis and Prediction (FAP)

The Framework for Analysis and Prediction (FAP), a free and open-source packages written in Java, is intended for practitioners and researchers working with time-series data [31, 103, 115, 122]. It aims to provide a comprehensive and extensible toolkit for various tasks related to temporal data. The framework implements a diverse set of techniques used for mining and analysing time-series data including various representations, similarity measures, classification algorithms, validation methods and preprocessing procedures. The FAP library served as the common tool for all the experiments in this book.

In its current version, in addition to common similarity measures such as distances based on L_p norms and the elastic measures presented in this chapter, FAP also includes the Canberra [123, 124] and Lorentzian [124, 125] distances, as well as more complex approaches like Swale [17], Spline [122] and Time Warp Edit Distance (TWED) [126]. In addition, elastic similarity measures are implemented with different warping window constraints (Itakura and Sakoe-Chiba), as well as without constraints. The library also includes classification algorithms like the nearest neighbor (1NN) method and its majority-voting extension (kNN) with various weighting schemes presented in Chap. 3.

FAP provides various performance evaluation algorithms too, like the holdout technique, the leave-one-out (LOO) method and (stratified) k-fold cross-validation (SCV) (they are described in more detail in Chap. 3). The behavior of these classifier evaluation methods is controlled through different parameters. The holdout method can be adjusted by setting the percentage of the provided dataset used for classifier training, the random seed intended to control partitioning of the data into test and training subsets and a Boolean value indicating whether to use a stratified or random split. The cross-validation method, in addition to the options defined for the holdout technique, can also be parameterized by specifying the number of folds. The leave-one-out method does not require any parameter setting.

FAP also offers methods for performing preprocessing transformations on input data. These techniques are often essential to clean and prepare the dataset prior to its utilization in various tasks of data mining. Some common approaches to data preprocessing are: data integration, handling missing data, reduction, cleaning, discretization and transformation. The FAP preprocessing component implements the following transformations [6, 127, 128]: z-score and min–max normalization, shifting, (decimal) scaling and linear equiscaling. Naturally, this list is not final; it is expected that new methods will be implemented and integrated in the future.

FAP also supports various time-series representations which are usually necessary or preferred in time-series analysis since storing raw data often requires significant storage capacity and makes various tasks more difficult to perform. Time-series representation techniques do not memorize the exact values of individual time-series data points. Instead, they calculate the simplified version of time series with significantly reduced dimensionality, while the general shape and trend of temporal data is preserved. FAP supports the following representation techniques: Symbolic Aggregate Approximation (SAX) [129], Piecewise Linear Approximation (PLA) [130], Adaptive Piecewise Constant Approximation (APCA) [131], Piecewise Aggregate Approximation (PAA) [132] and Spline [122]. Again, this list is not final and it is open to further extensions.

The need for a framework for analysis and prediction arose during the development of a new time-series representation for the CuBaGe system [122]. Initially implementing fundamental classes, the framework gradually expanded to include new capabilities. Subsequent research on the impact of global constraints on time-series distance measures and the potential enhancement of classifiers led to the addition of classes to FAP in order to implement these functionalities. During this development, FAP was extended to include a range of distance measures, global constraints, classifiers, evaluation techniques, representation techniques and preprocessing methods.

With these extensions, FAP has proven to be adaptable and suitable for multidisciplinary research areas [104, 105], including: emotion recognition [133], medicine [134] and psychology [72].

The positive experiences with FAP in research projects motivated the decision to release it as an open-source library. The framework was found to be a valuable tool for students enrolled in Informatics programs, particularly those interested in machine learning and data mining. The usage of FAP facilitated the realization of several seminar and thesis projects. Additionally, students could incorporate FAP into their undergraduate and master thesis experiments across different domains.

During its history and different development phases FAP showed its versatility and adaptability for different research and practical applications. Students and researchers interested in mining and analysing time series can utilize the library to explore and compare different techniques in temporal data analysis, making it a valuable resource in the field.

2.5.2 Distributed Distance Matrix Generation

In this subsection we will consider a rather powerful solution to deal with time-consuming calculations of similarities in time series datasets based on agent technology and distributed programming.

As we already mentioned there are a lot of domains where time series data are generated like medical observations, social and natural phenomena and results of a range of economic, engineering and scientific observations and experiments. Almost all these tasks rely on determining the similarity between available time series data in various datasets. For these calculations we can select different similarity measures and approaches. However, it is expected that the chosen similarity measure should reflect the underlying (dis)similarity of the data represented by the time series [6]. The main problem during performing extensive experiments with big datasets of time series is time and, in some cases, memory space. This is especially important for small institutions and researchers that lack high-performance computers and limited hardware support.

On the other hand, thanks to the highly developed research area of agent technologies [135–138] it is rather easy and convenient to develop and implement distributed software solutions and services. Software agents [139, 140] are widely recognized as autonomous entities expressing proactive and goal-directed behavior. Mobile agents [141] as another kind of agents are self-contained and identifiable programs which move over the network and can perform different activities and tasks acting on behalf of the user or another entity. However, in some situations it is better not to use mobile agents due to the additional communication cost over the network. They can execute some tasks on a computer in a network for a while before halting it and migrating to another computer in the network and continuing execution of already started task on the new computer.

Accordingly, our idea, which will be briefly presented in this subsection is to unify the advantages of these two powerful techniques to fulfill time and memory requirements in environments with limited computer network capacities. In order to produce a distance matrix for time series data supported by powerful agent technology we implemented software agents which deal with dynamic properties of the computational network. Bearing in mind that in a network we can suddenly encounter unavailability of several computers or need for load-balancing and task sharing we should take care of these sensitive properties to obtain continuation of already started calculations without data loss.

The local computer network of our Department consists of a limited number of computers of modest power and capacity. Furthermore, these computers are used for different purposes (including both theoretical and practical teaching) and we cannot utilize them continuously and in longer periods of time to calculate distance matrices for our experiments. Therefore, it was challenging to find a suitable solution for their effective engagement so as not to disrupt other activities. As a result of this effort, we proposed and implemented a unique approach based on agent technologies to achieve good performance with limited hardware resources in small local area networks.

The computer ecosystem of our Department includes several interconnected computer centers that can be employed for generating time series distance matrices. However, due to the intensive use of the computers of these centers for various purposes (educational and research), they may become (un)available at any time, depending on the activities of students and other researchers. To cope with the dynamic nature of the network and to minimize the amount of network traffic, one of the main tasks of our software agents was to maintain an adequate list of available computers [142].

We designed and implemented an agent-based system for distributed computing named ADiS (Agent-based Distributed computing System), with crucial functionality to speed-up the rather demanding distance matrix calculation processes relying on limited hardware equipment and resources. Our intention was to develop an extensible system that, through various plugins, Is capable to perform other tasks as well, not only distance matrix calculations. These plug-ins can be incorporated dynamically in other infrastructures developed by other researchers and for different purposes, to support any form of distributed computations [121]. As the job-distribution architecture of ADiS supports efficient load-balancing, it can be extremely useful for researchers in institutions which cannot afford high performance grid systems but possesses environments equipped with low performance multipurpose personal computers.

Based on the above, we proposed a time-efficient solution by introducing two types of agents, the so-called *system-level* and *middle* agents. One type is dedicated to building and managing dynamic networks, while the other is responsible for load balancing. In software architectures, system-level agents are very flexible and allow a new agent with new functionality to be easily incorporated into the architecture without disrupting other parts of the system.

Using agents for network management brings various benefits. They enable flexible, dynamic network building, can efficiently discover faulty nodes and offer remote maintenance. The agent-based approach that we proposed for building and maintaining dynamic networks confirms that it is simple to implement, effective concerning the network traffic overhead and reliable as it operates regardless of the number of failed nodes in the network.

Our environment is specific since it consists of a number of lower-end personal computers interconnected with a stable, wired, high-speed network. It provides efficient load-balancing and job-distribution architecture. When a computer becomes available, the agent just divides its set of tasks into two distinct sub-sets. One sub-set is kept by the given agent while the other is sent to the new computer.

2.5.2.1 ADiS Architecture

The proposed architecture is fully distributed where all computers are equal, i.e. none of them is designated as a central component of the system, which results in high flexibility in selecting available computers. The essential components of ADiS are presented in Fig. 2.66 [142]:

- **FAP**—Framework for Analysis and Prediction,
- **DIMAG**—Distance Matrix Generator component,
- **DIMAGAgent**—a simple agent that controls DIMAG in performing computational activities,
- **HBAgent**—an agent for establishing and maintaining a dynamic network of available computers,
- **PD Module**—a platform-dependent component of the system.

The rest of this subsection is dedicated to presenting the functionality of main ADiS components in more detail:

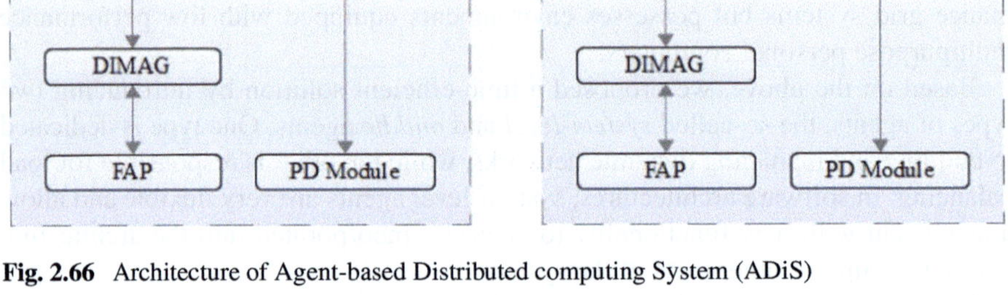

Fig. 2.66 Architecture of Agent-based Distributed computing System (ADiS)

- **DIMAG** is the key component of the system responsible for generating distance matrices relying on the **FAP** library. It is also responsible for dividing larger tasks into smaller sub-tasks that can be distributed among the available computers included in the system in order to speed up calculations.

  DIMAG is controlled by a set of instructions (called a program) for computing (partial) distance matrices. Each instruction includes the following information:

  - The name of the dataset whose (partial) distance matrix should be generated.
  - The name of the distance measure between time series.
  - A number indicating how many parts (of approximately the same size) the distance matrix should to be divided into.
  - Index of the current sub-matrix: determining which part of the distance matrix should be calculated by the current instruction.
  - A switch indicating whether to generate a full distance matrix or a diagonal one.
  - Additional parameters for initializing the distance measure.

  Each individual instruction controls the computation of one (partial) distance matrix. The task-sharing algorithm takes care of the allocation of available computers. When a computer becomes available and the DIMAGAgent receives a task-sharing request, its instruction set is split into two parts. One part remains on the given computer and the other is sent to one of the available computers.

  When the generation of all partial matrices is complete, DIMAG merges them into the final distance matrix.
- **HBAgent** is in charge of taking care of the system state by detecting changes in the computer network. This is of critical importance in a dynamic environment where the availability of individual computers depends on their engagement for other tasks and may become (un)available at any time. HBAgent maintains the correct view of the network state and it tries to connect to the first available computer when it joins the network. An important assumption during this process is that inter-computer connections will not break. When a connection is established, the agent becomes aware of all other computers in the network. Its subsequent activity is to exchange information with the agent on the target computer. The connection between HBAgents is based on a *heartbeat* connection principle and they "ping" each other at regular time intervals to indicate their presence. When an agent detects the availability of another one, it spreads this information to other remaining agents. As optimization of the whole process is of special importance the intention is to minimize network traffic as much as possible. Accordingly, the heartbeat connection is realized in a special way. Agent communication and activities within the network should support fault tolerant regime of work and to achieve this a specially designed algorithm is proposed. Its experimental validation is presented in detail in [143].
- **PD Module** is responsible for determining the computer's availability and the current CPU-load. If there is no input activity (e.g. mouse movement or keyboard

typing) on the given computer for a period of time, this module signals the appropriate HBAgent that the computer is available for computing. When a user occupies the computer and starts using some of its input devices, the PD Module will instruct the HBAgent to stop all calculations, freeing up system resources.

When a computer completes an assigned subtask, it propagates the resulting (partial) matrix over the network. That way, if a computer becomes unavailable for any reason, its calculation results will not be lost since they can be collected from any other computer connected to the system. In addition, its unfinished subtasks can easily be transferred to other available computers.

In addition to our initial intention to develop the ADiS system (i.e. accelerating the generation of distance matrices), it can easily be extended into a general-purpose architecture for distributed computing, with the following advantages:

- As it relies on agent technology, it is flexible and allows a high level of adaptivity to changes in dynamic networks.
- It has a well-defined plugin architecture and supports extensibility in performing arbitrary calculations for task distribution.

References

1. Shannon, C.E.: A mathematical theory of communication. Bell Syst. Tech. J. **27**, 379–423 (1948). https://doi.org/10.1002/j.1538-7305.1948.tb01338.x
2. Das, G., Gunopulos, D.: Time series similarity and indexing. In: Ye, N. (ed.) Handb. Data Min., pp. 279–304. Lawrence Erlbaum Associates, Mahwah, NJ (2003)
3. Mitsa, T.: Temporal data mining. Taylor & Francis (2010). http://books.google.rs/books?id=4P_7ydvW7cAC
4. Esling, P., Agon, C.: Time-series data mining. ACM Comput. Surv. **45**, 12:1–12:34 (2012). https://doi.org/10.1145/2379776.2379788
5. Laxman, S., Sastry, P.S.: A survey of temporal data mining. Sadhana **31**, 173–198 (2006). https://doi.org/10.1007/BF02719780
6. Han, J., Pei, J., Tong, H.: Data mining: concepts and techniques, 4th ed. Morgan Kaufmann (2022)
7. Shaukat, K., Alam, T.M., Luo, S., Shabbir, S., Hameed, I.A., Li, J., Abbas, S.K., Javed, U.: A review of time-series anomaly detection techniques: a step to future perspectives, pp. 865–877 (2021). https://doi.org/10.1007/978-3-030-73100-7_60
8. Blázquez-García, A., Conde, A., Mori, U., Lozano, J.A.: A review on outlier/anomaly detection in time series data. ACM Comput. Surv. **54**, 1–33 (2022). https://doi.org/10.1145/3444690
9. Javed, A., Lee, B.S., Rizzo, D.M.: A benchmark study on time series clustering. Mach. Learn. Appl. **1**, 100001 (2020). https://doi.org/10.1016/j.mlwa.2020.100001
10. Alqahtani, A., Ali, M., Xie, X., Jones, M.W.: Deep time-series clustering: a review. Electronics **10**, 3001 (2021). https://doi.org/10.3390/electronics10233001
11. Faouzi, J.: Time series classification: a review of algorithms and implementations. In: Kotecha, K. (ed.) Mach. Learn. (Emerging Trends Appl.). Proud Pen (2022). https://inria.hal.science/hal-03558165
12. Abanda, A., Mori, U., Lozano, J.A.: A review on distance based time series classification. Data Min. Knowl. Discov. **33**, 378–412 (2019). https://doi.org/10.1007/s10618-018-0596-4

13. Torres, J.F., Hadjout, D., Sebaa, A., Martínez-Álvarez, F., Troncoso, A.: Deep learning for time series forecasting: a survey. Big Data **9**, 3–21 (2021). https://doi.org/10.1089/big.2020.0159
14. Masini, R.P., Medeiros, M.C., Mendes, E.F.: Machine learning advances for time series forecasting. J. Econ. Surv. **37**, 76–111 (2023). https://doi.org/10.1111/joes.12429
15. Aßfalg, J.: Advanced analysis on temporal data. LMU München (2008) (PhD thesis). http://edoc.ub.uni-muenchen.de/8798/1/Assfalg_Johannes.pdf
16. Ding, H., Trajcevski, G., Scheuermann, P., Wang, X., Keogh, E.: Querying and mining of time series data: experimental comparison of representations and distance measures. In: Proc. VLDB Endow., VLDB Endowment, 2008, pp. 1542–1552. https://doi.org/10.14778/1454159.1454226
17. Morse, M.D., Patel, J.M.: An efficient and accurate method for evaluating time series similarity. In: Proc. 2007 ACM SIGMOD Int. Conf. Manag. Data—SIGMOD '07, pp. 569–580. ACM Press, New York, USA (2007). https://doi.org/10.1145/1247480.1247544
18. Hetland, M.L.: A survey of recent methods for efficient retrieval of similar time sequences. In: Last, M., Kandel, A., Bunke, H. (eds.) Data Min. Time Ser. Databases, pp. 23–42. World Scientific Publishing Company (2004). https://doi.org/10.1142/9789812565402_0002
19. Faloutsos, C., Ranganathan, M., Manolopoulos, Y.: Fast subsequence matching in time-series databases. ACM SIGMOD Rec. **23**, 419–429 (1994). https://doi.org/10.1145/191843.191925
20. Bai, B., Li, G., Wang, S., Wu, Z., Yan, W.: Time series classification based on multi-feature dictionary representation and ensemble learning. Expert Syst. Appl. **169**, 114162 (2021). https://doi.org/10.1016/j.eswa.2020.114162
21. Yin, N., Wang, S., Hong, S., Li, H.: A segment-wise method for pseudo periodic time series prediction. In: Luo, X., Yu, J.X., Li, Z. (eds.) Adv. Data Min. Appl., pp. 461–474. Springer, Cham (2014). https://doi.org/10.1007/978-3-319-14717-8_36
22. Wang, X., Mueen, A., Ding, H., Trajcevski, G., Scheuermann, P., Keogh, E.: Experimental comparison of representation methods and distance measures for time series data. Data Min. Knowl. Discov. **26**, 275–309 (2013). https://doi.org/10.1007/s10618-012-0250-5
23. Little, R., Rubin, D.: Statistical analysis with missing data, 3rd edn. Wiley (2019). https://doi.org/10.1002/9781119482260
24. Ciaburro, G., Iannace, G.: Machine learning-based algorithms to knowledge extraction from time series data: a review. Data **6**, 55 (2021). https://doi.org/10.3390/data6060055
25. Ratanamahatana, C.A., Lin, J., Gunopulos, D., Keogh, E., Vlachos, M., Das, G.: Mining time series data. In: Maimon, O., Rokach, L. (eds.) Data Min. Knowl. Discov. Handb., 2nd edn., pp. 1049–1077. Springer, Boston, MA (2009). https://doi.org/10.1007/978-0-387-09823-4_56
26. Fu, T.: A review on time series data mining. Eng. Appl. Artif. Intell. **24**, 164–181 (2011). https://doi.org/10.1016/j.engappai.2010.09.007
27. Fakhrazari, A., Vakilzadian, H.: A survey on time series data mining. In: 2017 IEEE Int. Conf. Electro Inf. Technol., pp. 476–481. IEEE (2017). https://doi.org/10.1109/EIT.2017.8053409
28. Mörchen, F.: Time Series Knowledge Mining. Philipps-University Marburg, Germany (2006) (PhD thesis). https://www.mybytes.de/papers/moerchen06tskm.pdf
29. Milanović, M., Stamenković, M.: Data mining in time series. Ekon. Horizonti. **13**, 5–25 (2011). http://www.horizonti.ekfak.kg.ac.rs/sites/default/files/Casopis/2011_1/1_Milanovic_Stamenkovic.pdf
30. Last, M., Kandel, A., Bunke, H.: Data mining in time series databases. World Scientific (2004)
31. Kurbalija, V.: Time series analysis and prediction using case based reasoning technology. University of Novi Sad (2009). (PhD thesis) https://doi.org/10.2298/NS20091005KURBALIJA
32. Parmezan, A.R.S., Souza, V.M.A., Batista, G.E.A.P.A.: Evaluation of statistical and machine learning models for time series prediction: identifying the state-of-the-art and the best conditions for the use of each model. Inf. Sci. (NY) **484**, 302–337 (2019). https://doi.org/10.1016/j.ins.2019.01.076

33. Hawkins, D.M.: Identification of outliers. Springer, Dordrecht (1980). https://doi.org/10.1007/978-94-015-3994-4
34. Zaki, M.J., Meira, W.J.: Data mining and machine learning: fundamental concepts and algorithms, 2nd edn. Cambridge University Press (2020)
35. Rakthanmanon, T., Campana, B., Mueen, A., Batista, G., Westover, B., Zhu, Q., Zakaria, J., Keogh, E.: Searching and mining trillions of time series subsequences under dynamic time warping. In: Proc. 18th ACM SIGKDD Int. Conf. Knowl. Discov. Data Min., pp. 262–270. ACM, New York, NY, USA (2012). https://doi.org/10.1145/2339530.2339576
36. Shasha, D., Zhu, Y.: High performance discovery in time series. Springer, New York, NY (2004). https://doi.org/10.1007/978-1-4757-4046-2
37. Singh, D., Singh, B.: Investigating the impact of data normalization on classification performance. Appl. Soft Comput. **97**, 105524 (2020). https://doi.org/10.1016/j.asoc.2019.105524
38. Paparrizos, J., Liu, C., Elmore, A.J., Franklin, M.J.: Debunking four long-standing misconceptions of time-series distance measures. In: Proc. 2020 ACM SIGMOD Int. Conf. Manag. Data, pp. 1887–1905. ACM, New York, NY, USA (2020). https://doi.org/10.1145/3318464.3389760
39. Anh Dau, H., Keogh, E., Kamgar, K., Michael Yeh, C.-C., Zhu, Y., Gharghabi, S., Ann Ratanamahatana, C., Chen, Y., Hu, B., Begum, N., Bagnall, A., Mueen, A., Batista, G.: Hexagon-ML, The UCR time series classification archive (2019). https://www.cs.ucr.edu/~eamonn/time_series_data_2018/
40. Paparrizos, J.: 2018 UCR Time-series archive: backward compatibility, missing values, and varying lengths (2019). https://github.com/johnpaparrizos/UCRArchiveFixes
41. Briandet, R., Kemsley, E.K., Wilson, R.H.: Discrimination of arabica and robusta in instant coffee by fourier transform infrared spectroscopy and chemometrics. J. Agric. Food Chem. **44**, 170–174 (1996). https://doi.org/10.1021/jf950305a
42. Petitjean, F., Inglada, J., Gancarski, P.: Satellite image time series analysis under time warping. IEEE Trans. Geosci. Remote Sens. **50**, 3081–3095 (2012). https://doi.org/10.1109/TGRS.2011.2179050
43. Olszewski, R.T.: Generalized feature extraction for structural pattern recognition in time-series data. Carnegie Mellon University (2001). (PhD thesis) https://www.cs.cmu.edu/~bobski/pubs/tr01108-twosided.pdf
44. Ratanamahatana, C.A., Keogh, E.: Everything you know about dynamic time warping is wrong. In: 3rd Work. Min. Temporal Seq. Data, Conjunction with 10th ACM SIGKDD Int. Conf. Knowl. Discov. Data Min. (KDD-2004), Seattle, WA, 2004
45. Wei, L., Keogh, E.: Semi-supervised time series classification. In: Proc. 12th ACM SIGKDD Int. Conf. Knowl. Discov. Data Min., pp. 748–753. ACM, New York, NY, USA (2006). https://doi.org/10.1145/1150402.1150498
46. Rath, T.M., Manmatha, R.: Word image matching using dynamic time warping. In: Comput. Vis. Pattern Recognition, 2003. Proceedings. 2003 IEEE Comput. Soc. Conf., 2003, vol. 2, pp. II-521–II-527. https://doi.org/10.1109/CVPR.2003.1211511
47. Ratanamahatana, C.A., Keogh, E.: Making time-series classification more accurate using learned constraints. In: Proc. 2004 SIAM Int. Conf. Data Min., pp. 11–22. Society for Industrial and Applied Mathematics, Philadelphia, PA (2004). https://doi.org/10.1137/1.9781611972740.2
48. Dau, H.A., Bagnall, A., Kamgar, K., Yeh, C.-C.M., Zhu, Y., Gharghabi, S., Ratanamahatana, C.A., Keogh, E.: The UCR time series archive. IEEE/CAA J. Autom. Sin. **6**, 1293–1305 (2019). https://doi.org/10.1109/JAS.2019.1911747
49. Keogh, E., Wei, L., Xi, X., Lonardi, S., Shieh, J., Sirowy, S.: Intelligent Icons: integrating lite-weight data mining and visualization into GUI operating systems. In: Sixth Int. Conf. Data Min., pp. 912–916. IEEE (2006). https://doi.org/10.1109/ICDM.2006.90
50. Guillame-Bert, M., Dubrawski, A.: Classification of time sequences using graphs of temporal constraints. J. Mach. Learn. Res. **18**, 1–34 (2017). http://jmlr.org/papers/v18/15-403.html

51. Gao, J., Giri, S., Kara, E.C., Bergés, M.: PLAID: a public dataset of high-resolution electrical appliance measurements for load identification research. In: Proc. 1st ACM Conf. Embed. Syst. Energy-Efficient Build., pp. 198–199. ACM, New York, NY, USA (2014). https://doi.org/10.1145/2674061.2675032
52. Huang, X., Ye, Y., Xiong, L., Lau, R.Y.K., Jiang, N., Wang, S.: Time series k-means: a new k-means type smooth subspace clustering for time series data. Inf. Sci. (NY) **367–368**, 1–13 (2016). https://doi.org/10.1016/j.ins.2016.05.040
53. Rebbapragada, U., Protopapas, P., Brodley, C.E., Alcock, C.: Finding anomalous periodic time series. Mach. Learn. **74**, 281–313 (2009). https://doi.org/10.1007/s10994-008-5093-3
54. Geurts, P.: Contributions to decision tree induction: bias/variance tradeoff and time series classification. University of Liège, Belgium (2002) (PhD thesis)
55. Liu, J., Zhong, L., Wickramasuriya, J., Vasudevan, V.: uWave: accelerometer-based personalized gesture recognition and its applications. Pervasive Mob. Comput. **5**, 657–675 (2009). https://doi.org/10.1016/j.pmcj.2009.07.007
56. Tan, P.-N., Steinbach, M., Kumar, V., Karpatne, A.: Introduction to data mining, 2nd ed. Pearson Education (2019)
57. Esposito, F., Malerba, D., Tamma, V., Bock, H.H.: Similarity and dissimilarity. In: Bock, H.-H., Diday, E. (eds.) Anal. Symb. Data, pp. 139–197. Springer, Berlin, Heidelberg (2000). https://doi.org/10.1007/978-3-642-57155-8_8
58. Simovici, D.: Mathematical analysis for machine learning and data mining. World Scientific (2018). https://doi.org/10.1142/10702
59. Bagnall, A., Lines, J., Bostrom, A., Large, J., Keogh, E.: The great time series classification bake off: a review and experimental evaluation of recent algorithmic advances. Data Min. Knowl. Discov. **31**, 606–660 (2017). https://doi.org/10.1007/s10618-016-0483-9
60. Su, H., Liu, S., Zheng, B., Zhou, X., Zheng, K.: A survey of trajectory distance measures and performance evaluation. VLDB J. **29**, 3–32 (2020). https://doi.org/10.1007/s00778-019-00574-9
61. Górecki, T., Piasecki, P.: A comprehensive comparison of distance measures for time series classification. In: Steland, A., Rafajłowicz, E., Okhrin, O. (eds.) Stoch. Model. Stat. Their Appl., pp. 409–428. Springer, Cham (2019). https://doi.org/10.1007/978-3-030-28665-1_31
62. Serrà, J., Arcos, J.L.: An empirical evaluation of similarity measures for time series classification. Knowl.-Based Syst. **67**, 305–314 (2014). https://doi.org/10.1016/j.knosys.2014.04.035
63. Giusti, R., Batista, G.E.A.P.A.: An empirical comparison of dissimilarity measures for time series classification. In: 2013 Brazilian Conf. Intell. Syst., pp. 82–88. IEEE (2013). https://doi.org/10.1109/BRACIS.2013.22
64. Jiang, W.: Time series classification: nearest neighbor versus deep learning models. SN Appl. Sci. **2**, 721 (2020). https://doi.org/10.1007/s42452-020-2506-9
65. Agrawal, R., Faloutsos, C., Swami, A.: Efficient similarity search in sequence databases. In: Lomet, D.B. (ed.) Proc. 4th Int. Conf. Found. Data Organ. Algorithms (FODO '93), pp. 69–84. Springer, Berlin, Heidelberg (1993). https://doi.org/10.1007/3-540-57301-1_5
66. Zhao, F., Gao, Y., Li, X., An, Z., Ge, S., Zhang, C.: A similarity measurement for time series and its application to the stock market. Expert Syst. Appl. **182**, 115217 (2021). https://doi.org/10.1016/j.eswa.2021.115217
67. Mirylenka, K., Dallachiesa, M., Palpanas, T.: Data series similarity using correlation-aware measures. In: Proc. 29th Int. Conf. Sci. Stat. Database Manag., pp. 1–12. ACM, New York, NY, USA (2017). https://doi.org/10.1145/3085504.3085515
68. Peng, B., Fatourou, P., Palpanas, T.: Fast data series indexing for in-memory data. VLDB J. **30**, 1041–1067 (2021). https://doi.org/10.1007/s00778-021-00677-2
69. Wang, Y., Wang, P., Pei, J., Wang, W., Huang, S.: A data-adaptive and dynamic segmentation index for whole matching on time series. Proc. VLDB Endow. **6**, 793–804 (2013). https://doi.org/10.14778/2536206.2536208
70. Batista, G.E.A.P.A., Wang, X., Keogh, E.J.: A complexity-invariant distance measure for time series. In: Proc. 2011 SIAM Int. Conf. Data Min., pp. 699–710. Society for Industrial and Applied Mathematics, Philadelphia, PA (2011). https://doi.org/10.1137/1.9781611972818.60

71. Bankó, Z., Abonyi, J.: Correlation based dynamic time warping of multivariate time series. Expert Syst. Appl. **39**, 12814–12823 (2012). https://doi.org/10.1016/j.eswa.2012.05.012
72. Kurbalija, V., Ivanović, M., von Bernstorff, C., Nachtwei, J., Burkhard, H.-D.: Matching observed with empirical reality—what you see is what you get? Fundam. Informaticae **129**, 133–147 (2014). https://doi.org/10.3233/FI-2014-965
73. Post, A.R., Harrison, J.H.: Temporal data mining. Clin. Lab. Med. **28**, 83–100 (2008). https://doi.org/10.1016/j.cll.2007.10.005
74. Keogh, E.: Efficiently finding arbitrarily scaled patterns in massive time series databases. In: Nada, L., Dragan, G., Ljupčo, T., Hendrik, B. (eds.) Knowl. Discov. Databases PKDD 2003, pp. 253–265. Springer, Berlin, Heidelberg (2003). https://doi.org/10.1007/978-3-540-39804-2_24
75. Das, G., Lin, K.-I., Mannila, H., Renganathan, G., Smyth, P.: Rule discovery from time series. In: Proc. Fourth Int. Conf. Knowl. Discov. Data Min., pp. 16–22. AAAI Press (1998)
76. Chen, S.-Y., Tseng, T.-T., Ke, H.-R., Sun, C.-T.: Social trend tracking by time series based social tagging clustering. Expert Syst. Appl. **38**, 12807–12817 (2011). https://doi.org/10.1016/j.eswa.2011.04.073
77. Goldin, D.Q., Kanellakis, P.C.: On similarity queries for time-series data: constraint specification and implementation. In: Montanari, U., Rossi, F. (eds.) Princ. Pract. Constraint Program.—CP '95, pp. 137–153. Springer, Berlin, Heidelberg (1995). https://doi.org/10.1007/3-540-60299-2_9
78. Kantardzic, M.: Data mining: concepts, models, methods, and algorithms, 3rd edn. Wiley-IEEE Press (2019)
79. Cassisi, C., Montalto, P., Aliotta, M., Cannata, A., Pulvirenti, A.: Similarity measures and dimensionality reduction techniques for time series data mining. In: Adv. Data Min. Knowl. Discov. Appl. InTech (2012). https://doi.org/10.5772/49941
80. Vlachos, M., Kollios, G., Gunopulos, D.: Discovering similar multidimensional trajectories. In: Proc. 18th Int. Conf. Data Eng., pp. 673–684. IEEE Comput. Soc. (2002). https://doi.org/10.1109/ICDE.2002.994784
81. Chen, L., Özsu, M.T., Oria, V.: Robust and fast similarity search for moving object trajectories. In: Proc. 2005 ACM SIGMOD Int. Conf. Manag. Data—SIGMOD '05, pp. 491–502. ACM Press, New York, USA (2005). https://doi.org/10.1145/1066157.1066213
82. Berndt, D., Clifford, J.: Using dynamic time warping to find patterns in time series. In: Usama, R.U., Fayyad, M. (eds.) Knowl. Discov. Databases Pap. from 1994 AAAI Work., pp. 359–370. AAAI Press, Seattle, Washington (1994). http://dblp.uni-trier.de/rec/bib/conf/kdd/BerndtC94
83. Buza, K.: Time series classification and its applications. In: Proc. 8th Int. Conf. Web Intell. Min. Semant., pp. 1–4. ACM, New York, NY, USA (2018). https://doi.org/10.1145/3227609.3227690
84. Li, H.: Multivariate time series clustering based on common principal component analysis. Neurocomputing **349**, 239–247 (2019). https://doi.org/10.1016/j.neucom.2019.03.060
85. Petitjean, F., Forestier, G., Webb, G.I., Nicholson, A.E., Chen, Y., Keogh, E.: Dynamic time warping averaging of time series allows faster and more accurate classification. In: 2014 IEEE Int. Conf. Data Min., pp. 470–479. IEEE (2014). https://doi.org/10.1109/ICDM.2014.27
86. Silva, D.F., Giusti, R., Keogh, E., Batista, G.E.A.P.A.: Speeding up similarity search under dynamic time warping by pruning unpromising alignments. Data Min. Knowl. Discov. **32**, 988–1016 (2018). https://doi.org/10.1007/s10618-018-0557-y
87. Li, G., Yuan, T., Qin, S.J., Chai, T.: Dynamic time warping based causality analysis for root-cause diagnosis of nonstationary fault processes. IFAC-PapersOnLine **48**, 1288–1293 (2015). https://doi.org/10.1016/j.ifacol.2015.09.146
88. Chen, Y., Nascimento, M.A., Ooi, B.C., Tung, A.K.H.: SpADe: on shape-based pattern detection in streaming time series. In: 2007 IEEE 23rd Int. Conf. Data Eng., pp. 786–795. IEEE (2007). https://doi.org/10.1109/ICDE.2007.367924
89. Bernecker, T., Houle, M.E., Kriegel, H.-P., Kröger, P., Renz, M., Schubert, E., Zimek, A.: Quality of similarity rankings in time series. In: Pfoser, D., Tao, Y., Mouratidis, K., Nascimento, M.A., Mokbel, M., Shekhar, S., Huang, Y. (eds.) Adv. Spat. Temporal Databases,

pp. 422–440. Springer, Berlin, Heidelberg (2011). https://doi.org/10.1007/978-3-642-22922-0_25
90. Vlachos, M., Gunopulos, D., Das, G.: Indexing time-series under conditions of noise. In: Last, M., Kandel, A., Bunke, H. (eds.) Data Min. Time Ser. Databases, pp. 67–100. World Scientific Publishing Company (2004). https://doi.org/10.1142/9789812565402_0004
91. Wagner, R.A., Fischer, M.J.: The string-to-string correction problem. J. ACM **21**, 168–173 (1974). https://doi.org/10.1145/321796.321811
92. Hirschberg, D.S.: A linear space algorithm for computing maximal common subsequences. Commun. ACM **18**, 341–343 (1975). https://doi.org/10.1145/360825.360861
93. Cormen, T.H., Leiserson, C.E., Rivest, R.L., Stein, C.: Introduction to algorithms, 4th ed. The MIT Press (2022)
94. Yazdani, N., Ozsoyoglu, Z.M.: Sequence matching of images. In: Proc. 8th Int. Conf. Sci. Stat. Data Base Manag., pp. 53–62. IEEE Comput. Soc. Press (1996). https://doi.org/10.1109/SSDM.1996.505915
95. Levenshtein, V.I.: Binary codes capable of correcting deletions, insertions, and reversals. Sov. Phys. Dokl. **10**, 707–710 (1966)
96. Ren, K., Qu, D., Xu, S., Jiao, X., Tai, L., Zhang, H.: Uncertainty visualization of transport variance in a time-varying ensemble vector field. ISPRS Int. J. Geo-Inf. **9**, 19 (2020). https://doi.org/10.3390/ijgi9010019
97. Toohey, K., Duckham, M.: Trajectory similarity measures. SIGSPATIAL Spec. **7**, 43–50 (2015). https://doi.org/10.1145/2782759.2782767
98. Nyirarugira, C., Kim, T.: Stratified gesture recognition using the normalized longest common subsequence with rough sets. Signal Process. Image Commun. **30**, 178–189 (2015). https://doi.org/10.1016/j.image.2014.10.008
99. Chen, L., Ng, R.: On the marriage of Lp-norms and edit distance. In: Nascimento, M.A., Özsu, M.T., Kossmann, D., Miller, R.J., Blakeley, J.A., Schiefer, K.B. (eds.) Proc. 2004 VLDB Conf., pp. 792–803. Elsevier (2004). https://doi.org/10.1016/B978-012088469-8.50070-X
100. Allison, L., Yee, C.: Minimum message length encoding and the comparison of macromolecules. Bull. Math. Biol. **52**, 431–453 (1990). https://doi.org/10.1016/S0092-8240(05)80219-8
101. Bandera, J.P., Marfil, R., Bandera, A., Rodríguez, J.A., Molina-Tanco, L., Sandoval, F.: Fast gesture recognition based on a two-level representation. Pattern Recognit. Lett. **30**, 1181–1189 (2009). https://doi.org/10.1016/j.patrec.2009.05.017
102. Koide, S., Xiao, C., Ishikawa, Y.: Fast subtrajectory similarity search in road networks under weighted edit distance constraints. Proc. VLDB Endow. **13**, 2188–2201 (2020). https://doi.org/10.14778/3407790.3407818
103. Kurbalija, V., Radovanović, M., Geler, Z., Ivanović, M.: A framework for time-series analysis. In: Dicheva, D., Dochev, D. (eds.) Artif. Intell. Methodol. Syst. Appl. SE-5, pp. 42–51. Springer, Berlin, Heidelberg (2010). https://doi.org/10.1007/978-3-642-15431-7_5
104. Kurbalija, V., Ivanović, M., Geler, Z., Radovanović, M.: Two faces of the framework for analysis and prediction, Part 1—Education. Inf. Technol. Control. **47**, 249–261 (2018). https://doi.org/10.5755/j01.itc.47.2.18746
105. Kurbalija, V., Ivanović, M., Geler, Z., Radovanović, M.: Two faces of the framework for analysis and prediction, Part 2—Research. Inf. Technol. Control. **47**, 489–502 (2018). https://doi.org/10.5755/j01.itc.47.3.18747
106. Keogh, E., Ratanamahatana, C.A.: Exact indexing of dynamic time warping. Knowl. Inf. Syst. **7**, 358–386 (2005). https://doi.org/10.1007/s10115-004-0154-9
107. Müller, M.: Music synchronization. In: Fundam. Music Process., pp. 115–166. Springer, Cham (2015). https://doi.org/10.1007/978-3-319-21945-5_3
108. Oregi, I., Pérez, A., Del Ser, J., Lozano, J.A.: On-line elastic similarity measures for time series. Pattern Recognit. **88**, 506–517 (2019). https://doi.org/10.1016/j.patcog.2018.12.007
109. Geler, Z., Kurbalija, V., Ivanović, M., Radovanović, M.: Elastic distances for time-series classification: Itakura versus Sakoe-Chiba constraints. Knowl. Inf. Syst. **64**, 2797–2832 (2022). https://doi.org/10.1007/s10115-022-01725-1

110. Geler, Z., Kurbalija, V., Ivanovic, M., Radovanovic, M., Dai, W.: Dynamic time warping: Itakura vs Sakoe-Chiba. In: 2019 IEEE Int. Symp. Innov. Intell. Syst. Appl., pp. 1–6. IEEE (2019). https://doi.org/10.1109/INISTA.2019.8778300
111. Sakoe, H., Chiba, S.: Dynamic programming algorithm optimization for spoken word recognition. IEEE Trans. Acoust. **26**, 43–49 (1978). https://doi.org/10.1109/TASSP.1978.1163055
112. Itakura, F.: Minimum prediction residual principle applied to speech recognition. IEEE Trans. Acoust. **23**, 67–72 (1975). https://doi.org/10.1109/TASSP.1975.1162641
113. Choi, W., Cho, J., Lee, S., Jung, Y.: Fast constrained dynamic time warping for similarity measure of time series data. IEEE Access **8**, 222841–222858 (2020). https://doi.org/10.1109/ACCESS.2020.3043839
114. Xi, X., Keogh, E., Shelton, C., Wei, L., Ratanamahatana, C.A.: Fast time series classification using numerosity reduction. In: Proc. 23rd Int. Conf. Mach. Learn.—ICML '06, pp. 1033–1040. ACM Press, New York, USA (2006). https://doi.org/10.1145/1143844.1143974
115. Geler, Z.: Role of similarity measures in time series analysis. University of Novi Sad, Serbia (2015). (PhD thesis) https://nardus.mpn.gov.rs/handle/123456789/1703
116. Geler, Z., Kurbalija, V., Radovanović, M., Ivanović, M.: Impact of the Sakoe-Chiba band on the DTW time series distance measure for kNN classification. In: Buchmann, R., Kifor, C.V., Yu, J. (eds.) 7th Int. Conf. Knowl. Sci. Eng. Manag. KSEM 2014, pp. 105–114. Springer, Cham (2014). https://doi.org/10.1007/978-3-319-12096-6_10
117. Kurbalija, V., Radovanović, M., Geler, Z., Ivanović, M.: The influence of global constraints on DTW and LCS similarity measures for time-series databases. In: Dicheva, D., Markov, Z., Stefanova, E. (eds.) Third Int. Conf. Software, Serv. Semant. Technol. S3T 2011 SE-10, pp. 67–74. Springer, Berlin, Heidelberg (2011). https://doi.org/10.1007/978-3-642-23163-6_10
118. Kurbalija, V., Radovanović, M., Geler, Z., Ivanović, M.: The influence of global constraints on similarity measures for time-series databases. Knowl.-Based Syst. **56**, 49–67 (2014). https://doi.org/10.1016/j.knosys.2013.10.021
119. Ratanamahatana, C.A., Keogh, E.: Three myths about dynamic time warping data mining. In: Proc. 2005 SIAM Int. Conf. Data Min., pp. 506–510. Society for Industrial and Applied Mathematics, Philadelphia, PA (2005). https://doi.org/10.1137/1.9781611972757.50
120. Dau, H.A., Silva, D.F., Petitjean, F., Forestier, G., Bagnall, A., Mueen, A., Keogh, E.: Optimizing dynamic time warping's window width for time series data mining applications. Data Min. Knowl. Discov. **32**, 1074–1120 (2018). https://doi.org/10.1007/s10618-018-0565-y
121. Mitrovic, D., Ivanović, M., Geler, Z.: Agent-based distributed computing for dynamic networks. Inf. Technol. Control. **43**, 88–97 (2014). https://doi.org/10.5755/j01.itc.43.1.4588
122. Kurbalija, V., Ivanović, M., Budimac, Z.: Case-based curve behaviour prediction. Softw. Pract. Exp. **39**, 81–103 (2009). https://doi.org/10.1002/spe.891
123. Kotsiantis, S.B.: Supervised machine learning: a review of classification techniques. In: Proc. 2007 Conf. Emerg. Artif. Intell. Appl. Comput. Eng. Real Word AI Syst. with Appl. EHealth, HCI, Inf. Retr. Pervasive Technol., pp. 3–24. IOS Press, Amsterdam, The Netherlands (2007). http://dl.acm.org/citation.cfm?id=1566770.1566773
124. Deza, M.M., Deza, E.: Encyclopedia of distances. Springer, Berlin, Heidelberg (2016). https://doi.org/10.1007/978-3-662-52844-0
125. Cha, S.-H.: Comprehensive survey on distance/similarity measures between probability density functions. Int. J. Math. Model. Methods Appl. Sci. **1**, 300–307 (2007). http://www.gly.fsu.edu/~parker/geostats/Cha.pdf
126. Marteau, P.-F.: Time warp edit distance with stiffness adjustment for time series matching. IEEE Trans. Pattern Anal. Mach. Intell. **31**, 306–318 (2009). https://doi.org/10.1109/TPAMI.2008.76
127. Larose, D.T., Larose, C.D.: Discovering knowledge in data, 2nd edn. Wiley (2014). https://doi.org/10.1002/9781118874059
128. Symeonidis, A.L., Mitkas, P.A.: Agent intelligence through data mining (multiagent systems, artificial societies, and simulated organizations). Springer, Secaucus, NJ, USA (2005)

129. Lin, J., Keogh, E., Wei, L., Lonardi, S.: Experiencing SAX: a novel symbolic representation of time series. Data Min. Knowl. Discov. **15**, 107–144 (2007). https://doi.org/10.1007/s10618-007-0064-z
130. Keogh, E.J., Chu, S., Hart, D., Pazzani, M.: Segmenting time series: a survey and novel approach. In: Last, M., Kandel, A., Bunke, H. (eds.) Data Min. Time Ser. Databases, pp. 1–22. World Scientific Publishing Company (2004)
131. Keogh, E., Chakrabarti, K., Pazzani, M., Mehrotra, S.: Locally adaptive dimensionality reduction for indexing large time series databases. In: Proc. 2001 ACM SIGMOD Int. Conf. Manag. Data, pp. 151–162. ACM, New York, NY, USA (2001). https://doi.org/10.1145/375663.375680
132. Keogh, E., Chakrabarti, K., Pazzani, M., Mehrotra, S.: Dimensionality reduction for fast similarity search in large time series databases. Knowl. Inf. Syst. **3**, 263–286 (2001). https://doi.org/10.1007/PL00011669
133. Kurbalija, V., Ivanović, M., Radovanović, M., Geler, Z., Dai, W., Zhao, W.: Emotion perception and recognition: an exploration of cultural differences and similarities. Cogn. Syst. Res. **52**, 103–116 (2018). https://doi.org/10.1016/j.cogsys.2018.06.009
134. Kurbalija, V., Radovanović, M., Ivanović, M., Schmidt, D., von Trzebiatowski, G.L., Burkhard, H.-D., Hinrichs, C.: Time-series analysis in the medical domain: a study of Tacrolimus administration and influence on kidney graft function. Comput. Biol. Med. **50**, 19–31 (2014). https://doi.org/10.1016/j.compbiomed.2014.04.007
135. Mitrović, D., Ivanović, M., Budimac, Z., Vidaković, M.: Radigost: interoperable web-based multi-agent platform. J. Syst. Softw. **90**, 167–178 (2014). https://doi.org/10.1016/j.jss.2013.12.029
136. Sredojević, D., Vidaković, M., Ivanović, M.: ALAS: agent-oriented domain-specific language for the development of intelligent distributed non-axiomatic reasoning agents. Enterp. Inf. Syst. **12**, 1058–1082 (2018). https://doi.org/10.1080/17517575.2018.1482567
137. Savaglio, C., Ganzha, M., Paprzycki, M., Bădică, C., Ivanović, M., Fortino, G.: Agent-based Internet of Things: State-of-the-art and research challenges. Futur. Gener. Comput. Syst. **102**, 1038–1053 (2020). https://doi.org/10.1016/j.future.2019.09.016
138. Bădică, C., Budimac, Z., Burkhard, H.-D., Ivanovic, M.: Software agents: languages, tools, platforms. Comput. Sci. Inf. Syst. **8**, 255–298 (2011). https://doi.org/10.2298/CSIS110214013B
139. Mitrović, D., Ivanović, M., Vidaković, M., Budimac, Z.: The Siebog multiagent middleware. Knowl.-Based Syst. **103**, 56–59 (2016). https://doi.org/10.1016/j.knosys.2016.03.017
140. Mitrovic, D., Ivanovic, M., Vidakovic, M., Budimac, Z.: Siebog: an enterprise-scale multi-agent middleware. Inf. Technol. Control. **45** (2016). https://doi.org/10.5755/j01.itc.45.2.12621
141. Mitrovic, D., Ivanovic, M., Budimac, Z., Vidakovic, M.: Supporting heterogeneous agent mobility with ALAS. Comput. Sci. Inf. Syst. **9**, 1203–1229 (2012). https://doi.org/10.2298/CSIS120102025M
142. Mitrović, D., Geler, Z., Ivanović, M.: Distributed distance matrix generator based on agents. In: Proc. 2nd Int. Conf. Web Intell. Min. Semant.—WIMS '12, pp. 1–6. ACM Press, New York, USA (2012). https://doi.org/10.1145/2254129.2254179
143. Mitrović, D., Budimac, Z., Ivanović, M., Vidaković, M.: Agent-based approaches to managing fault-tolerant networks of distributed multi-agent systems. Multiagent Grid Syst. **7**, 203–218 (2011). https://doi.org/10.3233/MGS-2011-0176

Chapter 3
Time Series Classification

Classification refers to the task of assigning predefined labels—*classes*—to individual data objects. It is performed based on one selected attribute (the *class label*), that has a fixed number of categorical values. The number of categories (classes) is known in advance, which is one of the main differences between clustering and classification: for clustering, the number of groups is usually not known, but rather revealed during the clustering process. Therefore, classification is an example of *supervised learning*, while clustering is considered *unsupervised learning*. Supervised learning algorithms infer functions that make predictions for previously unseen data examples based on examples from the training (labeled) data [1, 2]. Unsupervised learning algorithms, on the other hand, infer patterns (in the case of clustering, groups of data examples) by exclusively looking at unlabeled data examples [1, 2].

One of the basic data-mining tasks is training a classifier using available training data. The goal of this process is to infer models that perform accurate mapping of data objects (time series, in our case) onto predetermined classes. The classification workflow is usually comprised of the following steps [3, 4]:

1. **Classifier training**: the classifier is trained using a given set of known data examples (which constitute the *training set*) relying on a selected learning method. The value of the class attribute (i.e., the *label*) must be provided for each training example. Based on the training dataset, the learning algorithm derives the classification model, i.e., a function that assigns one of the predefined classes to each data instance.
2. **Classifier testing**: the classifier obtained in the first phase must be evaluated using a separate set of data examples (referred to as the *test set*) which have not been seen by the classifier during its training. The label must be known in advance also for the examples included in the test set. The *accuracy* (see Sect. 3.2) of the classifier is determined by applying it to the test examples and comparing the predicted and the actual class labels. If the classifier accuracy is not at an adequate level, the training process needs to be revisited. To improve the classifier,

Z. Gellér et al., *Recent Advances in Time-Series Classification—Methodology and Applications*, Intelligent Systems Reference Library 264,
https://doi.org/10.1007/978-3-031-77527-7_3

we can consider multiple (not necessarily mutually exclusive) possibilities: a larger training set may be needed, some key peculiarities of the problem may not have been accounted for, *hyperparameter* tuning may be needed, or perhaps the selected learner is not appropriate for the problem at hand.

3. **Classifier application**: If the trained classifier exhibits adequate accuracy on the test set, it can be deployed for classification of previously unseen data objects.

Classification of time series is a very well-researched topic in the time-series community [5–12]. Górecki and Luczak [13] developed a novel distance measure based on time-series shapes as opposed to point-to-point comparison. Jeong et al. [14] introduced the weighted version of the Dynamic Time Warping (DTW) similarity measure and proposed a modified logistic weight function. A consequence of the *curse of dimensionality*[1] called *hubness* (the tendency of some data examples to appear in k nearest neighbor lists of unexpectedly many other examples in a dataset) was studied by Radovanović et al. [15] in the context of nearest neighbor-based time-series classification. Spiegel et al. [16] presented a DTW variation based on a greedy method that evaluates the elements of the *warping matrix* (see Sect. 2.3.2) that most likely contributes to the *warping path*, thereby speeding up distance computations and, consequently, time-series classification. Xi et al. [17] suggested to combine constrained DTW (see Sect. 2.4) with *numerosity reduction* (sampling of the training data), with the goal of producing compact classifiers and preserving accuracy. To overcome some shortcomings of the nearest-neighbor classifier, Ye and Keogh [18] proposed a primitive called *time series shapelets*.[2]

Methods for classifier learning range from simpler statistical approaches, e.g., exponential smoothing [19] and ARIMA [20], to more complex machine-learning models like support vector machines [21, 22], genetic algorithms [23], fuzzy systems [24] and neural networks [25].

Autoregressive (AR) models are a common statistical approach for time series analysis: they predict future values by relying on a linear combination of past observations in the series. Such solutions are more naturally applicable to prediction of future time-series values (i.e., *forecasting*). However, interesting approaches exist in the domain of classification. The large margin autoregressive (LMAR) method introduced by Kini and Sekhar [26] builds an AR model for each class and employs a large margin method for parameter estimation. Turrado García et al. [27] described a system that groups time series with the same forecasting model in the supply-chain domain. Huan and Palaniappan [28] applied neural network-based classification of autoregressive features to electroencephalogram signals recorded during different mental tasks.

An established procedure of classifying time series relies on transforming them into feature vectors of fixed dimensionality. In such a setting, many classical machine-learning and data-mining approaches can then be used for classification or clustering. In their work on time series clustering, Zhang et al. [29] focused on the

[1] The curse of dimensionality covers issues that may arise when analyzing high-dimensional data (i.e. data with a large number of features) [66].

[2] Time series segments that capture the key characteristics of a specific class [18].

orthogonal wavelet transform. They employed the coefficients of the Haar wavelet to create feature vectors for clustering. With the aim of reducing the Euclidean distance between signals belonging to the same class, Eruhimov et al. [30] utilized DTW to transform the time axis. Adaptations of segment-based representations for extracting features from time series [31–33] also need to be mentioned. Other works in the area of transforming time series into vector representations relies on extraction of meaningful patterns [34] and application of segment-base features [35].

Another common technique in the field of time series classification is to establish a distance measure for quantifying the (dis)similarity between time series. This measure is then leveraged by distance-based classification algorithms for data categorization [9]. The key challenge lies in chosing an appropriate measure that effectively captures the shapes of the time series data. Fundamental similarity measures used for this purpose are overviewed in the previous chapter. In addition to them, a number of other measures have been proposed, such as, for example, the Time Warp Edit Distance (TWED)—an alignment-based distance measure introduced by Marteau [36]. Górecki and Luczak [13] proposed a novel method for time series distance calculation that leverages derivatives. This approach goes beyond point-to-point comparisons, enabling the consideration of the overall shape of the time series data. Weighted DTW (WDTW), a generalization of DTW by Jeong et al. [14], penalizes time-series points by phase difference between a reference point of the first time series and the test point of the second time series. This strategy can alleviate some imbalanced alignments where one point from the first series gets mapped onto many points from the second series.

In this book, we focus on the approach to time-series classification that employs time-series distance measures in combination with common basic distance-based classifiers—the 1NN rule and its extensions. This was motivated by numerous papers that compare various classification algorithms, where the simple nearest-neighbor approach proved to be competitive [5, 6, 12, 37, 38]. An additional advantage of the nearest neighbor classifier is that its accuracy directly reflects the suitability of the used (dis)similarity measure [39], providing the groundwork for our examination of the influence of elastic distance measure constraints and facilitating practical illustrations of the findings.

3.1 The Nearest Neighbor Rule

Let $Q = (q_1, q_2, \dots, q_n)$ be a time series with n values representing the evolution of a phenomenon at consecutive fixed time intervals [40–42]. The objective of classification is to assign a class label to new (unlabeled) time series based on the information contained within a provided set of labeled time series (i.e., the training set): $T = \{(T_1, T_1^c), (T_2, T_2^c), \dots, (T_m, T_m^c)\}$ [41, 43, 44]. In the formulas and their descriptions in the rest of this section, the set of labels associated with the members of the training set T is denoted by C:

$$C = \{T_1^c, T_2^c, \ldots, T_{|C|}^c\}$$

where T_i^c is the class of time series T_i.

When it comes to classifying time series data, the simple nearest-neighbor (1NN) method [45] has emerged as a prominent technique [5, 37, 46]. It assigns the class label of the closest training example (most similar time series) to new unlabeled time series. Generally for the 1NN classifier, Cover and Hart [45] demonstrated that the asymptotic misclassification rate R is bounded as follows:

$$R^* \leq R \leq R^* \times \left(2 - R^* \times \frac{|C|}{|C| - 1}\right)$$

where R^* is the Bayes error.

The kNN rule was first formulated by by Fix and Hodges [47], among others. Let $C = \{c_1, c_2\}$ denote a set containing two class labels and let the class c_i $(i = 1,2)$ in the training set T be represented by n_i instances. Accoding to Fix and Hodges, class c_1 is assigned to an unlabeled time series Q if the condition $k_1/n_1 > k_2/n_2$ is satisfied, where k_i is the number of representatives of class c_1 in the subset of the training set T containing the $k = k_1 + k_2$ nearest neighbors of Q. The normalization of nearest neighbor counts using class frequencies aims to address potential bias introduced by class imbalance. This approach ensures that the prevalence of a particular class among the nearest neighbors of a new time series (Q) is proportionate to its representation within the training dataset.

The kNN algorithm, which leverages majority voting among the k nearest neighbors, provides an even more straightforward generalization of the 1NN classifier: a given unlabeled time series Q is assigned the most frequent class among its k nearest neighbors in T (without normalization using class frequencies). This procedure can be formulated as follows:

$$Q^C = \underset{c \in C}{arg\ max} \left\{ \sum_{i=1}^{k} E\left(c = T_i^c\right) \right\}$$

where $E(\bullet)$ is the indicator function and T_i^c is the ith nearest neighbor's class.

Wu et al. [48] discuss some challenges associated with selecting the number of considered nearest neighbors. With small k, the kNN rule is more likely to be misled by points representing noise. A large k value can lead to the inclusion of neighbors from diverse classes, compromising the local nature of the neighborhood. Another issue may arise from treating all nearest neighbors equally, regardless of their distance from Q. One way to mitigate this is by weighing neighbors' votes based on their distance from the query object in the following manner:

$$Q^C = \underset{c \in C}{arg\ max} \left\{ \sum_{i=1}^{k} w_i \times E\left(c = T_i^c\right) \right\}$$

where w_i represents the ith nearest neighbor's weight.

The subsections that follow present various weighting functions, laying the foundation for the in-depth analysis of their impact on classification accuracy in Chaps. 5 and 6.

3.1.1 Dudani's Weighting Functions

A distance-based weighting scheme for kNN voting was first proposed by Dudani in [49] (the method will be henceforth denoted "DN"). It assigns weights to neighbors between 0 and 1: the first neighbor is weighted by 1 and the farthest by 0. The weights of the neighbors between the first and the last are determined by linear scaling as follows:

$$w_i = \begin{cases} 1 & d_k = d_1 \\ \frac{d_k - d_i}{d_k - d_1} & d_k \neq d_1 \end{cases}$$

where d_i is the distance between Q and its ith nearest neighbor.

Providing an example from the field of three-class problems (where the a priori probabilities of the classes were equal), Dudani showed that in the case of a small-sized training set, the DN weighting function produces a lower error probability than the majority-voting kNN [49]. The likelihood of error was assessed by Monte Carlo analysis, using Euclidean neighbor distances. The results illustrated in Fig. 3.1 (from [49]) demonstrate that the error probability of the weighting function introduced by Dudani is lower than in the case of the majority rule. Furthermore, Dudani noted that the high probability of error for the majority rule with $k = 2$ was caused by many ties.

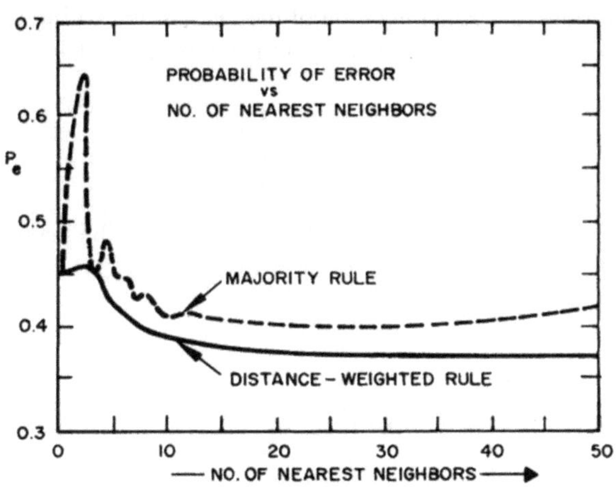

Fig. 3.1 Probability of error in relation to the number of nearest neighbors [49]

Dudani's opinion is that DN is appropriate for small and moderately-sized training sets [49], with one of the reasons for its superiority being the lower probability of ties.

Dudani also proposed two other weighting schemes [49]—the inverse distance weight (denoted "IN"):

$$w_i = \frac{1}{d_i}$$

and the rank weight (denoted "RK"):

$$w_i = k - i + 1$$

The inverse of the *squared* distance [48, 50, 51] (denoted "IS") can be used instead of the inverse distance. However, division by zero is possible in both cases, which is usually addresses by the addition of a small constant ε to the denominator [52, 53] (in the experiments whose results are presented in Chaps. 5 and 6, we used 0.001) as in

$$w_i = \frac{1}{d_i^2 + \varepsilon}$$

3.1.2 Macleod's Weighting Function

Dudani's weighting scheme completely ignores the last (kth) neighbor in the (quite usual) case when $d_k \neq d_1$, since $d_k - d_i = 0$ for $i = k$. In [54], Macleod et al. proposed a generalization of the Dudani weighting through the introduction of two parameters: $s \geq k$ and $\alpha \geq 0$ (henceforth denoted "MC"):

$$w_i = \begin{cases} 1 & d_s = d_1 \\ \frac{(d_s - d_i) + \alpha \times (d_s - d_1)}{(1+\alpha) \times (d_s - d_1)} & d_s \neq d_1 \end{cases}$$

When $s = k$ and $\alpha = 0$, MC reverts to the original Dudani weighting. Among the parameter combinations investigated in [54], we will utilize $s = k$ with $\alpha = 1$ in Chap. 5.

3.1.3 The Fibonacci Weighting Function

The application of the Fibonacci sequence for the purposes of nearest neighbor weighting was proposed by Pao et al. [55]:

Table 3.1 Accuracies of various weighting techniques used in combination with three kNN variants [55]

	Dudani (%)	Inverse (%)	Rank (%)	Fibonacci (%)
WKNN	75.6	74.2	73.8	76.1
WCAP	74.2	73.6	73.1	74.5
WDKNN	78.7	79.5	81.2	81.4

$$w_i = w_{i+1} + w_{i+2}$$

$$w_{k-1} = w_k = 1$$

They compared the Fibonacci method (denoted "FB") with the three of Dudani's weighting schemes (DN, IN and RK) defined in the domain of emotion recognition from Mandarin speech. In addition to majority-vote and distance-weighted kNN (WKNN) rules, two additional kNN variants were also included in their experiments: *Categorical Average Patterns* (WCAP) [56] and *Weighted Discrete* kNN (WDKNN) [57]. The results revealed that the FB function achieved better performance than all other weighting methods across the tested classifiers.

Pao et al. [55] enlisted 16 female and 18 male subjects to express five emotions through speech (neutral, happy, angry, sad and bored). To ensure high-quality data for the experiments, only recordings with a strong consensus (above 80% agreement) among ten independent human listeners were retained. The dataset prepared in this manner contained 116 neutral, 96 happy, 151 angry, 124 sad and 83 bored recordings.

During the experiments, the authors extracted three acoustic features from spoken utterances: *Linear Predictive Cepstral Coefficients* (LPCC), *Linear Predictive Coefficients* (LPC) and *Mel-Frequency Cepstral Coefficients* (MFCC). Min–max normalization was applied to the data in the preprocessing stage.

To identify the optimal value of k (the number of nearest neighbors) between 1 and 15, Pao et al. employed the *leave-one-out* (LOO) procedure (described in Sect. 3.2) and the majority-voting classification technique. After that, a comparative analysis of weighting schemes and different classifiers was conducted using the established optimal value of k. The obtained results presented in Table 3.1 (from [55]) indicated that the Fibonacci weighting scheme produced the best results.

3.1.4 The Uniform and Dual-Uniform Weighting Functions

Gou et al. [58] proposed a weighting method expressed as the inverse value of the neighbors' rank (denoted "UN"):

$$w_i = \frac{1}{i}$$

Table 3.2 The UCI datasets used by Gou et al. in their experiments [58]

Dataset	Attributes	Instances	Classes
Pendigits	16	10,992	10
Oppdigits	64	5,620	10
Ionosphere	34	351	2
Glass	10	214	7
Landsat Satellite	36	6,435	7
Libras Movement	90	360	15
Wine Quality-Red	11	1,599	11
Zoo	17	101	7
Vehicle	18	946	4
Wine Quality-White	11	4,898	11
Letter	16	20,000	26
Image Segmentation	19	2,310	7

and an extension of Dudani's scheme (denoted "DU"):

$$w_i = \begin{cases} 1 & d_k = d_1 \\ \frac{d_k - d_i}{d_k - d_1} \times \frac{1}{i} & d_k \neq d_1 \end{cases}$$

The authors denoted the weighted kNN classifiers employing the UN, DU and DN weighting functions, respectively, by UWKNN, DWKNN and WKNN. The conclusion of comparing their accuracies with the performance of the majority-voting kNN was that DU outperformed all the other examined methods.

In addition to artificially generated data, the experiments were also performed on 12 real-world UCI [59] datasets (Table 3.2) with numerical attributes only. Comparison of dataset elements was realized relaying on the Euclidean distance, and the optimal neighborhood size (k) was chosen by the LOO algorithm.

The first phase of the experiments involved 600 elements of a synthetic dataset and the value of k was varied from 1 to 50. The observed dependence of classification accuracy on the number of nearest neighbors is illustrated in Fig. 3.2 taken from [58].

In the second phase, the number of elements in the synthetic dataset was gradually increased from 100 to 2000 in steps of 100 but the number of neighbors was fixed at 15. Figure 3.3 taken from [58] shows classification accuracy as a function of the amount of available data.

The findings suggested that the DU weighting scheme (DWKNN) outperformed the alternatives in both phases of the experiments. It is worth mentioning that the majority-voting kNN was outperformed by the other classifiers in both experimental setups (which will be confirmed Chaps. 5 and 6 as well).

The lowest obtained error rates and the corresponding optimal k values are shown in Table 3.3 (from [58]). For 7 datasets out of 12, the best performance was exhibited

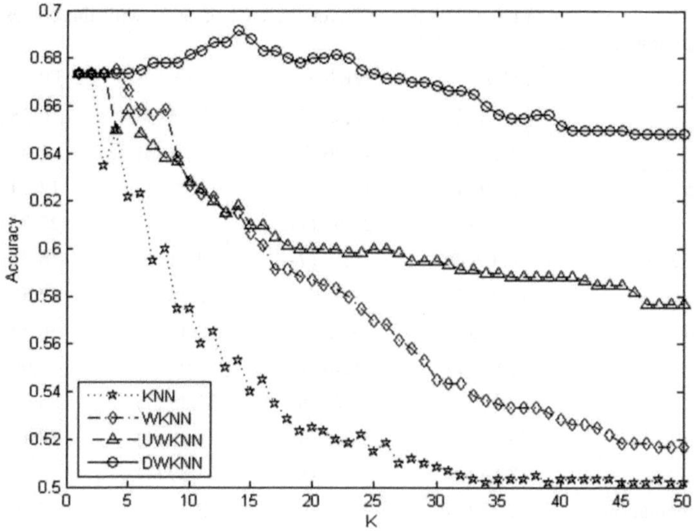

Fig. 3.2 Dependence of classification accuracy on the number of considered nearest neighbors [58]

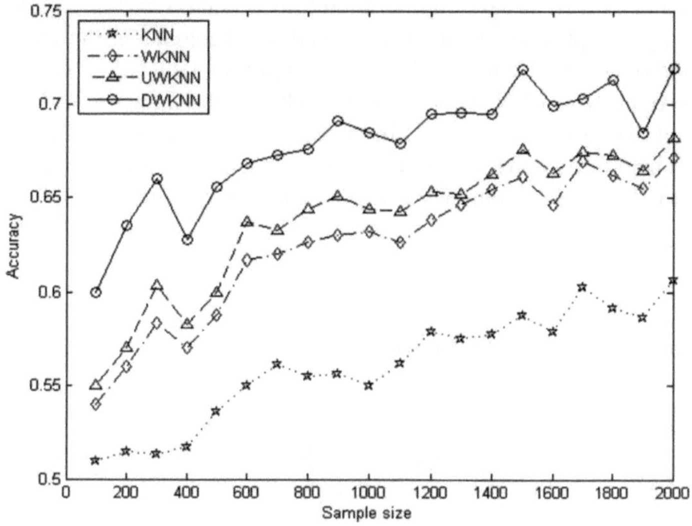

Fig. 3.3 Impact of the amount of available data on classification accuracy [58]

by the DU weighting scheme (DWKNN). Moreover, among the classifiers examined, the DU weighting method achieved the lowest average classification error.

Based on the results shown in Table 3.3, it can be also seen that, when using kNN, WKNN and UWKNN, the optimal k-value was low (< 10) in the case of each considered dataset (for most of them it was even equal to 1). Contrary to them, DWKNN

Table 3.3 The lowest error rates (%) achieved over the UCI datasets with the corresponding k-values in parentheses [58]

Datasets	kNN	WKNN	UWKNN	DWKNN
Pendigits	0.58 (3)	**0.53 (6)**	0.55 (4)	0.55 (22)
Oppdigits	1.00 (4)	**0.96 (7)**	1.01 (8)	1.03 (32)
Ionosphere	13.39 (1)	13.39 (1)	13.39 (1)	**12.54 (14)**
Glass	26.64 (1)	26.64 (1)	26.64 (1)	**25.23 (8)**
Landsat Satellite	8.44 (4)	8.45 (7)	**8.34 (6)**	**8.34 (6)**
Libras Movement	12.78 (1)	12.78 (1)	12.78 (1)	**12.50 (6)**
Wine Quality-Red	38.46 (1)	38.15 (6)	38.46 (1)	**36.52 (44)**
Zoo	**1.98 (1)**	**1.98 (1)**	**1.98 (1)**	**1.98 (1)**
Vehicle	33.45 (5)	**32.74 (6)**	33.45 (7)	33.92 (30)
Wine Quality-White	38.38 (1)	38.36 (4)	38.38 (1)	**38.10 (17)**
Letter	3.64 (4)	**3.29 (8)**	3.39 (7)	3.32 (50)
Image segmentation	3.33 (1)	**3.12 (6)**	3.33 (1)	3.29 (9)
Average error	15.17	15.03	15.14	**14.78**

often preferred a much larger neighborhood (the *Zoo* and *Landsat Sattelite* datasets were the only exceptions). We also confirmed this behavior of the DU scheme in Chaps. 5 and 6 over the UCR [60] datasets. In addition, Gou et al. [58] reported that when classifying with kNN, WKNN and UWKNN, the increase in the number of nearest neighbors was accompanied by a decrease in accuracy, DWKNN's performance remained fairly consistent (and even exhibited growth in the case of some of the datasets). Again, DU demonstrated this feature also with the UCR data sets in Chap. 5.

3.1.5 Zavrel's Weighting Function

Zavrel [52] proposed another weighting method, comparing it with DN and IN weighting schemes, as well as baseline 1NN and kNN classifiers. This scheme (denoted "ZA") relies on the following exponential function:

$$w_i = e^{-\alpha d_i^{\beta}}$$

Within the above equation, α and β represent constant values. Alpha (α) dictates the slope of the exponential decay function and beta (β) governs the power of the decay. In accordance with Zavrel's recommendations, in the experiments presented in Chap. 5, we set the value of both of these parameters to 1.

Zavrel's survey included one linguistic dataset (*PP-attachment*) [61] and several datasets from the UCI repository [59]. Table 3.4 (source: [52]) provides an overview of the datasets used in this study.

In the experiments, Zavrel used the cosine distance [62]. In the case of datasets marked with 10CV in the "Error estimation" column of Table 3.4, the training set comprised 90% of the available data and the test set the remaining 10%. In addition, these experiments were repeated 10 times, with random shuffling of the datasets before each evaluation. With the rest of the datasets, the original test partition (or a separate test dataset) defined in the original papers was used. The optimal value of the parameter k was chosen from the interval $[1, 100]$ so that the accuracy on the test set was maximal.

According to Zavrel's findings (Table 3.5), weighting the nearest neighbor votes improved the performance of the kNN classifier. In doing so, DN outperformed the other two schemes (which was also confirmed over the UCR [60] datasets in Chap. 5). The k-values that provided the best results are listed in parentheses. Dashes indicate that the performance of the given method did not exceed that of 1NN.

Looking at the results, it can be seen that 1NN outperformed the other variants only on the *Segmentation* dataset and that in most cases the DN scheme proved to be the most successful. Moreover, Zavrel's findings showed that, with the increase in the number of considered neighbors, after reaching the optimal accuracy, the performance turns into an almost monotonous decrease. The conclusion of the study [52] was that the application of a weighting function can potentially increase the robustness of kNN in choosing the value of the parameter k and lead to improved accuracy. In our extensive experiments (Chaps. 5 and 6), we made similar findings for time-series mining.

Table 3.4 UCI datasets used by Zavrel [52]

Dataset	Attributes	Instances	Classes	Error estimation
PP-attachment	100	20,801	2	Test set 3,097 cases
Glass	9	214	6	10CV
Wine	13	178	3	10CV
Sonar	60	208	2	10CV
Letter	16	16,000 train	26	Test set 4,000 cases
Isolet	617	6,238 train	26	Test set 1,559 cases
Vowel	10	517 train	11	Test set 473 cases
Segmentation	19	210 train	7	Test set 2,100 cases
Ionosphere	34	351	2	10CV
Diabetes (pima)	8	768	2	10CV
Breast cancer (wpbc)	32	198	2	10CV
Breast cancer (wdbc)	30	569	2	10CV
Cleveland heart	13	303	5	10CV

Table 3.5 Overview of Zavrel's results [52]

Dataset	1NN	kNN	Inverse	Dudani	Zavrel
PP-attachment	80.1	83.4 (13)	83.7 (13)	**84.2 (30)**	84.0 (35)
Glass	76.4	**77.3 (2)**	–	**77.3 (5)**	76.8 (3)
Wine	96.7	**97.8 (3)**	**97.8 (3)**	**97.8 (7)**	**97.8 (3)**
Sonar	82.5	–	83.1 (7)	**85.0 (9)**	–
Letter	95.6	–	–	**96.0 (5)**	–
Isolet	88.6	91.9 (13)	92.4 (13)	**92.9 (15)**	92.4 (3)
Vowel	52.6	–	55.6 (7)	**55.8 (15)**	55.0 (7)
Segmentation	**90.9**	–	–	–	–
Ionosphere	90.0	–	–	**90.6 (5)**	–
Diabetes (pima)	66.1	70.1 (3)	69.7 (80)	**70.3 (100)**	70.1 (3)
Breast cancer (wpbc)	69.0	79.5 (11)	**81.0 (9)**	79.5 (21)	79.5 (11)
Breast cancer (wdbc)	89.1	**93.3 (5)**	93.2 (5)	92.8 (30)	**93.3 (5)**
Cleveland heart	57.0	**62.7 (2)**	58.7 (11)	58.7 (9)	59.3 (100)

3.1.6 *The Dual Distance-Weighted Function*

As an extension of Dudani's weighting scheme, Gou et al. [63] introduced the dual distance-weighted function (denoted "DD") in the following manner:

$$w_i = \begin{cases} 1 & d_k = d_1 \\ \frac{d_k - d_i}{d_k - d_1} \times \frac{d_k + d_1}{d_k + d_i} & d_k \neq d_1 \end{cases}$$

This novel weighting adapts the DN scheme by weighting the nearest and farthest neighbors the same way but assigning smaller values to neighbors in between. This approach was proposed with the aim to enhance kNN's accuracy in the presence of outliers, as well as when the dataset exhibits an imbalanced class distribution.

In [63], the performance of the DD method was compared with 1NN, kNN and the DN weighting function utilizing 12 datasets available through the UCI Machine Learning Repository [59] (Table 3.6). Evaluation was performed using 20-fold cross-validation (see Sect. 3.2) relying on the Euclidean distance, with the number of neighbors (k) being taken from the interval [1,15].

As can be seen from Table 3.7 (the optimal numbers of nearest neighbors are given in parentheses), DD outperformed every other kNN variant considered on every dataset, without exception. Furthermore, compared to other methods, DD relied on a slightly larger neighborhood in order to achieve its best performance. In Chaps. 5 and 6 DD emerged as one of the best-performing weighting schemes on time-series data as well.

A close examination of the results provided by Gou et al. indicated that both the newly introduced and the DN weighting scheme noticeably outperformed the

Table 3.6 The UCI datasets that served as the benchmark for the experiments introducing the DD weighting scheme [63]

Dataset	Attributes	Instances	Classes	Training samples	Testing samples
Glass	10	214	7	140	74
Wine	13	178	3	100	78
Sonar	60	208	2	120	88
Parkin	22	195	2	120	75
Iono	34	351	2	200	151
Musk	166	476	2	276	200
Vehicle	18	846	4	500	346
Image	19	2,310	7	1,310	1,000
Cardio	21	2,126	10	1,126	1,000
Opt	64	5,620	10	3,000	2,620
Landsat	36	6,435	7	3,435	3,000
Letter	16	20,000	26	10,000	10,000

Table 3.7 Classification accuracies reported in [63]

Dataset	1NN	kNN	DN	DD
Glass	69.86 ± 1.35	69.86 ± 1.35(1)	69.86 ± 1.56 (1)	**70.14 ± 1.35 (5)**
Wine	71.15 ± 1.81	71.15 ± 1.81(1)	71.47 ± 1.54 (4)	**71.99 ± 1.38 (4)**
Sonar	80.62 ± 1.62	80.62 ± 1.62 (1)	81.59 ± 1.81 (4)	**82.05 ± 1.81 (5)**
Parkin	82.67 ± 2.05	83.00 ± 1.80 (4)	83.53 ± 1.65 (8)	**83.93 ± 1.86 (8)**
Iono	84.01 ± 1.36	84.01 ± 1.36 (1)	84.27 ± 1.24 (7)	**84.44 ± 1.24 (8)**
Musk	83.98 ± 0.94	83.98 ± 1.06 (1)	84.77 ± 0.87 (6)	**85.10 ± 0.94 (7)**
Vehicle	63.16 ± 0.79	63.76 ± 1.24 (3)	63.96 ± 1.02 (8)	**64.34 ± 1.17 (9)**
Image	95.19 ± 0.31	95.19 ± 0.31 (1)	95.19 ± 0.33 (1)	**95.21 ± 0.31 (4)**
Cardio	69.84 ± 0.49	69.84 ± 0.49 (1)	70.12 ± 0.48 (5)	**70.30 ± 0.45 (6)**
Opt	98.43 ± 0.12	98.52 ± 0.12 (4)	98.64 ± 0.13 (7)	**98.65 ± 0.14 (7)**
Landsat	89.89 ± 0.21	90.35 ± 0.27 (4)	90.63 ± 0.25 (10)	**90.65 ± 0.26 (11)**
Letter	94.34 ± 0.086	94.38 ± 0.10 (4)	94.89 ± 0.092 (9)	**94.93 ± 0.088 (9)**

majority-voting rule (kNN) for most *k*-values. This is clearly recognizable in the three selected examples of datasets shown in Fig. 3.4 (they were taken from [63] which contains plots also the other datasets included in the research). In Chaps. 5 and 6 we detected similar behavior also with the UCR [60] datasets: the application of weights can have a favorable effect on the kNN's classification performance both over the simple 1NN rule and the majority-voting kNN.

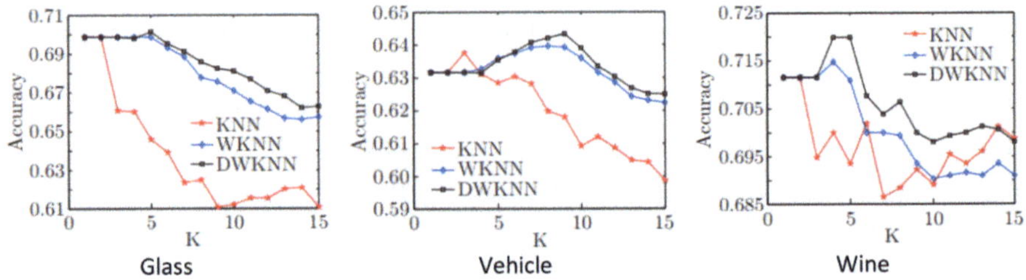

Fig. 3.4 Relationship between classification accuracy and the number of nearest neighbors [63]

3.2 Classifier Evaluation Methods

The effectiveness of a classifier can be assessed by counting how many test examples it classifies correctly or incorrectly. The percentage of correctly classified examples of the test set represents the accuracy of the classifier (A) and the percentage of wrongly classified examples represents its error rate (E). Formally:

$$A = \frac{\text{number of correctly classified test examples}}{\text{total number of test examples}}$$

$$E = \frac{\text{number of misclassified test examples}}{\text{total number of test examples}}$$

Accuracy and error rate are closely related measures of performance: $A = 1-E$. A more comprehensive discussion of these and other classifier performance indicators can be found in [3, 39, 43, 64].

In the following, we provide brief descriptions of the most commonly used techniques for dividing the available labeled data into disjoint training and test sets.

Holdout [3, 39, 43]. This method is based on the simple strategy of randomly dividing the initial labeled dataset into two completely separated subsets—one of which is then used to train the classifier and the other one to test its performance. Commonly used ratios between the numbers of test and training examples are 50%–50%, 20%–80%, 10%–90% and 1/3–2/3—but generally it can be chosen arbitrarily. To get a better estimate of the performance, the process can be repeated several times and the average of the obtained accuracies/error rates could be reported. This technique is known as *random subsampling*.

Cross-Validation [2, 3, 39, 43, 51, 64]. After randomly splitting the available labeled dataset into k approximately equal sized disjoint subsets (called *folds*), the performance evaluation is conducted in k iterations: one (not yet utilized) fold is reserved for testing in each iteration and the classifier is trained with the union of the remaining $k-1$ folds. This approach ensures that every object of the initial dataset is evaluated once as the test case and that it participates in training the classifier $k-1$ times. The overall classification accuracy is computed by averaging the obtained k accuracies.

k-fold cross-validation indicates that the performance evaluation included exactly k folds. To achieve a more reliable assessment, *n runs of k-fold cross-validation* repeats the entire process n times whereby the final assessment is made by averaging the $n \times k$ intermediate results. Random shuffling of the data is applied before each run. In data mining, a popular approach to estimate classification accuracy is 10 runs of tenfold cross-validation (CV10x10). This technique is believed to provide a reliable assessment of the classifier's true performance [3, 43]. Aside from SCV10x10, researchers often use 5 runs of twofold (CV5x2) [64] or a single ninefold (CV1x9) [65] configuration for performance evaluation.

There are two categories of cross-validation techniques, distinguished by how they split the initial data into separate folds (disjoint subsets) for training and testing purposes:

- *Random split cross-validation* randomly divides the data into folds without considering the distribution of class labels.
- *Stratified cross-validation* (SCV) aims to maintain the original class distribution in the test and training subsets. It ensures that each fold has a similar proportion of each class as the entire dataset, addressing potential bias caused by imbalanced data.

Ding et al. employed a unique variation of cross-validation [38]. Instead of the typical approach, they used one of the k disjoint folds for training and the union of the remaining $k - 1$ folds for testing.

Leave-One-Out [2, 3, 39, 43, 64]. The leave-one-out (LOO) method is a special implementation of the k-fold cross-validation algorithm where each object of the initial dataset (D) acts as its own fold for testing while all others are used for training, i.e., $k = |D|$. By utilizing all available examples except one for training, this method offers an advantage in terms of efficient use of available data. LOO does have some limitations, however. It can be computationally expensive due to the numerous training iterations required. Additionally, the test sets, containing only a single example each, are not good representatives of the considered problem.

References

1. Jones, M.T.: Artificial intelligence: a aystems approach. Infinity Science Press (2008). http://books.google.rs/books?id=40fuRAAACAAJ
2. Mohri, M., Rostamizadeh, A., Talwalkar, A.: Foundations of machine learning. The MIT Press (2012)
3. Abonyi, J.: Adatbányászat a hatékonyság eszköze, 1st edn. ComputerBooks, Budapest (2006)
4. Kotsiantis, S.B.: Supervised machine learning: a review of classification techniques. In: Proc. 2007 Conf. Emerg. Artif. Intell. Appl. Comput. Eng. Real Word AI Syst. with Appl. EHealth, HCI, Inf. Retr. Pervasive Technol., pp. 3–24. IOS Press, Amsterdam, The Netherlands (2007). http://dl.acm.org/citation.cfm?id=1566770.1566773

5. Ruiz, A.P., Flynn, M., Large, J., Middlehurst, M., Bagnall, A.: The great multivariate time series classification bake off: a review and experimental evaluation of recent algorithmic advances. Data Min. Knowl. Discov. **35**, 401–449 (2021). https://doi.org/10.1007/s10618-020-00727-3
6. Bagnall, A., Lines, J., Bostrom, A., Large, J., Keogh, E.: The great time series classification bake off: a review and experimental evaluation of recent algorithmic advances. Data Min. Knowl. Discov. **31**, 606–660 (2017). https://doi.org/10.1007/s10618-016-0483-9
7. Lee, Y.-H., Wei, C.-P., Cheng, T.-H., Yang, C.-T.: Nearest-neighbor-based approach to time-series classification. Decis. Support. Syst. **53**, 207–217 (2012). https://doi.org/10.1016/j.dss.2011.12.014
8. Buza, K.: Time series classification and its applications. In: Proc. 8th Int. Conf. Web Intell. Min. Semant., pp. 1–4. ACM, New York, NY, USA (2018). https://doi.org/10.1145/3227609.3227690
9. Abanda, A., Mori, U., Lozano, J.A.: A review on distance based time series classification. Data Min. Knowl. Discov. **33**, 378–412 (2019). https://doi.org/10.1007/s10618-018-0596-4
10. Faouzi, J.: Time series classification: a review of algorithms and implementations. In: Kotecha, K. (ed.) Mach. Learn. Emerging Trends Appl. Proud Pen (2022). https://inria.hal.science/hal-03558165
11. Ismail Fawaz, H., Forestier, G., Weber, J., Idoumghar, L., Muller, P.-A.: Deep learning for time series classification: a review. Data Min. Knowl. Discov. **33**, 917–963 (2019). https://doi.org/10.1007/s10618-019-00619-1
12. Jiang, W.: Time series classification: nearest neighbor versus deep learning models. SN Appl. Sci. **2**, 721 (2020). https://doi.org/10.1007/s42452-020-2506-9
13. Górecki, T., Łuczak, M.: Using derivatives in time series classification. Data Min. Knowl. Discov. **26**, 310–331 (2013). https://doi.org/10.1007/s10618-012-0251-4
14. Jeong, Y.-S., Jeong, M.K., Omitaomu, O.A.: Weighted dynamic time warping for time series classification. Pattern Recognit. **44**, 2231–2240 (2011). https://doi.org/10.1016/j.patcog.2010.09.022
15. Radovanović, M., Nanopoulos, A., Ivanović, M.: Time-series classification in many intrinsic dimensions. In: Proc. 2010 SIAM Int. Conf. Data Min., pp. 677–688. Society for Industrial and Applied Mathematics, Philadelphia, PA (2010). https://doi.org/10.1137/1.9781611972801.59
16. Spiegel, S., Jain, B.-J., Albayrak, S.: Fast time series classification under lucky time warping distance. In: Proc. 29th Annu. ACM Symp. Appl. Comput., pp. 71–78. ACM, New York, NY, USA (2014). https://doi.org/10.1145/2554850.2554885
17. Xi, X., Keogh, E., Shelton, C., Wei, L., Ratanamahatana, C.A.: Fast time series classification using numerosity reduction. In: Proc. 23rd Int. Conf. Mach. Learn.—ICML '06, pp. 1033–1040. ACM Press, New York, NY, USA (2006) https://doi.org/10.1145/1143844.1143974
18. Ye, L., Keogh, E.: Time series shapelets: a new primitive for data mining. In: Proc. 15th ACM SIGKDD Int. Conf. Knowl. Discov. Data Min., pp. 947–956. ACM, New York, NY, USA (2009). https://doi.org/10.1145/1557019.1557122
19. Chatfield, C., Koehler, A.B., Ord, J.K., Snyder, R.D.: A new look at models for exponential smoothing. J. R. Stat. Soc. Ser. D (The Stat.) **50**, 147–159 (2001). https://doi.org/10.1111/1467-9884.00267
20. Box, G.E.P., Jenkins, G.M., Reinsel, G.C., Ljung, G.M.: Time series analysis: forecasting and control, 5th edn. Wiley, Hoboken, New Jersey (2015)
21. Grabockal, J., Bedallig, E., Schmidt-Thieme, L.: Efficient classification of long time-series. ICT Innov Secur. Intell. Syst. **207**(2013), 47–57 (2012)
22. Wu, Y., Chang, E.Y.: Distance-function design and fusion for sequence data. In: Proc. Thirteen. ACM Conf. Inf. Knowl. Manag.—CIKM '04, pp. 324–333. ACM Press, New York, NY, USA (2004). https://doi.org/10.1145/1031171.1031238
23. Eads, D.R., Hill, D., Davis, S., Perkins, S.J., Ma, J., Porter, R.B., Theiler, J.P.: Genetic algorithms and support vector machines for time series classification. Proc. SPIE **4787**, 74–85 (2002). https://doi.org/10.1117/12.453526

24. Petridis, V., Kehagias, A.: Predictive modular fuzzy systems for time-series classification. Fuzzy Syst. IEEE Trans. **5**, 381–397 (1997). https://doi.org/10.1109/91.618275
25. Nanopoulos, A., Alcock, R., Manolopoulos, Y.: Feature-based classification of time-series data. Int. J. Comput. Res. **10**, 49–61 (2001)
26. Kini, V.B., Sekhar, C.C.: Large margin mixture of AR models for time series classification. Appl. Soft Comput. **13**, 361–371 (2013)
27. Turrado García, F., García Villalba, L.J., Portela, J.: Intelligent system for time series classification using support vector machines applied to supply-chain. Expert Syst. Appl. **39**, 10590–10599 (2012). https://doi.org/10.1016/j.eswa.2012.02.137
28. Huan, N.-J., Palaniappan, R.: Neural network classification of autoregressive features from electroencephalogram signals for brain–computer interface design. J. Neural Eng. **1**, 142–150 (2004). http://stacks.iop.org/1741-2552/1/i=3/a=003
29. Zhang, H., Ho, T.B., Zhang, Y., Lin, M.S.: Unsupervised feature extraction for time series clustering using orthogonal wavelet transform. Informatics **30**, 305–319 (2006)
30. Eruhimov, V., Martyanov, V., Tuv, E.: Constructing high dimensional feature space for time series classification. In: Kok, J.N., Koronacki, J., de Mántaras, R.L., Matwin, S., Mladenic, D., Skowron, A. (eds.) Knowl. Discov. Databases PKDD 2007, 11th Eur. Conf. Princ. Pract. Knowl. Discov. Databases, pp. 414–421. Springer (2007)
31. Geurts, P., Wehenkel, L.: Segment and combine approach for non-parametric time-series classification. In: Proc. 9th Eur. Conf. Princ. Pract. Knowl. Discov. Databases, pp. 478–485. Springer, Berlin, Heidelberg (2005). https://doi.org/10.1007/11564126_48
32. Lee, J.-G., Han, J., Li, X., Gonzalez, H.: TraClass: trajectory classification using hierarchical region-based and trajectory-based clustering. Proc. VLDB Endow. **1**, 1081–1094 (2008). http://dl.acm.org/citation.cfm?id=1453856.1453972
33. Megalooikonomou, V., Wang, Q., Li, G., Faloutsos, C.: A multiresolution symbolic representation of time series. In: Data Eng. 2005. ICDE 2005. Proceedings. 21st Int. Conf., 2005, pp. 668–679. https://doi.org/10.1109/ICDE.2005.10
34. Zhang, X., Wu, J., Yang, X., Ou, H., Lv, T.: A novel pattern extraction method for time series classification. Optim. Eng. **10**, 253–271 (2009). https://doi.org/10.1007/s11081-008-9056-0
35. Zhang, Z., Cheng, J., Li, J., Bian, W., Tao, D.: Segment-based features for time series classification. Comput. J. **55**, 1088–1102 (2012). https://doi.org/10.1093/comjnl/bxs029
36. Marteau, P.-F.: Time warp edit distance with stiffness adjustment for time series matching. IEEE Trans. Pattern Anal. Mach. Intell. **31**, 306–318 (2009). https://doi.org/10.1109/TPAMI.2008.76
37. Lima, F.T., Souza, V.M.A.: A large comparison of normalization methods on time series. Big Data Res. **34**, 100407 (2023). https://doi.org/10.1016/j.bdr.2023.100407
38. Ding, H., Trajcevski, G., Scheuermann, P., Wang, X., Keogh, E.: Querying and mining of time series data: experimental comparison of representations and distance measures. In: Proc. VLDB Endow., VLDB Endowment, pp. 1542–1552 (2008). https://doi.org/10.14778/1454159.1454226
39. Tan, P.-N., Steinbach, M., Kumar, V., Karpatne, A.: Introduction to data mining, 2nd edn. Pearson Education (2019)
40. Chen, L., Ng, R.: On the marriage of Lp-norms and edit distance. In: Nascimento, M.A., Özsu, M.T., Kossmann, D., Miller, R.J., Blakeley, J.A., Schiefer, K.B. (eds.) Proc. 2004 VLDB Conf., pp. 792–803. Elsevier (2004). https://doi.org/10.1016/B978-012088469-8.50070-X
41. Esling, P., Agon, C.: Time-series data mining. ACM Comput. Surv. **45**, 12:1–12:34 (2012). https://doi.org/10.1145/2379776.2379788
42. Zhang, H., Ho, T.B., Lin, M.S.: A non-parametric wavelet feature extractor for time series classification. In: Dai, H., Srikant, R., Zhang, C. (eds.) Adv. Knowl. Discov. Data Min. SE-71, pp. 595–603. Springer, Berlin, Heidelberg (2004). https://doi.org/10.1007/978-3-540-24775-3_71
43. Han, J., Pei, J., Tong, H.: Data mining: concepts and techniques, 4th edn. Morgan Kaufmann (2022)

44. Laxman, S., Sastry, P.S.: A survey of temporal data mining. Sadhana **31**, 173–198 (2006). https://doi.org/10.1007/BF02719780
45. Cover, T., Hart, P.: Nearest neighbor pattern classification. IEEE Trans. Inf. Theory **13**, 21–27 (1967). https://doi.org/10.1109/TIT.1967.1053964
46. Górecki, T., Łuczak, M., Piasecki, P.: An exhaustive comparison of distance measures in the classification of time series with 1NN method. J. Comput. Sci. **76**, 102235 (2024). https://doi.org/10.1016/j.jocs.2024.102235
47. Fix, E., Hodges, J.L.: Discriminatory analysis. Nonparametric discrimination: consistency properties. Int. Stat. Rev. / Rev. Int. Stat. **57**, 238–247 (1989). https://doi.org/10.2307/1403797
48. Wu, X., Kumar, V., Ross Quinlan, J., Ghosh, J., Yang, Q., Motoda, H., McLachlan, G.J., Ng, A., Liu, B., Yu, P.S., Zhou, Z.-H., Steinbach, M., Hand, D.J., Steinberg, D.: Top 10 algorithms in data mining. Knowl. Inf. Syst. **14**, 1–37 (2008). https://doi.org/10.1007/s10115-007-0114-2
49. Dudani, S.A.: The distance-weighted k-nearest-neighbor rule. IEEE Trans. Syst. Man. Cybern. SMC-6, 325–327 (1976). https://doi.org/10.1109/TSMC.1976.5408784
50. Larose, D.T., Larose, C.D.: Discovering knowledge in data, 2nd edn. Wiley (2014). https://doi.org/10.1002/9781118874059
51. Mitchell, T.M.: Machine learning. McGraw-Hill Inc., New York, NY, USA (1997)
52. Zavrel, J.: An empirical re-examination of weighted voting for k-NN. in: Proc. 7th Belgian-Dutch Conf. Mach. Learn., 1997, pp. 139–148
53. Xu, J.: An empirical comparison of weighting functions for multi-label distance weighted K-nearest neighbour method, pp. 13–20 (2011). https://doi.org/10.5121/csit.2011.1302
54. Macleod, J., Luk, A., Titterington, D.: A re-examination of the distance-weighted k-nearest neighbor classification rule. IEEE Trans. Syst. Man Cybern. **17**, 689–696 (1987). https://doi.org/10.1109/TSMC.1987.289362
55. Pao, T.-L., Chen, Y.-T., Yeh, J.-H., Cheng, Y.-M., Lin, Y.-Y.: A comparative study of different weighting schemes on KNN-based emotion recognition in mandarin speech. In: Huang, D.-S., Heutte, L., Loog, M. (eds.) Adv. Intell. Comput. Theor. Appl. With Asp. Theor. Methodol. Issues, pp. 997–1005. Springer, Berlin, Heidelberg (2007). https://doi.org/10.1007/978-3-540-74171-8_101
56. Takigawa, Y., Hott, S., Kiyasu, S., Miyahara, S.: Pattern classification using weighted average patterns of categorical k-nearest neighbors. In: Proc. 1th Int. Work. Camera-Based Doc. Anal. Recognit., pp. 111–118 (2005)
57. Pao, T.-L., Chen, Y.-T., Yeh, J.-H., Chang, Y.-H.: Emotion recognition and evaluation of mandarin speech using weighted D-KNN classification. In: Proc. 17th Conf. Comput. Linguist. Speech Process. ROCLING 2005, Taiwan, ROC, 2005, pp. 203–212. Association for Computational Linguistics and Chinese Language Processing (ACLCLP), Taiwan (2005) http://aclweb.org/anthology/O/O05/O05-1013.pdf
58. Gou, J., Xiong, T., Kuang, Y.: A novel weighted voting for K-nearest neighbor rule. J. Comput. **6**, 833–840 (2011). https://doi.org/10.4304/jcp.6.5.833-840
59. Kelly, M., Longjohn, R., Nottingham, K.: UCI machine learning repository (n.d.). https://archive.ics.uci.edu. Accessed 7 Mar 2024
60. Anh Dau, H., Keogh, E., Kamgar, K., Michael Yeh, C.-C., Zhu, Y., Gharghabi, S., Ann Ratanamahatana, C., Chen, Y., Hu, B., Begum, N., Bagnall, A., Mueen, A., Batista, G.: Hexagon-ML, the UCR time series classification archive (2019). https://www.cs.ucr.edu/~eamonn/time_series_data_2018/
61. Ratnaparkhi, A., Reynar, J., Roukos, S.: A maximum entropy model for prepositional phrase attachment. In: Proc. Work. Hum. Lang. Technol., pp. 250–255. Association for Computational Linguistics, Stroudsburg, PA, USA (1994). https://doi.org/10.3115/1075812.1075868
62. Deza, M.M., Deza, E.: Encyclopedia of distances. Springer, Berlin, Heidelberg (2016). https://doi.org/10.1007/978-3-662-52844-0
63. Gou, J., Du, L., Zhang, Y., Xiong, T.: A new distance-weighted k-nearest neighbor classifier. J. Inf. Comput. Sci. **9**, 1429–1436 (2012)
64. Alpaydin, E.: Introduction to machine learning. MIT Press (2004)

65. Thang, L.Q., Temiyasathit, C.: Investigation of regularization theory for four-class classification in brain-computer interface, pp. 275–285 (2014). https://doi.org/10.1007/978-3-319-12778-1_21
66. Radovanović, M.: High-dimensional data representations and metrics for machine learning and data mining. University of Novi Sad, Serbia (2010). (PhD thesis). https://nardus.mpn.gov.rs/handle/123456789/1652

Chapter 4
The Impact of Global Constraints on the Accuracy of Elastic Similarity Measures

The direct dependence of the efficiency of the 1NN classifier on the underlying distance measure makes it ideally suited for investigating the performance of different methods proposed for estimating the similarity of time series [1–10]. Although its simplicity and the absence of parameters that should be optimized by training are in favor of the 1NN classifier, the classification of a new time series requires comparing it with each individual time series of the training dataset in order to choose the one that is the most similar to it, which can be both time and computationally demanding [1]. When it comes to elastic distance measures, one of the possibilities to speed up the process of finding the nearest neighbor is to reduce the number of time series points that are compared by applying global constraints such as the Sakoe-Chiba band (SC) [11] and the Itakura parallelogram (IT) [12].

Relying on the 1NN classifier, in this chapter we present the results of the analysis of the impact of global constraints on the four basic elastic measures presented in the second chapter (their concise overview is given in Appendix A): Dynamic Time Warping (DTW) [13], Longest Common Subsequence (LCS) [14], Edit distance with Real Penalty (ERP) [15] and Edit Distance on Real sequence (EDR) [16]. In the first phase of the experiments, we investigated the intensity of changes in the NN graph generated by constraining the elasticity. In the second phase, we compared the accuracies obtained with different warping-window widths in order to get an initial assessment of the effect of employing global constraints. We verified the preliminary findings of the first two phases in the third phase by comparing the performance of the unconstrained measures with their SC and IT constrained counterparts using optimal warping-window widths.

The analysis performed on the datasets provided by the UCR repository [17] included all widths of the warping window from the range between 0% and 50% of the time series' length in steps of one percentage point and from the range between 50% and 100% in steps of 5 percentage points, i.e. the value of the parameter r was chosen from the set $R = \{0,\ 1,\ 2, \ldots, 48,\ 49,\ 50,\ 55,\ 60,\ \ldots, 90,\ 95,\ 100\}$. Since the time series were z-normalized before the experiments using Paparizzos' script

Z. Gellér et al., *Recent Advances in Time-Series Classification—Methodology and Applications*, Intelligent Systems Reference Library 264,
https://doi.org/10.1007/978-3-031-77527-7_4

[18], the value of the matching threshold parameter ε used by the LCS and EDR measures was fixed at 0.25 as recommended in [16]. Therefore, when interpreting their results, it should be borne in mind that this parameter was not optimized (so it would not be fair to compare them with the performance of other distance measures).

The experiments implemented for the purposes of this chapter were based on our previous findings [19–23] and they were conducted employing our *Framework for Analysis and Prediction* [24] (except for statistical analyses), which has been successfully applied both for educational purposes [25] and in various research involving working with time series [26] for more than a decade.

4.1 Evaluation Methodology

In this section, we present the methodology for assessing the impact of global constraints on the accuracy of elastic similarity measures that will be applied in the remainder of this chapter.

4.1.1 Changes in the NN Graph

The *nearest neighbor graph* (NN graph) of a dataset X is a directed graph in which each time series $A \in X$ is connected to a time series $B \in X$ where $B \neq A$ and $d(A, B) \leq d(A, C)$ for each $C \in X/\{A\}$, with respect to a given distance measure d [27]. As the 1NN classifier assigns to each yet unclassified time series A the class of the time series B (from the training set) that is the most similar to A with respect to the chosen distance measure d, changes in the NN graph directly affect its classification performance (when it is evaluated on X relying on the leave-one-out method). Based on this consideration, in the first phase of the examination of the effect of employing global constraints on the accuracy of the 1NN classifier, we examined the intensity of the changes they generated in the nearest-neighbor graphs of the four elastic measures discussed in this book. For each considered width of the warping window (r), we examined the percentage of time series whose first (closest) neighbors were replaced as a result of constraining the warping window (Figs. 4.5, 4.17, 4.29 and 4.41) and the proportion of time series whose first neighbors were assigned a different class label compared to the one in the NN graph constructed using the unconstrained measure (Figs. 4.6, 4.18, 4.30 and 4.42). The motivation behind the first phase of the experiments was based on the assumption that the more changes generated by the global constraints in the NN graph, the more the constrained measures will differ from the unconstrained one in classification accuracy.

As an illustration, Fig. 4.1 shows the NN graphs of the *BeetleFly* (8) dataset corresponding to the unconstrained DTW ($r = 100\%$) and the (squared) Euclidean distance ($r = 0\%$). The *BeetleFly* (8) dataset contains 40 time series grouped into two classes that are marked in green and yellow. It is a *balanced* dataset, 20 time series

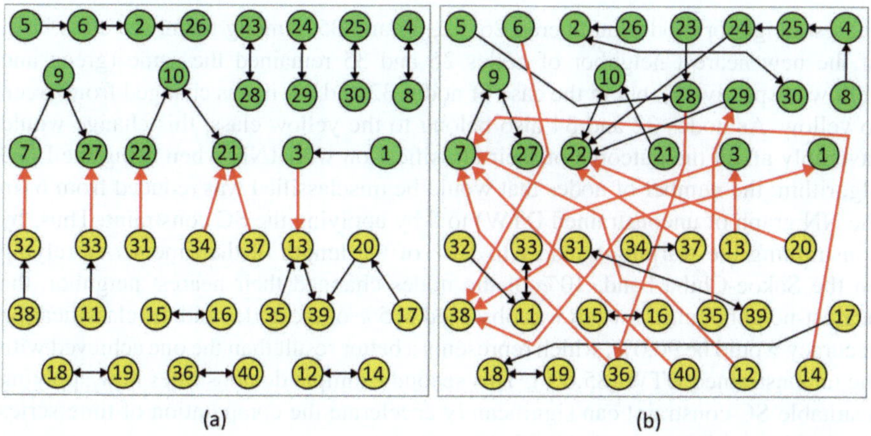

Fig. 4.1 Nearest neighbor graphs of the *BeetleFly* (2) dataset obtained using the unconstrained DTW ($r = 100\%$) (**a**) and the (squared) Euclidean distance ($r = 0\%$) (**b**); a red edge indicates that the nearest neighbor does not belong to the same class as the given node

belong to both classes. The red color of the edges indicates that the nearest neighbor does not belong to the same class as the given node. The time series represented by such a node would be misclassified by the 1NN classifier (in combination with the LOO evaluation method). In combination with the unconstrained DTW there would be only 6 such time series (nodes numbered 3, 31, 32, 33, 34 and 37) and the classification accuracy would be 85.0%.

As can be seen in Fig. 4.1b, disallowing any elasticity (i.e. constraining the warping-window to the diagonal) generated significant rearrangements in the NN graph. Under the influence of such a strict constraint, 27 nodes changed their nearest neighbor,[1] which amounts to 67.5% of the time series. In 15 time series,[2] this led to the situation that the class of the new nearest neighbor differs from the class of the old one (the one in the NN graph generated by the unconstrained DTW), which means that the change of the class of the nearest neighbor was present in 37.5% of the time series. Since the number of time series that do not belong to the same class to which their nearest neighbor belongs is 13,[3] by applying a global constraint of width $r = 0\%$, the accuracy of the 1NN classifier calculated with LOO would be 67.5%, which is noticeably less than 85.0% that would be obtained with the unconstrained DTW. This simple example clearly demonstrates the potential harm of lack of elasticity on classification accuracy and the advantage of DTW over the Euclidean distance.

Let us now consider the NN graph of the *BeetleFly* (8) dataset in Fig. 4.2a acquired by utilizing the Sakoe-Chiba band of width $r = 10\%$. With respect to the NN graph produced by the unconstrained DTW in Fig. 4.1a, 4 time series have changed their

---

[1] Nodes 1–6, 10, 12–14, 17, 20, 24, 25, 28–40.

[2] Nodes 1, 6, 12–14, 20, 28, 31–36, 38, 40.

[3] Nodes 1, 3, 6, 12–14, 20, 28, 35–38, 40.

nearest neighbor: nodes numbered 26, 32, 34 and 35. Among them, the class label of the new nearest neighbor of nodes 26 and 35 remained the same (green and yellow, respectively), but in the case of nodes 32 and 34, it was changed from green to yellow. As nodes 32 and 34 also belong to the yellow class, this change would favorably affect the outcome of their classification with 1NN when using the LOO algorithm: the number of nodes that would be misclassified was reduced from 6 (in the NN graph of unconstrained DTW) to 4 by applying the SC constraint. Thus, by constraining the warping-window to 10% of the length of the time series relying on the Sakoe-Chiba band, 10% of the nodes changed their nearest neighbor, the nearest-neighbor class switch was observed in 5% of the nodes and the classification accuracy would be 90.0%, which represents a better result than the one achieved with the unconstrained DTW (85.0%). This second example demonstrates how applying a suitable SC constraint can significantly accelerate the computation of time-series similarities, with improved classification accuracy.

On the example of the *BeetleFly* (8) dataset, the application of the Itakura parallelogram of width $r = 27\%$ shows that the classification accuracy of the unconstrained DTW may be achieved by comparing a smaller number of time series points. By inspecting the NN graphs shown in Figs. 4.1a and 4.2b, it can be seen that the application of the IT constraint caused 9 time series to change their nearest neighbor (nodes numbered 3, 11, 12, 14, 17, 26, 32, 34 and 35) thereby affecting 22.5% of the dataset. Among them, only in the case of nodes 12, 14, 32 and 34 did the class of the nearest neighbor change, which represents 10% of nodes. Each of the nodes 12, 14, 32 and 34 belongs to the yellow class. At the same time, nodes 12 and 14 replaced their yellow nearest neighbors with a green one and nodes 32 and 34 changed their green ones with yellow ones. In this way, the number of nearest neighbors that do not belong to the same class to which the given time series belongs remained unchanged

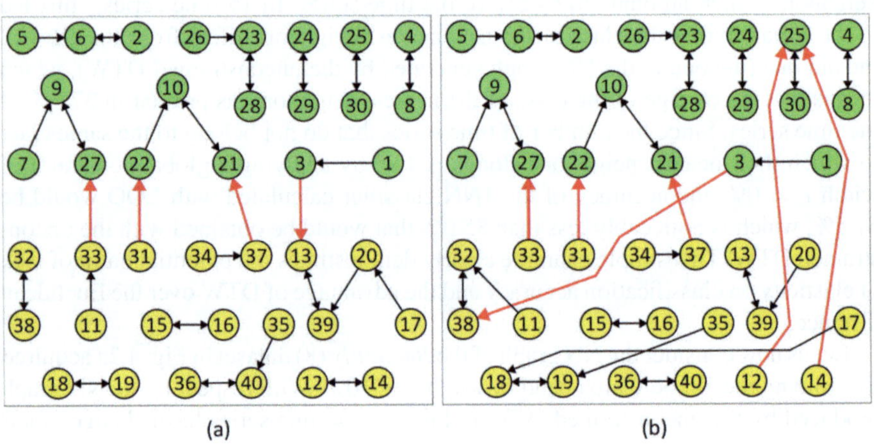

Fig. 4.2 Nearest neighbor graphs of the *BeetleFly* (2) dataset obtained using the Sakoe-Chiba constrained DTW ($r = 10\%$) (**a**) and the Itakura constrained DTW ($r = 27\%$) (**b**); a red edge indicates that the nearest neighbor does not belong to the same class as the given node

in relation to the NN graph of the unconstrained DTW and the same classification accuracy would be achieved using the 1NN classifier (in combination with the LOO evaluation technique).

4.1.2 Classification Accuracy—Individual Warping-Window Widths

With the aim of obtaining a preliminary assessment of the performance of the constrained measures, following the reports in [28–31], the focus of the second phase of our study was on examining the classification accuracy (averaged across all datasets) of SC and IT with different warping-window widths, looking at individual values of the parameter r (Figs. 4.7, 4.19, 4.31 and 4.43). Special attention was paid to the exploration of the differences between $r = 0\%$ (no elasticity), $r = 100\%$ (unconstrained elasticity) and the intermediate values. In addition, we examined the number of datasets for which a given unconstrained measure performed better, worse or the same as the unconstrained variants (Figs. 4.8, 4.20, 4.32 and 4.44).

For performance evaluation, we used ten times repeated tenfold stratified cross-validation (SCV10x10), a standard approach for estimating classifier efficiency in data mining [32]. We divided each of the UCR datasets into 10 disjoint stratified folds of approximately the same size. We used one of them as the test set and the pooling of the others as the training set (Fig. 4.3). After using each of the 10 folds for testing, we shuffled the elements of the dataset and repeated the described procedure 9 more times. In this way, we obtained 100 different splits into testing and training subsets and thus calculated the average accuracy for the given dataset. For each value of the parameter r, we used the same 100 splits into testing and training sets.

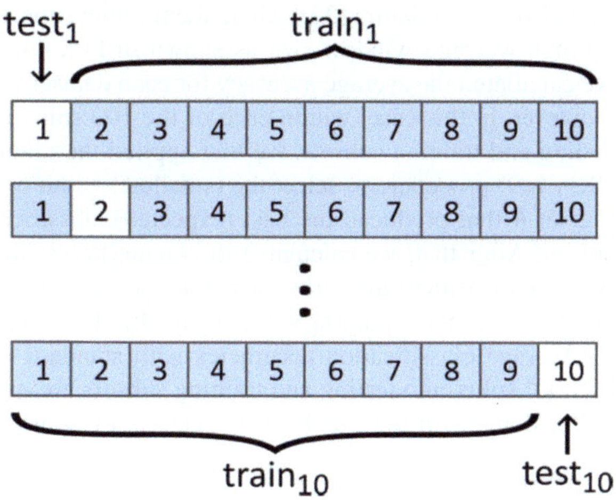

Fig. 4.3 Calculating classification accuracy for individual values of the parameter r applying tenfold cross-validation on the entire dataset

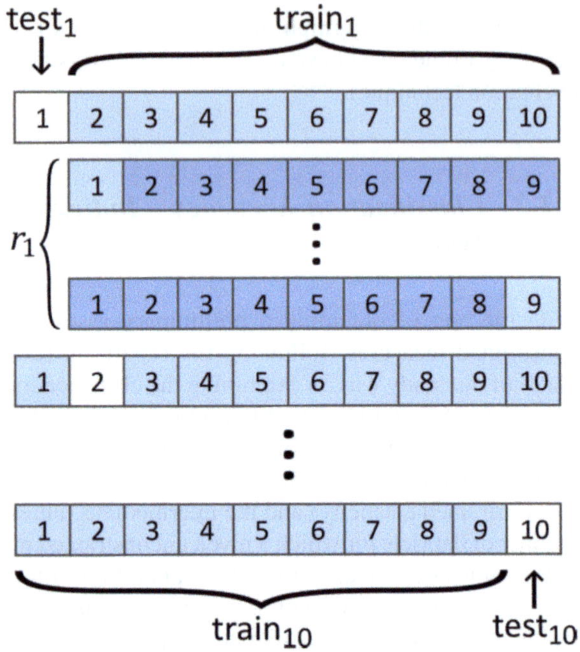

Fig. 4.4 Calculating classification accuracy using tenfold cross-validation where the optimal value of the parameter r is chosen using ninefold cross-validation on the training subset

4.1.3 Classification Accuracy—Optimal Warping-Window Widths

The initial observations we reached in the first two phases of the experiments were verified in the third phase by a in-depth analysis of the influence of global constraints on the classification accuracy of elastic measures based on the 1NN classifier and nested cross-validation [32], where the training subsets were employed to obtain the optimal warping window size as shown in Fig. 4.4. Similar to the second phase, we calculated the average accuracy for each dataset using the SCV10x10 technique. However, in this case, within each of the 100 splits of a given dataset into disjoint testing and training subsets, we first applied the ninefold stratified cross-validation (SCV1x9) procedure to select the (smallest) width of the warping window (r) from the set R that produced the best (expected) classification accuracy on the training subset. After that, we calculated the (actual) classification accuracy by comparing the classes of the time series of the testing subset with the classes of their nearest neighbors from the training subset using the chosen optimal warping-window width.

Average classification accuracies with standard deviations calculated over the 12,800[4] splits into testing and training subsets are given in Tables 4.1, 4.4, 4.7 and 4.10, along with the numbers of datasets on which different variants of a given

[4] Each of the 128 datasets was split 100 times into testing and training subsets by the SCV10x10 procedure.

elasticity measure produced the best results and their mean ranks. These tables report both the average value and standard deviations of the optimal warping windows widths of the corresponding constrained measures.

To gain insight into which window widths were preferred by the Sakoe-Chiba and Itakura constraints, we created histograms that for each warping-window width show how many times out of 12,800 splits into testing and training subsets was a given width chosen as optimal by the SCV1x9 algorithm on the training set (Figs. 4.9a, 4.21a, 4.33a and 4.45a). In addition, we also prepared accumulative charts showing the percentages of 12,800 splits in which the optimal warping-window width was less than a given value of the parameter r (Figs. 4.9b, 4.21b, 4.33b and 4.45b). The x-axis of these graphs shows the values of the parameter $r \in R$ and the y-axis shows the percentages of splits in which the optimal warping-window width was less than or equal to the given value. To explore the differences in optimal window widths also at the level of individual datasets, Figs. 4.10, 4.22, 4.34 and 4.46 visualize their average r-values.

Figures 4.11, 4.23, 4.35 and 4.47 compare the classification accuracies of the three variants of the considered elastic measures in pairs, in the form of scatter plots. If we label one variant with A and the other one with B, each of the 128 sets is depicted by one point whose x-coordinate represents the classification accuracy obtained with A and y-coordinate corresponds to the classification accuracy achieved with B. Points positioned below the diagonal represent datasets on which A achieved better accuracy than B and points above the diagonal portray datasets on which B outperformed A. The closer the points are to the diagonal, the smaller the difference in performance between A and B. In Figs. 4.12, 4.24, 4.36 and 4.48, we express the distances of the points from the diagonal through the average differences in classification accuracy between the different variants of the analyzed measures over the datasets where the performance of the given variant was better and worse than the other variant (see Appendix C for details). To help identify the datasets where the performance of A and B differ the most, Figs. 4.13, 4.25, 4.37 and 4.49 show the differences in their classification accuracies across datasets. In these figures, datasets are represented by their numerical identifiers.

As recommended in [33], we examined the statistical significance of the difference in the performance of different variants of elastic measures using the Friedman test [34, 35] followed by Nemenyi's post-hoc test [36] for determining pairs of variants whose results differ significantly (we presented these findings in the form of critical difference diagrams in Figs. 4.14, 4.26, 4.38 and 4.50). Guided by the suggestions in [37], we also examined their pairwise differences using the Wilcoxon signed-rank test [38] adjusted with the Holm-Bonferroni correction [39] (Tables 4.2, 4.5, 4.8 and 4.11). In addition, we employed the corrected resampled t-test [40] to identify statistically significant wins and losses (Tables 4.3, 4.6, 4.9 and 4.12). All the mentioned statistical tests were performed with a significance level of $\alpha = 0.01$.

Besides the number of datasets on which the accuracy of method A was higher, lower or equal to the result of method B, as a more realistic indicator of the advantage of A over B, the authors of the paper [41] proposed comparing the expected and actual accuracy gains for each dataset using a scatter plot called the *Texas Sharpshooter*

(TS) plot (Figs. 4.15, 4.27, 4.39 and 4.51). The accuracy gain of method A over method B is obtained by dividing the accuracy of A by the accuracy of B. If the gain is > 1, A performed better than B and if the gain is < 1, it performed worse. The expected gain in accuracy is computed using the training set and the actual gain is determined by utilizing the test set. In TS plots, each dataset is displayed as a point whose x coordinate represents the expected and y coordinate the actual accuracy gain. Depending on whether the coordinates are < or > 1, four cases can be distinguished:

- $x > 1$ and $y > 1$—Based on the results on the training set, we expected A to be better than B and it really is. This is a *true positive* (TP) case.
- $x < 1$ and $y < 1$—The results on the training set indicated that A is worse than B and it really is worse. This is a *true negative* (TN) case.
- $x > 1$ and $y < 1$—Based on the training set, we expected A to be better than B, but it turned out to be worse. This is a *false positive* (FP) case, we opted for the worse solution.
- $x < 1$ and $y > 1$—The results on the training set told us that A performs worse than B, but it actually performs better. This is a *false negative* (FN) case, we missed an opportunity to improve the classification accuracy (but at least we didn't make it worse like in the FP case).

To examine on which datasets the choice of method A over method B turned out to be the wrong decision to the greatest extent, in Figs. 4.16, 4.28, 4.40 and 4.52 we present the difference between the expected and actual accuracy gains of those datasets that were included in the FP region of the respective TS plots.

Several things should be kept in mind when interpreting the TS plots in this chapter:

- Since we employed the SCV10x10 method to calculate classification accuracy, we divided each dataset into disjoint testing and training subsets 100 times. For each of those 100 splits, we calculated the expected and actual accuracy gain. Then, we obtained the expected and actual accuracy gain of the entire dataset as the average value of those 100 gains.
- Since SC and IT take $r = 100\%$ into account when choosing the optimal warping-window width, their accuracy on the training subsets is never less than the accuracy of the corresponding unconstrained measure. It follows that their expected gain is never < 1, i.e. the TN and FN regions are always empty when comparing SC and IT with the associated unconstrained measure.
- The DTW variants misclassified each time series of the testing subset in 3 out of 100 splits in the case of the *Beef* (7) dataset, so expected and actual gains were calculated based on only 97 splits.
- The unconstrained ERP misclassified each time series of the testing subset in 5 out of 100 splits in the case of the *Beef* (7) dataset, therefore expected and actual gains were calculated based on only 95 splits (when comparing SC and IT with ERP).

- The SC constrained ERP misclassified each time series of the testing subset in 2 out of 100 splits in the case of the *Beef* (7) dataset, hence expected and actual gains were calculated based on only 98 splits (when comparing IT with SC).

The remainder of this chapter is devoted to the presentation and interpretation of the results of the analysis of the influence of global constraints on the accuracy of the four basic elastic distance measures discussed in this book.

4.2 Dynamic Time Warping (DTW)

4.2.1 Changes in the NN Graph

Based on Fig. 4.5, it is clear that, in the case of the DTW measure, the application of the Itakura parallelogram (IT) affected the graph of the nearest neighbors to a much greater extent than the Sakoe-Chiba band (SC). At a warping level of $r = 0\%$ (no warping), averaged across all datasets, approximately 74% of time series replaced their nearest neighbors as a response to the absence of any elasticity. This value was the highest among the four examined elastic measures at $r = 0\%$. On the other hand, the standard deviation (SD) [20.6 percentage points (pp)] was the smallest here. It should be noted that, as the constrained DTW with $r = 0\%$ represents the (squared) Euclidean distance, this is also the average difference between the NN graphs of the Euclidean distance and the unconstrained DTW. By increasing the width of the warping window, the changes in the NN graphs decreased much faster in the case of SC than when using IT. When it comes to SC, the average value became < 25 as early as $r = 8\%$, and when applying IT, this limit was crossed only with $r = 34\%$.

On average, SC generated the most prominent changes in the NN graphs of the *SmallKitchenAppliances* (104), *LargeKitchenAppliances* (66) and *Computers* (17)

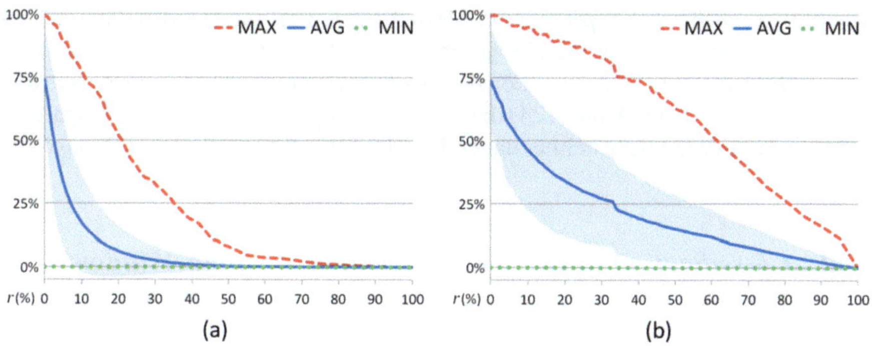

Fig. 4.5 DTW—proportion of time series that replaced their nearest neighbors as a result of applying the Sakoe-Chiba band (**a**) and the Itakura parallelogram (**b**)

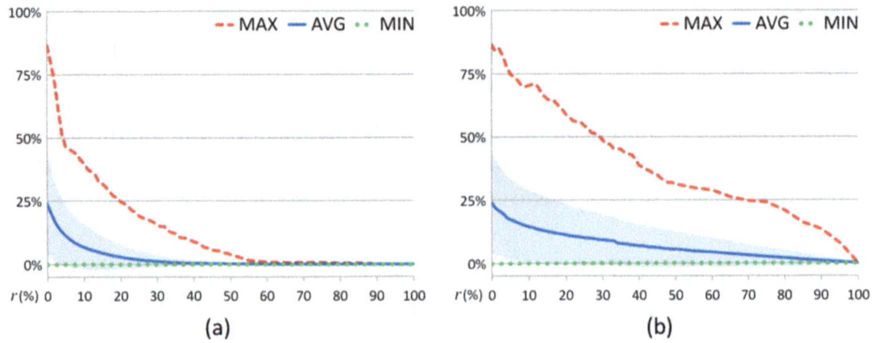

Fig. 4.6 DTW—proportion of time series for which a nearest neighbor class change was detected as a result of applying the Sakoe-Chiba band (**a**) and the Itakura parallelogram (**b**)

datasets, and IT in the graphs of the *Earthquakes* (29), *HouseTwenty* (60) and *ScreenType* (97) datasets. Regarding the smallest average changes, there was no difference between SC and IT: the three least affected datasets were *Meat* (70), *ChlorineConcentration* (14) and *Wine* (124) in both cases. Moreover, in the NN graph of the *Wine* (124) dataset, no time series changed its nearest neighbor for any of the warping-window widths considered, neither when using the Sakoe-Chiba nor when applying the Itakura constraint.

Figure 4.6 illustrates the impact of limiting DTW's elasticity through the proportion of those time series whose nearest neighbors altered their class. Comparing it with Fig. 4.5, it is easy to see that the changes were much less intense, but they were again more pronounced in the case of the Itakura parallelogram than in combination with the Sakoe-Chiba band. The maximum of the average value was close to 24% (with a standard deviation of about 19.7 pp) for $r = 0\%$. While when applying the SC constraint, the average value was > 10% only for $r < 7\%$, with IT it did not go below that limit as long as $r < 26\%$.

In this respect, SC generated the biggest changes in the case of the *ScreenType* (97), *SmallKitchenAppliances* (104) and *GestureMidAirD3* (49) datasets, and IT had the most impact on the NN graphs of the *PigCVP* (88), *Phoneme* (84) and *GestureMidAirD3* (49) datasets. For the *TwoLeadECG* (116), *ChlorineConcentration* (14), *Meat* (70) and *Wine* (124) datasets, no changes were observed compared to the unconstrained measure neither under the influence of the SC nor the IT constraint.

4.2.2 *Classification Accuracy—Individual Warping-Window Widths*

Looking at the classification accuracy calculated with SCV10x10 over the entire dataset for different warping-window widths, the average value reached its maximum

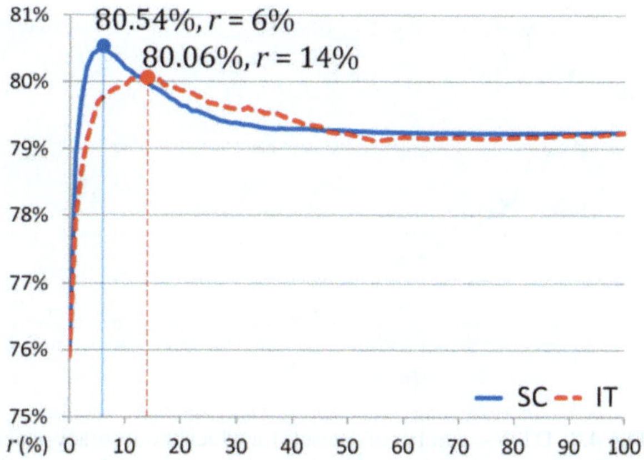

Fig. 4.7 DTW—average classification accuracy across all datasets for different warping-window widths

at $r = 6\%$ (80.54%) for SC and at $r = 14\%$ (80.06%) for IT (Fig. 4.7), suggesting that both constrained versions of DTW could outperform the unconstrained one. Also, these results indicate that a very strict restriction of their elasticity (close to $r = 0\%$) can have a negative impact on the performance of the classifier. As the Euclidean distance is a special case of the constrained DTW with a warping window of width $r = 0\%$, it can also be seen from Fig. 4.7 that the unconstrained DTW ($r = 100\%$) outperformed the Euclidean distance regarding the accuracy of the 1NN classifier.

From Fig. 4.8, it can be seen that the count of datasets for which the constrained DTW variants produced better classification accuracy than the unconstrained one (denoted by DTW<SC and DTW<IT in the figure) after a short period of growth turns into a decline as the value of the parameter r increases from 0% to 100%. It is also noticeable that both growth and decline are visibly sharper with SC than with IT. On the other hand, the number of datasets for which the unconstrained DTW outperformed the constrained variants (denoted by DTW>SC and DTW>IT in the figure) was for almost all values of the parameter r less than the number of datasets for which SC and IT outperformed the unconstrained DTW and shows a stronger downward trend for SC than for IT. A smaller deviation between these two numbers in the case of IT may indicate a potential SC advantage.

4.2.3 Classification Accuracy—Optimal Warping-Window Widths

Average classification accuracies (AV) over 128 UCR datasets and standard deviations (SD) of different DTW variants calculated using nested stratified tenfold cross-validation (with optimal warping-window widths chosen over the corresponding training sets using SCV1x9) are shown in Table 4.1.

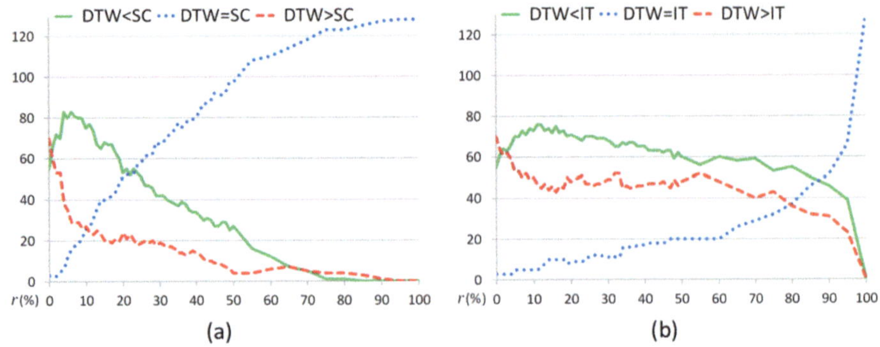

Fig. 4.8 DTW—number of datasets for which the accuracy generated by the unconstrained measure was lower (<), equal (=) and higher (>) than the result obtained by limiting the elasticity with the Sakoe-Chiba (**a**) and the Itakura (**b**) constraints

Table 4.1 DTW—classification accuracy and optimal warping-window width (across 12,800 splits): average value (AV) and standard deviation (SD); the number of datasets for which a given variant produced the best result (WIN) and the mean rank (MR)

	DTW	SC	r_{SC}	IT	r_{IT}
AV (%)	79.23	81.59	6.50	81.27	19.79
SD (pp)	19.45	19.19	10.56	18.84	28.38
WIN	29	69		41	
MR	2.42	1.64		1.94	

In terms of average accuracy, the best result was achieved by SC (81.59%) and the worst by the unconstrained DTW (79.23%). The deviation was the largest in the case of unconstrained DTW (19.45 pp) and the smallest when using the IT constraint (18.84 pp).

The average value of the optimal warping-window width and the standard deviation were significantly smaller with SC (r_{SC}) than with IT (r_{IT}). Specifically, IT produced approximately three times greater AV and approximately 2.7 times greater SD than SC.

Out of a total of 128 datasets, the best accuracy (WIN) most often came from SC (69), followed by IT (41) and least frequently from the unconstrained DTW (29), which is also reflected in the mean values of their ranks (MR).

Regarding the frequency of different warping-window widths chosen by the SCV1x9 method over the training sets, the higher concentration of smaller values in the case of SC highlights the difference between the global constraints: while SC preferred values closer to zero, IT relied more on larger values of the parameter r, even those close to 100% (Fig. 4.9a). Considering warping-window widths greater than 0% of the length of the time series, the standard deviation of the percentages of splits in which these widths appeared as optimal was higher when using the SC than the IT constraint (2.75 vs. 1.61), suggesting that IT more evenly selected different

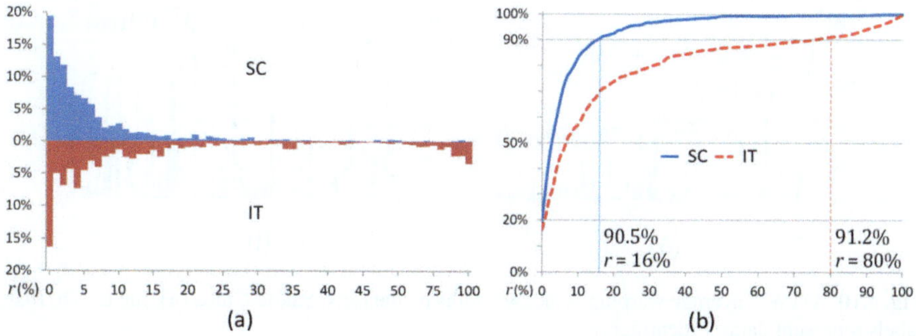

Fig. 4.9 DTW—frequencies (in percentage relative to the 12,800 splits) of different warping-window widths selected by the SCV1x9 algorithm over the training sets when evaluating classification accuracy using SCV10x10 (**a**); cumulative display of frequencies of optimal warping-window widths (**b**)

values of r. The cumulative plot in Fig. 4.9b highlights this difference even more clearly, namely, when using the SC constraint, in 90.5% of the splits, the optimal width of the warping window was not greater than 16% of the total width of the time series, but when it comes to the IT constraint, the value of 91.2% was reached only with $r = 80\%$.

It is also noticeable that both types of constraints strongly relied on $r = 0\%$ (Euclidean distance): SC in almost every fifth (19.4%) and IT in slightly more than 16% of the total of 12,800 iterations. Furthermore, for several datasets the optimal warping-window width was $r = 0\%$ in each of the 100 splits into test and train subsets. With SC, the *Crop* (21), *ItalyPowerDemand* (65), *Meat* (70), *MelbournePedestrian* (72), *MiddlePhalanxOutlineCorrect* (74) and *NonInvasive-FetalECGThorax2* (80) datasets had this feature, and with IT only *Crop* (21) and *MelbournePedestrian* (72). When using the Sakoe-Chiba band, the average value of r was the highest (66.04%) in the case of the *LargeKitchenAppliances* (66) dataset (Fig. 4.10). Averages higher than 20% were observed only with *Computers* (17), *GestureMidAirD2* (48), *HouseTwenty* (60), *MoteStrain* (78), *PLAID* (89), *ShakeGestureWiimoteZ* (101) and *Worms* (126). The largest average optimal warping-window widths, when relying on the Itakura parallelogram, were greater than 90% of the time series length for the following datasets: *LargeKitchenAppliances* (66), *MoteStrain* (78), *PigArtPressure* (87), *PigCVP* (88), *ShakeGestureWiimoteZ* (101), *StarLightCurves* (108).

The comparison of the different DTW variants in terms of the number of datasets for which one variant outperformed the other is given in Fig. 4.11. Out of a total of 128 datasets, SC generated better results than the unconstrained DTW in 95 datasets and underperformed in 30 of them, making the win-loss ratio between SC and DTW slightly better than 3:1. This ratio between IT and DTW was exactly 2:1 (84:42). In the competition between SC and IT, SC emerged as the winner, with a win-loss ratio of approximately 1.6:1. These findings further reinforce the advantage of constrained

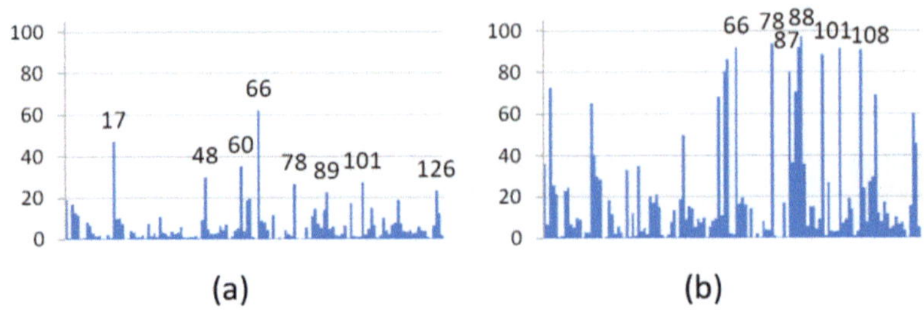

Fig. 4.10 DTW—average warping-window widths by datasets: Sakoe-Chiba (**a**), Itakura (**b**) (data labels represent dataset identifiers)

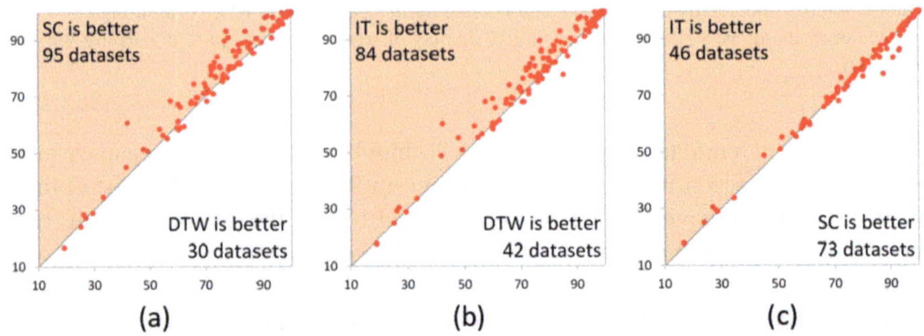

Fig. 4.11 DTW—comparison of similarity measure variants based on the number of datasets where they achieved higher classification accuracy: Sakoe-Chiba versus unconstrained DTW (**a**), Itakura versus unconstrained DTW (**b**), Itakura versus Sakoe-Chiba (**c**)

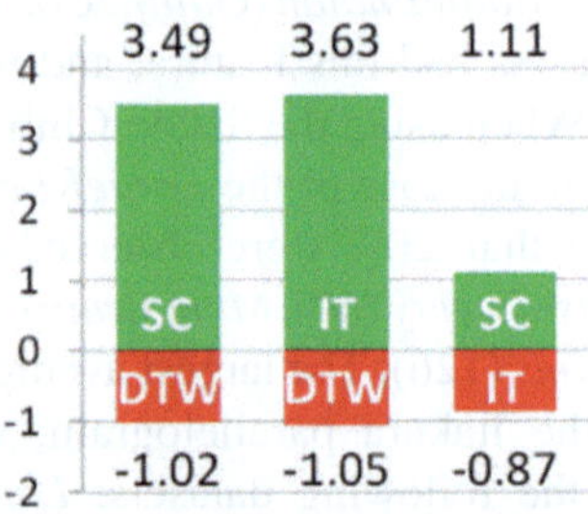

Fig. 4.12 DTW—average difference in classification accuracy of different DTW variants over datasets where the performance of a given variant was better (green) and worse (red) than the other one

variants over unconstrained DTW, at least as far as the accuracy of the 1NN classifier is concerned. In addition, they also give a slight advantage to SC over IT.

According to Fig. 4.12, the average advantage of SC and IT over DTW was almost 3.5 times higher than vice versa, which represents the highest detected advantage of the constrained variants in relation to the corresponding unconstrained measure

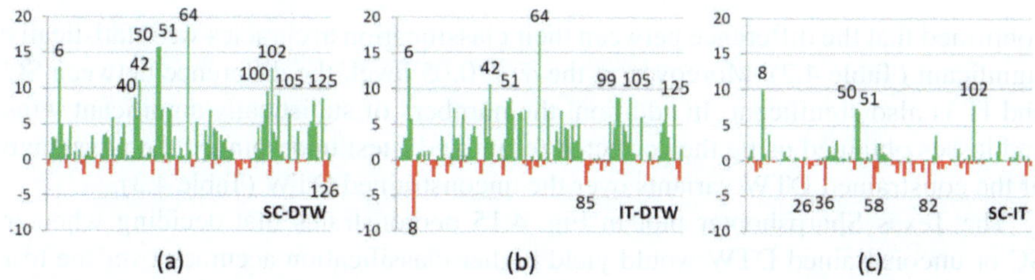

Fig. 4.13 DTW—differences between classification accuracies by datasets: Sakoe-Chiba minus unconstrained DTW (**a**), Itakura minus unconstrained DTW (**b**), Sakoe-Chiba minus Itakura (**c**) (data labels represent dataset identifiers)

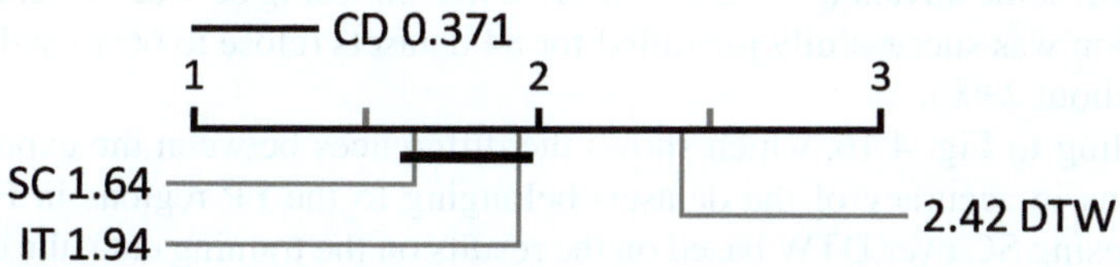

Fig. 4.14 DTW—critical difference diagram comparing average accuracy ranks based on Nemenyi's test: DTW variants connected by a solid bar did not differ statistically significantly

among the four considered elastic distances. The average difference between the classification accuracies of SC and IT over datasets where SC performed better was nearly 1.3 times higher than over datasets where IT outperformed SC, further supporting the previously noted subtle dominance of SC over IT. In addition, the difference between the average advantage of SC over IT and vice versa was the largest in the case of DTW.

The advantage of SC and IT over the unconstrained DTW is much more pronounced than vice versa (Fig. 4.13). Among the datasets where the classification accuracy of DTW was less than that of the constrained variants by more than 10 percentage points were *InsectWingbeatSound* (64), *ArrowHead* (6) and *FordA* (42). In the case of SC, *GesturePebbleZ2* (51), *GesturePebbleZ1* (50) and *ShapeletSim* (102) also belong to that list. The advantage of DTW over SC was more than 3 percentage points only in combination with the *Worms* (126) dataset, and when it comes to IT, there were two such datasets: *BeetleFly* (8) and *PigAirwayPressure* (85). The classification accuracy of SC was more than 5 percentage points higher than that of IT on the following datasets: *BeetleFly* (8), *ShapeletSim* (102), *GesturePebbleZ1* (50) and *GesturePebbleZ2* (51). An advantage of slightly more than 3 percentage points of IT over SC was observed only with *DodgerLoopDay* (26), *OSULeaf* (82), *Haptics* (58) and *EthanolLevel* (36).

The difference in performance between the three variants of DTW was, according to the Friedman test ($\chi^2(2) = 40.60, p < .001$), statistically significant. As detected by Nemenyi's post-hoc test (Fig. 4.14), SC and IT performed significantly better than DTW. The results of the Holm-Bonferroni-adjusted Wilcoxon signed-rank tests also confirmed that the difference between their classification accuracies was statistically significant (Table 4.2). Moreover, at the $\alpha = 0.05$ level, the difference between SC and IT is also significant. In addition, the numbers of statistically significant wins and losses obtained using the corrected resampled t-test underpinned the advantage of the constrained DTW variants over the unconstrained DTW (Table 4.3).

The Texas Sharpshooter plot in Fig. 4.15 demonstrates that deciding whether SC or unconstrained DTW would yield higher classification accuracy over the test set based on their accuracy over the training set was unsuccessful in approximately 18% of cases (23/128) and in about 77% of cases (99/128) it was successful. When choosing between IT and the unconstrained DTW, the failure rate was about 29% (37/128) and the success rate was slightly more than 66% (85/128). These results also confirm some advantage of SC over IT. When choosing between IT and SC, the better option was successfully identified for 84 datasets (close to 66%) and for 30 it was not (about 23%).

According to Fig. 4.16, which shows the differences between the expected and actual gains in accuracy of the datasets belonging to the FP regions in Fig. 4.15, when choosing SC over DTW based on the results on the training data, the error was the highest in the case of the *PigAirwayPressure* (86) and when preferring IT over DTW with the *BeetleFly* (8) dataset. When contrasting SC and DTW, a difference greater than 0.03 was also observed in datasets marked with numbers 17, 29, 48, 59, 87, 88, 126, i 127. The expected gain was more than 0.03 greater than the actual gain for datasets 9, 17, 29, 48, 85, 86, 113 and 127 when comparing IT to DTW. The overestimation of IT over SC was not greater than 0.02 in any dataset.

Table 4.2 DTW—Holm-Bonferroni-adjusted p-values of the pairwise Wilcoxon signed-rank test on classification accuracy (datasets)

A		81.59	81.27	79.23
		SC	IT	DTW
81.59	SC		.025	< .001
81.27	IT			< .001
79.23	DTW			

Table 4.3 DTW—statistically significant wins (W) and losses (L) reported by the corrected resampled t-test on classification accuracy (across 12,800 splits)

	SC	IT	DTW
W	105	80	37
L	32	50	140
W-L	73	30	−103

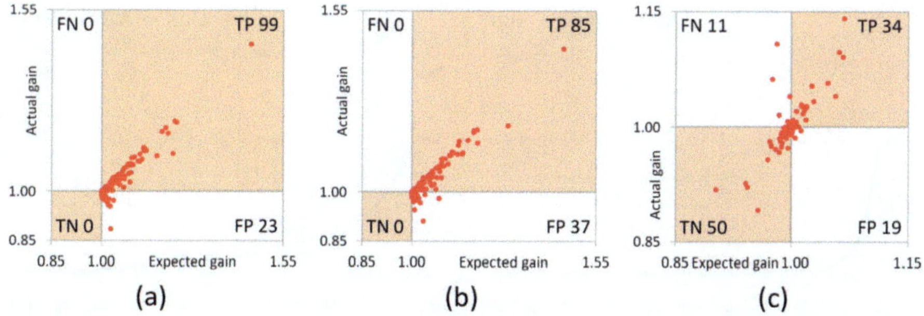

Fig. 4.15 DTW—comparing the actual and expected gain in classification accuracy: Sakoe-Chiba versus unconstrained DTW (**a**), Itakura versus unconstrained DTW (**b**), Itakura versus Sakoe-Chiba (**c**)

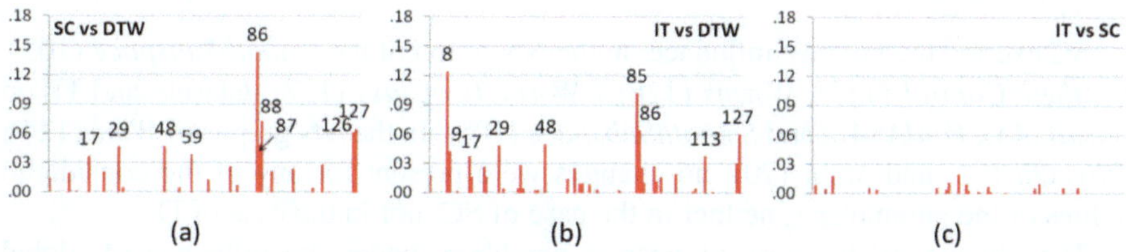

Fig. 4.16 DTW—differences between expected and actual accuracy gains of datasets belonging to the FP region: Sakoe-Chiba versus unconstrained DTW (**a**), Itakura versus unconstrained DTW (**b**), Itakura versus Sakoe-Chiba (**c**) (data labels represent dataset identifiers)

4.3 Longest Common Subsequence (LCS)

4.3.1 Changes in the NN Graph

Besides the visibly lower intensity, the most noticeable difference between the effect of global constraints on the structure of NN graphs generated with LCS (Fig. 4.17) and DTW (Fig. 4.5) is the stepped shape of the line representing the maximum change detected with different widths of the warping window. Alternating periods of stagnation followed by sharp declines resulting in a stepped pattern came from the shortest dataset of the UCR repository: *SmoothSubspace* (105).

By disabling any elasticity (i.e. for $r = 0\%$), the percentage of time series with changed nearest neighbors averaged about 49% (and the standard deviation was slightly greater than 29 percentage points (pp)). With the expansion of the warping window, the rate of decline was noticeably lower when using the Itakura parallelogram (IT) than when applying the Sakoe-Chiba band (SC). With IT, the average value became less than 25% at $r = 9\%$, but with SC it was below 25% even at $r = 3\%$. In addition, when relying on SC, for $r \geq 40\%$ none of the datasets showed any change in the NN graph compared to the unconstrained LCS.

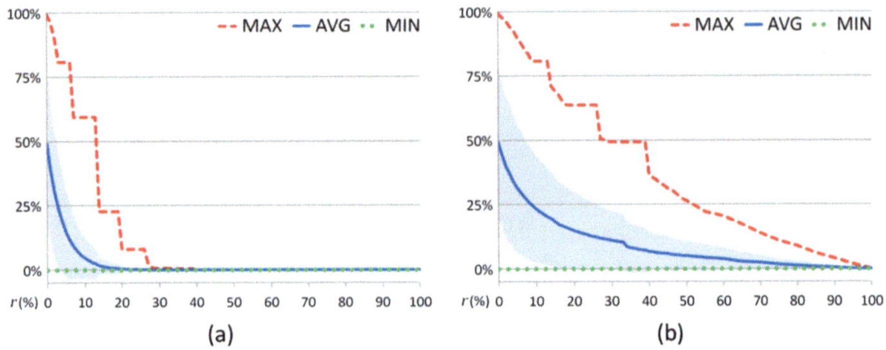

Fig. 4.17 LCS—proportion of time series that replaced their nearest neighbors as a result of applying the Sakoe-Chiba band (**a**) and the Itakura parallelogram (**b**)

SC exerted the greatest influence on the NN graph of the *SmoothSubspace* (105), *SyntheticControl* (112), *Worms* (126) i *WormsTwoClass* (127) datasets and IT on *FordB* (43), *FordA* (42) and *SmoothSubspace* (105). In the NN graphs of *Wine* (124), *OliveOil* (81) and *Meat* (70), no changes were present for any of the considered values of the parameter r, neither in the case of SC, nor in the case of IT.

The change of the class of nearest neighbors under the influence of global constraints was also much more moderate with SC than with IT (Fig. 4.18). By linear matching of time series points (i.e. when $r = 0\%$), the maximum value of the average was about 17.5%, with an approximate standard deviation of 18.1 pp. Average values greater than 10% in the case of the Sakoe-Chiba band were present only for $r = 0\%$ and $r = 1\%$, but the Itakura parallelogram generated averages above that limit up to $r = 4\%$. While with IT changes were present for all considered values of the parameter r, SC did not generate any changes for $r \geq 40\%$.

The three datasets with the largest changes compared to unconstrained LCS were *PigCVP* (88), *GestureMidAirD3* (49) and *Phoneme* (84) when applying SC and

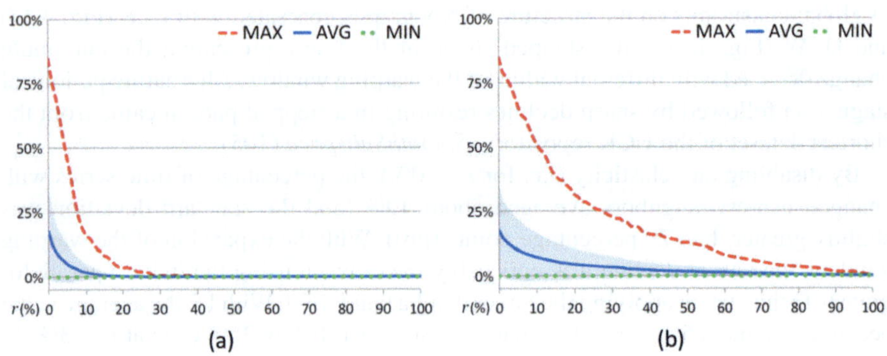

Fig. 4.18 LCS—proportion of time series for which a nearest neighbor class change was detected as a result of applying the Sakoe-Chiba band (**a**) and the Itakura parallelogram (**b**)

PigArtPressure (87), *PigCVP* (88) and *Phoneme* (84) with the IT constraint. In the NN graphs of the *Wine* (124), *OliveOil* (81), *Meat* (70), *Fungi* (46), *ProximalPhalanxOutlineAgeGroup* (92) and *Coffee* (16) datasets, not a single nearest neighbor of any time series changed its class either under the influence of SC or IT.

4.3.2 Classification Accuracy—Individual Warping-Window Widths

According to Fig. 4.19, the preliminary examination of the performance of the constrained LCS variants based on their classification accuracies obtained by applying SCV10x10 over the entire dataset for each individual value of the parameter r suggests their inability to outperform the unconstrained LCS. In the case of the Itakura parallelogram, the average accuracy increased almost monotonically between $r = 0\%$ (71.56%) and $r = 100\%$ (77.59%). The performance difference between the unconstrained and the fully constrained variant (i.e. when the warping window was reduced to the diagonal) was the largest among all four elasticity measures precisely in the case of LCS. The best result of the Sakoe-Chiba band was achieved with $r = 19\%$ (77.60%), but it differs from the result of the unconstrained LCS only at the second decimal place. When applying SC with $r \geq 40\%$ the results coincided with the accuracy of unconstrained LCS.

The fact that for $r \geq 40\%$ there was no difference in the performance of SC and LCS is clearly visible in Fig. 4.20 which shows the numbers of datasets for which the classification accuracy of the unconstrained LCS was lower, higher or equal to the results of the constrained variants. In the $r < 4\%$ region LCS dominated to a large extent and some advantage of SC was present only in the interval $7 \leq r \leq 9\%$. In this respect, IT was outperformed by the unconstrained LCS over virtually the entire range of r-values considered. Comparing Figs. 4.8, 4.20, 4.32 and 4.44, it can

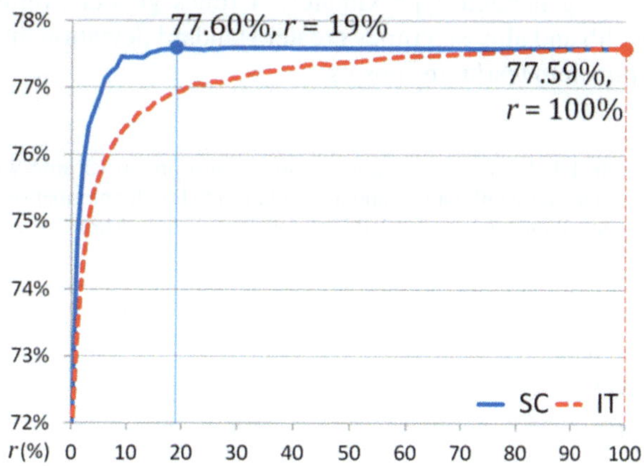

Fig. 4.19 LCS—average classification accuracy across all datasets for different warping-window widths

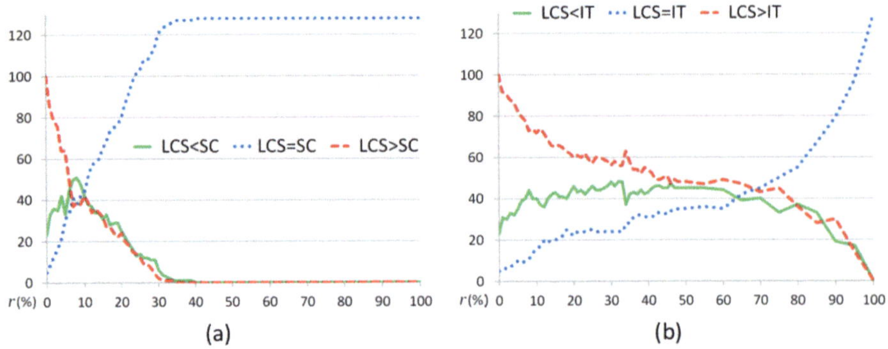

Fig. 4.20 LCS—number of datasets for which the accuracy generated by the unconstrained measure was lower (<), equal (=) and higher (>) than the result obtained by limiting the elasticity with the Sakoe-Chiba (**a**) and the Itakura (**b**) constraints

also be seen that the changes in the NN graph induced by $r = 0\%$ were the most unfavorable precisely in the case of LCS and EDR measures.

4.3.3 Classification Accuracy—Optimal Warping-Window Widths

Table 4.4 shows the average value (AV) and standard deviation (SD) of the classification accuracies of different LCS variants obtained by applying the SCV10x10 method, where the optimal warping-window width in each of the 100 iterations was chosen by the SCV1x9 algorithm on the corresponding training set. The difference between the best (SC) and worst (IT) results was only approximately 0.13 percentage points and the standard deviation values were also very close.

IT generated approximately 4 times greater average optimal warping-window width and almost 5 times greater standard deviation than SC (considering all 12,800 splits into test/train subsets).

Table 4.4 LCS—classification accuracy and optimal warping-window width (across 12,800 splits): average value (AV) and standard deviation (SD); the number of datasets for which a given variant produced the best result (WIN) and the mean rank (MR)

	LCS	SC	r_{SC}	IT	r_{IT}
AV (%)	77.59	77.67	6.01	77.54	24.80
SD (pp)	20.76	20.66	6.24	20.67	29.93
WIN	65	46		31	
MR	1.79	1.96		2.25	

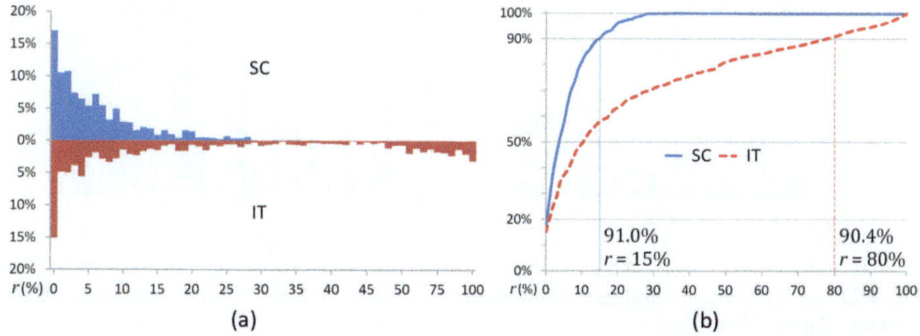

Fig. 4.21 LCS—frequencies (in percentage relative to the 12,800 splits) of different warping-window widths selected by the SCV1x9 algorithm over the training sets when evaluating classification accuracy using SCV10x10 (**a**); cumulative display of frequencies of optimal warping-window widths (**b**)

Among the 128 datasets, the best result (WIN) came from the unconstrained LCS 65 times, SC won 46 times and IT achieved the best accuracy in the case of 31 datasets.

The frequencies of the different optimal values of the parameter r depicted in Fig. 4.21a show a noticeable similarity to those in Fig. 4.9a generated by DTW. SC preferred narrow warping windows in this case as well and IT visibly relied also on values close to 100% of the time series length. Out of a total of 12,800 splits into training and testing subsets, the optimal value of the parameter r in combination with the Sakoe-Chiba constraint was not greater than 15% of the length of the time series in 91% of cases (Fig. 4.21b), but when using the Itakura parallelogram to reach the threshold of 90.4% of the splits it was necessary to take into account all widths up to $r = 80\%$, similar to the case of DTW. It is also apparent that when applying the SC constraint, the optimal value of r was not higher than 36% in any of the 12,800 splits.

The percentages of splits in which, on the training set, the highest classification accuracy was obtained with $r = 0\%$ were slightly lower compared to DTW: 17.0% (SC) and 15.1% (IT). When it comes to *FreezerSmallTrain* (45), *Meat* (70), *MelbournePedestrian* (72), *OliveOil* (81) and *Wine* (124) datasets, the optimal value of the parameter r was 0% of the length of the time series in each of the 100 iterations of the SCV10x10 procedure for both SC and IT and in the case of SC, this group also included *Chinatown* (13), *FreezerRegularTrain* (44) and *Strawberry* (109). The possibility of non-linear matching of time series points in these datasets did not lead to more accurate classification.

FordA (47) stood out in that the optimal warping-window width was 1% of the length of the time series in each of the 100 cross-validation splits when using both types of global constraints. When applying the Sakoe-Chiba constraint, this peculiarity was also observed in the case of the *FordB* (48) dataset and when it came to *Crop* (21), the optimal value of the parameter r was 3% in each iteration.

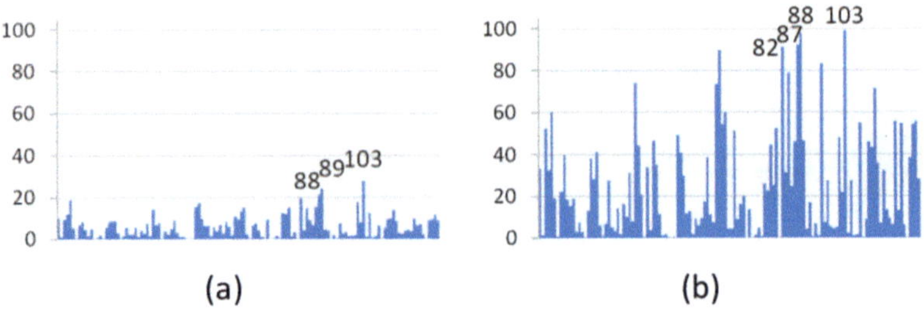

Fig. 4.22 LCS—average warping-window widths by datasets: Sakoe-Chiba (**a**), Itakura (**b**) (labels represent dataset identifiers)

Using the SC constraint, average *r*-values were greater than 20% only with *PigCVP* (88), *PLAID* (89) and *ShapesAll* (103) (Fig. 4.22). For many datasets, IT generated a significantly higher average value of *r* than SC, with *OSULeaf* (82), *PigArtPressure* (87), *PigCVP* (88) and *ShapesAll* (103), it was even > 90%.

Based on Fig. 4.23 which compares the performance of the considered LCS variants in terms of the number of datasets for which the classification accuracy of a given variant was higher or lower than the accuracy of the other two, it can be seen that the unconstrained LCS outperformed both SC and IT. The win-loss ratio between LCS and SC was 72 to 50 (approximately 1.4:1) and between LCS and IT it was 78 to 44 (roughly 1.8:1). IT remained inferior to SC in a similar proportion as compared to LCS: while SC produced better results in the case of 74 datasets, IT was only able to prevail in 43 cases (i.e. the ratio was about 1.7:1). However, the closeness of the points to the diagonal suggests that the classification accuracies were not significantly different.

The average difference in performance was approximately 1.7 and 1.6 times smaller on the datasets where LCS generated higher accuracy than SC and IT, respectively, than the other way around (Fig. 4.24). This means that the constrained measures achieved a higher average advantage over the unconstrained LCS on fewer

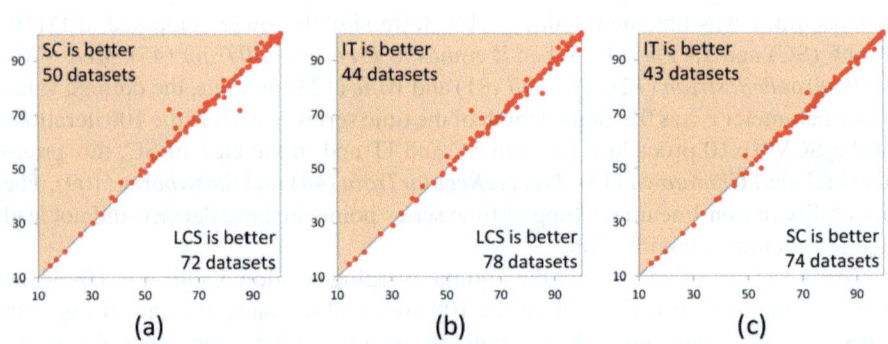

Fig. 4.23 LCS—comparison of similarity measure variants based on the number of datasets where they achieved higher classification accuracy: Sakoe-Chiba versus unconstrained LCS (**a**), Itakura versus unconstrained LCS (**b**), Itakura versus Sakoe-Chiba (**c**)

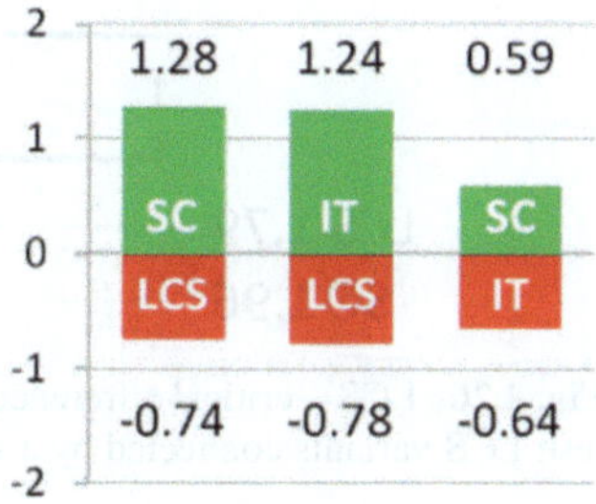

Fig. 4.24 LCS—average difference in classification accuracy of different LCS variants over datasets where the performance of a given variant was better (green) and worse (red) than the other one

datasets and LCS achieved a smaller average advantage over SC and IT on a larger number of datasets. In this regard, the results of SC and IT were approximately equal, i.e. the points above and below the diagonal in Fig. 4.23 are approximately at the same average distance from the diagonal.

The high level of uniformity of the results of LCS, SC and IT is also confirmed by Fig. 4.25, which shows the differences between their classification accuracies by individual datasets. The advantage of SC and IT over the unconstrained LCS was greater than 10 percentage points only in the case of the *FordA* (42) dataset. In addition, the constrained variants surpassed the unconstrained one by more than 5 percentage points only on the *FordB* (43) dataset. The advantage of LCS was greater than 10 percentage points over SC on the *BeetleFly* (8) and slightly more than 3 percentage points on the *Rock* (96) dataset. The only datasets where LCS beat IT by more than 3 percentage points were *BeetleFly* (8), *BirdChicken* (9) and *WormsTwoClass* (127). SC outperformed IT by more than 3 percentage points only on the *Herring* (59) dataset and IT prevailed by slightly more than 5 percentage points on the *BeetleFly* (8) and *DodgerLoopDay* (26) datasets.

According to the Friedman test, the difference in the performance of the analyzed LCS variants was statistically significant ($\chi^2(2) = 14.91$, $p < .001$) despite very close average accuracies and standard deviations in Table 4.4. Nemenyi's test assigned statistical significance to the difference between LCS and IT (Fig. 4.26). However, based on the Holm-Bonferroni-adjusted Wilcoxon signed-rank test, the

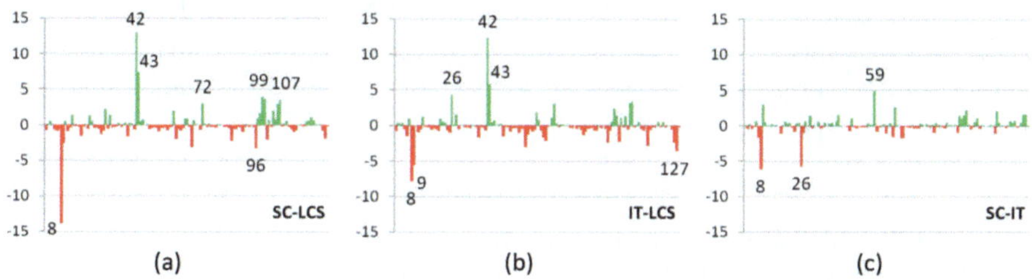

Fig. 4.25 LCS—differences between classification accuracies by datasets: Sakoe-Chiba minus unconstrained LCS (**a**), Itakura minus unconstrained LCS (**b**), Sakoe-Chiba minus Itakura (**c**) (labels represent dataset identifiers)

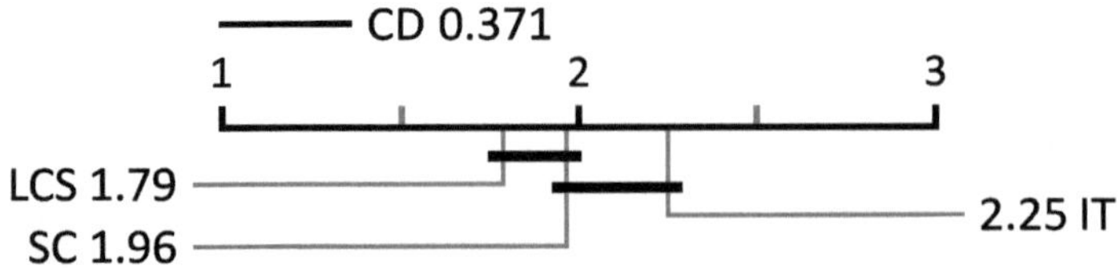

Fig. 4.26 LCS—critical difference diagram comparing average accuracy ranks based on Nemenyi's test: LCS variants connected by a solid bar did not differ statistically significantly

observed statistical significance was primary due to the difference between the Sakoe-Chiba and Itakura constraints (Table 4.5). The numbers of statistically significant wins and losses in Table 4.6 point primarily to IT's inferiority to SC and LCS.

When evaluating classification accuracy on the test set relying solely on classification accuracy on the training set, the prediction that SC would perform better than LCS was correct in approximately 40% (51/128) of cases, but in almost 52% (66/128) of the datasets it was wrong (Fig. 4.27). The advantage of IT over LCS was correctly estimated in about 37% (47/128) of cases and incorrectly in about 58% (74/128). The percentage of datasets for which we opted for the better variant when choosing between SC and IT amounted to approximately 66% (85/128), but we were wrong in about 22% (28/128) of them—which almost matches the result when considering DTW (Fig. 4.15).

In the FP region, the error in estimating the advantage of the constrained variants of LCS over the unconstrained one was the largest on the *BeetleFly* (8) dataset (Fig. 4.28). The difference between the expected and actual accuracy gain of SC over LCS was greater than 0.03 also for the following datasets: 26, 49, 67, 85, 96, 126 and 127. Among the datasets where the advantage of IT over LCS was overestimated by more than 0.03 were those marked with numbers 9, 49, 58, 59, 68, 85, 86, 96, 126

Table 4.5 LCS—Holm-Bonferroni-adjusted p-values of the pairwise Wilcoxon signed-rank test on classification accuracy

A		77.67	77.59	77.54
		SC	LCS	IT
77.67	SC		.377	.011
77.59	LCS			.063
77.54	IT			

Table 4.6 LCS—statistically significant wins (W) and losses (L) reported by the corrected resampled t-test on classification accuracy (across 12,800 splits)

	LCS	SC	IT
W	78	59	34
L	56	46	69
W-L	22	13	−35

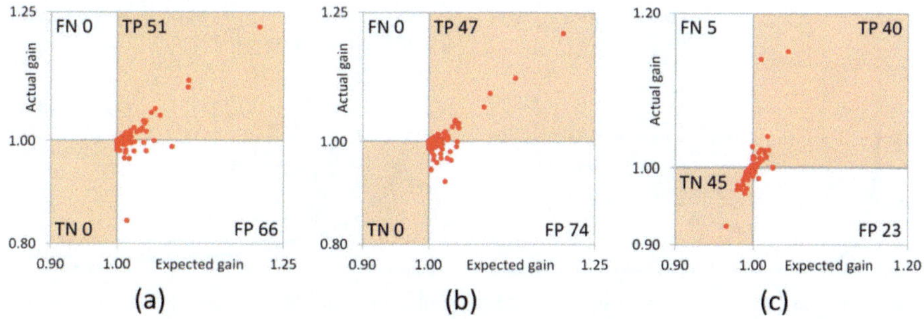

Fig. 4.27 LCS—comparing the actual and expected gain in classification accuracy: Sakoe-Chiba versus unconstrained LCS (**a**), Itakura versus unconstrained LCS (**b**), Itakura versus Sakoe-Chiba (**c**)

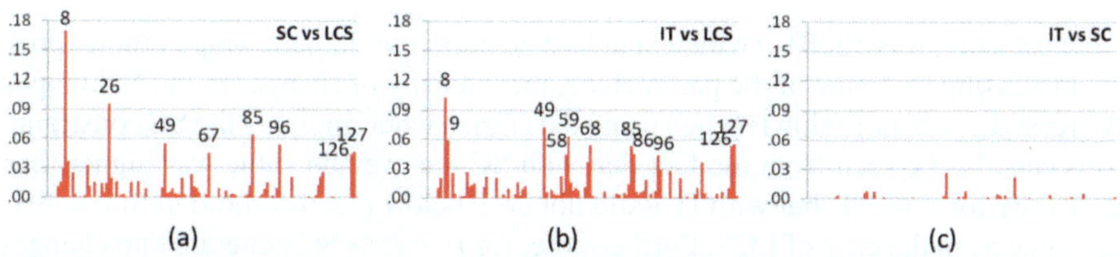

Fig. 4.28 LCS—differences between expected and actual accuracy gains of datasets belonging to the FP region: Sakoe-Chiba versus unconstrained LCS (**a**), Itakura versus unconstrained LCS (**b**), Itakura versus Sakoe-Chiba (**c**) (labels represent dataset identifiers)

and 127 as well. When comparing IT and SC, the error did not reach 0.03 for any of the 128 UCR dataset.

4.4 Edit Distance with Real Penalty (ERP)

4.4.1 Changes in the NN Graph

The general traits of the impact of global constraints on the nearest neighbor graph of the ERP measure presented in Fig. 4.29 show a high degree of similarity with the results obtained for LCS (Fig. 4.17). Alternating periods of constant values followed by sharp drops in the percentage of time series that changed their nearest neighbors under the influence of the Sakoe-Chiba band (SC) and the Itakura parallelogram (IT) compared to the unconstrained ERP are clearly visible in this case as well. The stepped shape of the line that represents the maximum value on the charts here also came from the shortest member of the UCR archive (i.e. *SmoothSubspace* (105)).

When the warping window was reduced to the diagonal (i.e., without any elasticity), the average value of the percentage of time series whose nearest neighbors

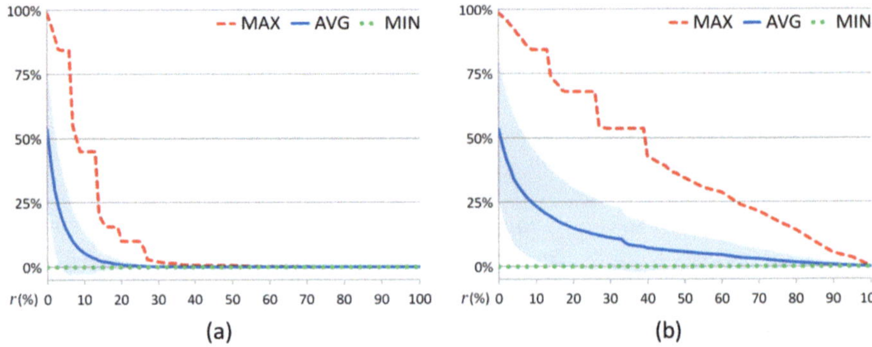

Fig. 4.29 ERP—proportion of time series that replaced their nearest neighbors as a result of applying the Sakoe-Chiba band (**a**) and the Itakura parallelogram (**b**)

were replaced was 53.4% (with a standard deviation of 28 percentage points (pp)). By increasing the value of the parameter r, the intensity of changes in the NN graphs decreased much more slowly when using the IT than when applying the SC constraint. This can also be seen from the fact that with SC the average value was higher than 25% only for $r \leq 3\%$, but with IT it did not drop below that threshold until $r = 8\%$, the same as in the case of LCS. Furthermore, for $r \geq 75\%$ SC generated no changes at all in the NN graphs of any dataset (when SC was applied to LCS, changes were absent as soon as $r \geq 40\%$).

Observing the average intensity of changes for all considered values of the parameter r, the Sakoe-Chiba band generated the greatest changes in the NN graphs of the *SmoothSubspace* (105), *PickupGestureWiimoteZ* (85) and *TwoPatterns* (117) datasets and the Itakura parallelogram had the greatest influence in the case of the *FordB* (43), *FordA* (42) and *TwoPatterns* (105) datasets. Analogous to LCS, neither SC nor IT generated any changes in the NN graphs of the *Wine* (124), *OliveOil* (81) and *Meat* (70) datasets.

Comparing Figs. 4.29 and 4.30, it is easy to see that under the influence of global constraints, only a small part of the new nearest neighbors altered their class in relation to the NN graph associated with the unconstrained measure. Similar to LCS, the maximum average value obtained with $r = 0\%$ was slightly higher than 17.2%, with a standard deviation of about 18.2 pp. As with LCS, when using the SC constraint, the average value was above 10% only for $r = 0\%$ and $r = 1\%$. Relying on the Itakura parallelogram, it was greater than 10% for $r \leq 5\%$. When the warping-window width reached 65% of the length of the time series, SC no longer generated any changes compared to the unconstrained ERP.

In terms of classes, the Sakoe-Chiba band produced the largest changes in the NN graphs of the *Phoneme* (84), *PickupGestureWiimoteZ* (85) and *PigCVP* (88) datasets, and the Itakura parallelogram in the case of the *PigArtPressure* (87), *PigCVP* (88) and *Phoneme* (84) datasets. The datasets numbered 13, 14, 44, 45, 46, 70, 81 and 124 were not affected by the application of global constraints.

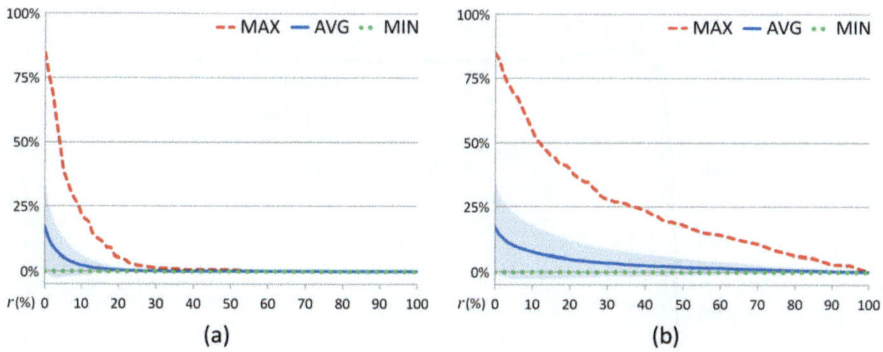

Fig. 4.30 ERP—proportion of time series for which a nearest neighbor class change was detected as a result of applying the Sakoe-Chiba band (**a**) and the Itakura parallelogram (**b**)

4.4.2 Classification Accuracy—Individual Warping-Window Widths

The average classification accuracies across all datasets obtained for different warping-window widths using 10 times repeated tenfold stratified cross-validation on the entire dataset presented in Fig. 4.31 anticipate a potential slight advantage of the Sakoe-Chiba constrained variant of ERP over the unconstrained one. Nevertheless, comparing the results obtained with $r = 0\%$ (76.97%) and $r = 100\%$ (81.48%) it is apparent that prohibiting any elasticity leads to performance degradation. SC reached its maximum with $r = 8\%$ (81.68%) and IT with $r = 85\%$ (81.49%), but compared to their results, the unconstrained ERP was only slightly inferior.

With very narrow warping windows ($r \leq 2\%$ for SC and $r \leq 7\%$ for IT) the number of datasets for which the classification accuracy of the unconstrained ERP outperformed the performance of the constrained variants is visibly greater than the number of datasets for which SC and IT surpassed ERP (Fig. 4.32). This further confirms the inferiority of SC and IT compared to the unconstrained ERP near $r = 0\%$, which is also noticeable in Fig. 4.31. The dominance of IT over ERP was constant for $r > 7\%$ and reached its peak with $r = 85\%$. In this regard, the greatest advantage of SC over ERP was around $r = 6\%$. The growth in the number of datasets for which the performance of the constrained and unconstrained ERP was equal was noticeably faster when using SC than in the case of IT. For $r \geq 70\%$, there was no difference between SC and ERP.

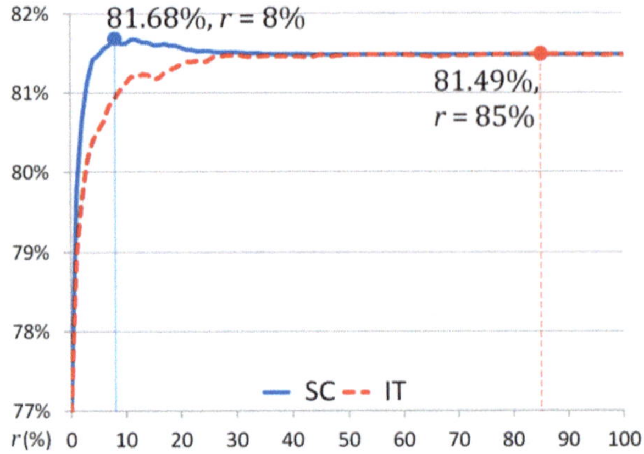

Fig. 4.31 ERP—average classification accuracy across all datasets for different warping-window widths

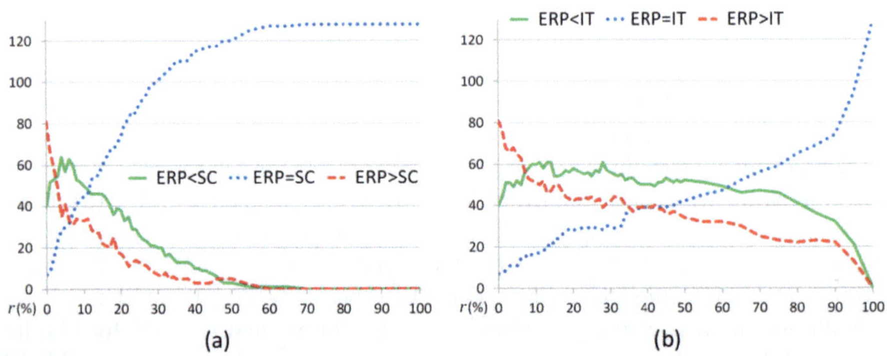

Fig. 4.32 ERP—number of datasets for which the accuracy generated by the unconstrained measure was lower (<), equal (=) and higher (>) than the result obtained by limiting the elasticity with the Sakoe-Chiba (**a**) and the Itakura (**b**) constraints

4.4.3 Classification Accuracy—Optimal Warping-Window Widths

In terms of average classification accuracy (calculated using the SCV10x10 method, where the optimal value of the parameter r was chosen with SCV1x9 on the training set in each of the 100 splits into testing and training subsets), the best result was achieved by the SC variant (82.04%), followed by IT (81.91%) and the unconstrained ERP (81.48%), as shown in Table 4.7. It should be noted that the difference between the best and the worst results did not even reach 0.6 percentage points. The performance of the three examined ERP variants was also very similar when it comes to standard deviation.

Table 4.7 ERP—classification accuracy and optimal warping-window width (across 12,800 splits): average value (AV) and standard deviation (SD); the number of datasets for which a given variant produced the best result (WIN) and the mean rank (MR)

	ERP	SC	r_{SC}	IT	r_{IT}
AV (%)	81.48	82.04	5.07	81.91	17.86
SD (pp)	17.97	17.70	6.54	17.71	25.31
WIN	50	61		44	
MR	2.15	1.86		2.00	

The optimal warping-window width was, on average (over a total of 12,800 test/train splits), more than 3.5 times larger in the case of the Itakura parallelogram than when using the Sakoe-Chiba constraint. Moreover, the standard deviation was approximately 3.9 times higher in combination with IT.

At the level of individual datasets, the highest classification accuracy was most often achieved by SC (61), followed by ERP (50) and finally IT (44), with mean ranks (MR) closely clustered around 2.

Based on the frequencies of the different optimal values of the parameter r shown in Fig. 4.33a, it can be evidently concluded that SC preferred narrower warping-windows and IT relied also on values close to the length of the time series (i.e., close to $r = 100\%$). Comparing Figs. 4.9, 4.21, 4.33 and 4.45, it can also be seen that the frequency of linear matching of time series points (i.e. $r = 0\%$) was the highest in the case of ERP for both types of global constraints—in more than 20% of the 12,800 splits into testing and training subsets.

According to Fig. 4.33b, optimal elasticity was achieved in about 90% of cases with $r \leq 13\%$ when using the Sakoe-Chiba band and with $r \leq 60\%$ when applying the Itakura parallelogram. It is also worth noting that widths greater than 55% of the length of the time series never emerged as optimal with SC. Furthermore, in the case

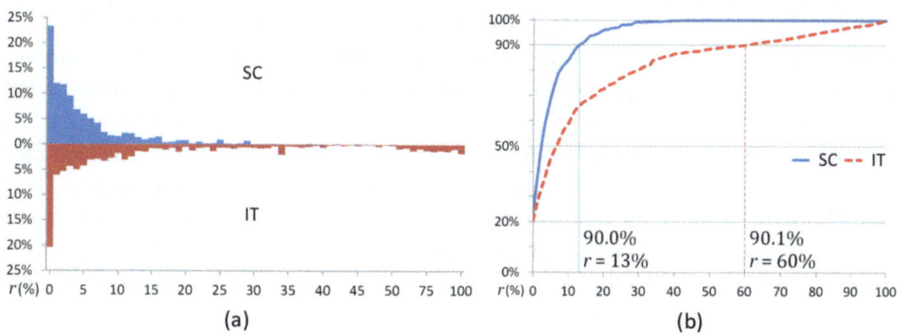

Fig. 4.33 ERP—frequencies (in percentage relative to the 12,800 splits) of different warping-window widths selected by the SCV1x9 algorithm over the training sets when evaluating classification accuracy using SCV10x10 (**a**); cumulative display of frequencies of optimal warping-window widths (**b**)

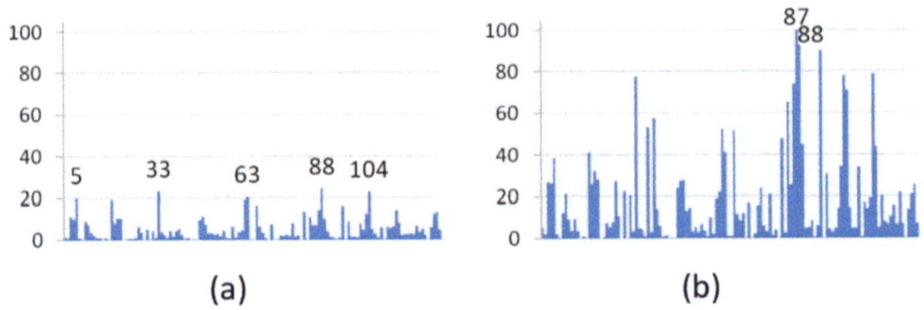

Fig. 4.34 ERP—average warping-window widths by datasets: Sakoe-Chiba (**a**), Itakura (**b**) (labels represent dataset identifiers)

of the *Crop* (21), *DodgerLoopWeekend* (28), *Fungi* (46), *Meat* (70), *Melbourne-Pedestrian* (72), *OliveOil* (81) and *Wine* (124) datasets the optimal value of the parameter r was 0% in each iteration of the SCV10x10 method with both SC and IT. This trait was also present in the results obtained with IT for *ECG200* (30) and with SC for *ItalyPowerDemand* (65). Elasticity did not appear to improve classification accuracy on these datasets. Applying the Sakoe-Chiba constraint, *Adiac* (2), *CBF* (12) and *FordB* (43) stood out in that their optimal warping-window width was 1% in each of the 100 splits into the testing and training subsets.

Figure 4.34a indicates that when relying on the SC constraint, the average of the optimal values of the parameter r was greater than 20% only for *AllGestureWiimoteZ* (5), *ElectricDevices* (33), *InsectEPGSmallTrain* (63), *PigCVP* (88) and *SmallKitchenAppliances* (104). In combination with the Itakura parallelogram, an average greater than 90% was observed only with *PigArtPressure* (87) and *PigCVP* (88) (Fig. 4.34b).

Out of the 128 UCR datasets, the unconstrained ERP was outperformed by SC and IT in 67 and 70 cases, respectively, and it was superior to them in 52 and 47 cases, respectively. Thus, the win-loss ratio between ERP and SC was approximately 1.3:1, and between ERP and IT it was around 1.5:1. Nonetheless, SC outperformed IT in a ratio of about 1.4:1 (67:45). Albeit the proximity of the points to the diagonal in Fig. 4.35 suggests that the differences in performance of the three ERP variants were not excessively large, on the datasets where SC and IT outperformed ERP, the average difference in classification accuracy was nearly 3 and close to 2 times higher, respectively, compared to the average difference on the datasets where ERP was better (Fig. 4.36). The difference between SC and IT was, in this respect, very small, both the points representing datasets where SC outperformed IT and the points denoting datasets where IT beat SC are very close to the diagonal in Fig. 4.35: their average distance from the diagonal was only approximately 0.42 (SC) and 0.37 (IT).

The slight advantage of SC and IT over ERP and their mutual similarity can be seen even more clearly in Fig. 4.37 which shows the difference between the generated classification accuracies over individual datasets. In contrasting SC and IT with ERP, the advantage of the constrained measures on no dataset was greater than 10 percentage points. An advantage greater than 5 percentage points was detected in the

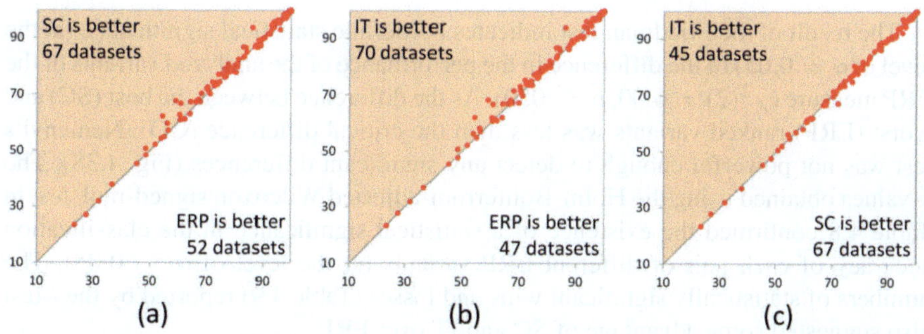

Fig. 4.35 ERP—comparison of similarity measure variants based on the number of datasets where they achieved higher classification accuracy: Sakoe-Chiba versus unconstrained ERP (**a**), Itakura versus unconstrained ERP (**b**), Itakura versus Sakoe-Chiba (**c**)

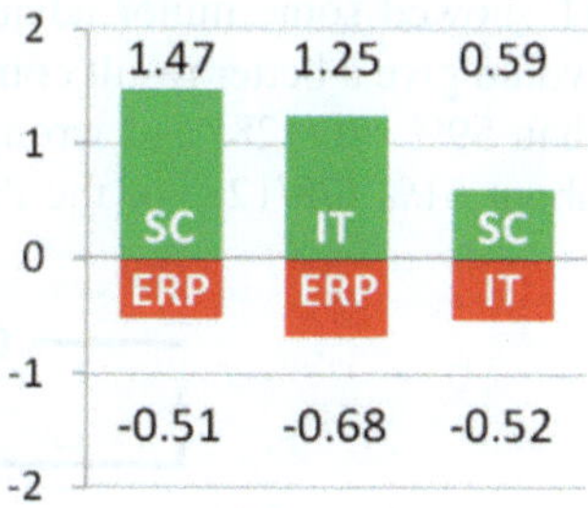

Fig. 4.36 ERP—average difference in classification accuracy of different ERP variants over datasets where the performance of a given variant was better (green) and worse (red) than the other one

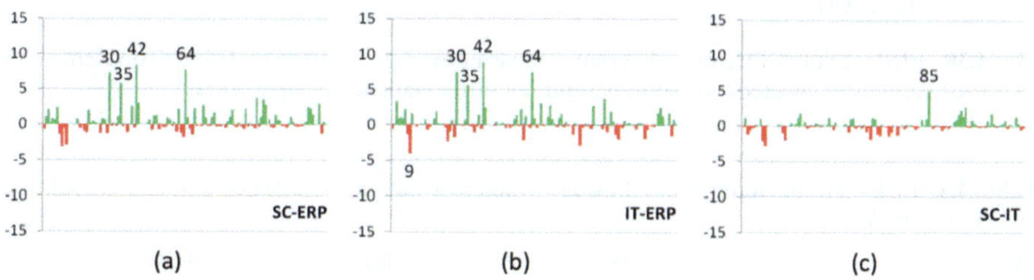

Fig. 4.37 ERP—differences between classification accuracies by datasets: Sakoe-Chiba minus unconstrained ERP (**a**), Itakura minus unconstrained ERP (**b**), Sakoe-Chiba minus Itakura (**c**) (labels represent dataset identifiers)

same four members of the UCR repository in both cases: *FordA* (42), *InsectWingbeatSound* (64), *ECG200* (30) and *EOGVerticalSignal* (35). A difference greater than 3 percentage points in favor of the unconstrained ERP was observed only with respect to IT and only in the case of the *BirdChicken* (9) dataset. The largest advantage of the Sakoe-Chiba band over the Itakura parallelogram was detected on the *PickupGestureWiimoteZ* (85) dataset and it amounted to 4.9 percentage points (rounded to one decimal place). The difference between the classification accuracy of IT and SC did not reach 3 percentage points on any dataset where IT outperformed SC.

The result of the Friedman test indicates borderline statistical significance (at the level of $\alpha = 0.05$) of the difference in the performance of the analyzed variants of the ERP measure ($\chi^2(2) = 6.00, p = .050$). As the difference between the best (SC) and worst (ERP) ranked variants was less than the critical difference (CD), Nemenyi's test was not powerful enough to detect any significant differences (Fig. 4.38). The p-values obtained using the Holm-Bonferroni-adjusted Wilcoxon signed-rank test in Table 4.8 confirmed the existence of a statistical significance in the classification accuracy of each pair of different ERP variants (at the level of $\alpha = 0.05$). The numbers of statistically significant wins and losses (Table 4.9) reported by the t-test also suggested some advantage of SC and IT over ERP.

In contrasting SC with ERP, in just over 35% (45 out of 128) of the datasets the prediction that SC would outperform ERP on the test set based on higher classification accuracy over the training set was incorrect (Fig. 4.39a). In slightly more than half of the datasets (71 out of 128, i.e. approximately 55%), this assessment was successful. IT showed some minor advantage over SC in this regard: the expectation that IT would give a better result compared to ERP on the test set was correct in little more than 59% (76/128) and wrong in about 31% (40/128) of the cases (Fig. 4.39b). In about 61% (78/128) of the datasets, it was successfully recognized whether IT or

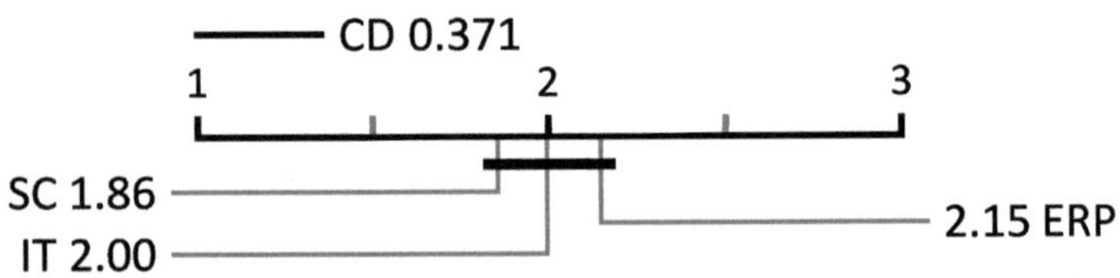

Fig. 4.38 ERP—critical difference diagram comparing average accuracy ranks based on Nemenyi's test: ERP variants connected by a solid bar did not differ statistically significantly

Table 4.8 ERP—Holm-Bonferroni-adjusted p-values of the pairwise Wilcoxon signed-rank test on classification accuracy

A		82.04	81.91	81.48
		SC	IT	ERP
82.04	SC		.036	.006
81.91	IT			.021
81.48	ERP			

Table 4.9 ERP—statistically significant wins (W) and losses (L) reported by the corrected resampled t-test on classification accuracy (across 12,800 splits)

	SC	IT	ERP
W	67	53	40
L	36	45	79
W-L	31	8	−39

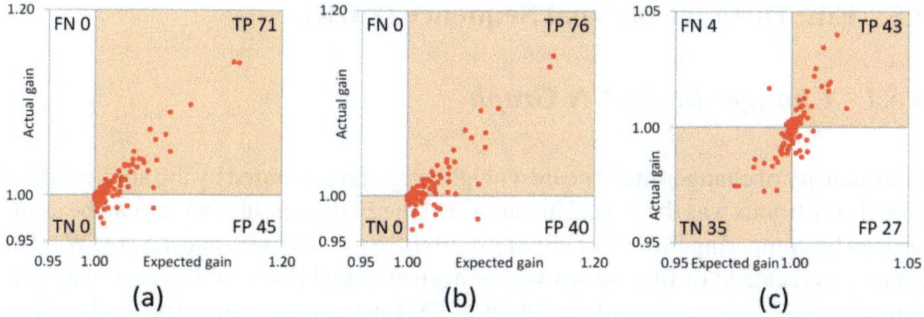

Fig. 4.39 ERP—comparing the actual and expected gain in classification accuracy: Sakoe-Chiba versus unconstrained ERP (**a**), Itakura versus unconstrained ERP (**b**), Itakura versus Sakoe-Chiba (**c**)

SC represented a better option (Fig. 4.39c), but the decision based on the expected performance led to worse actual results in approximately 24% (31/128) of them, which is somewhat behind the success rate seen with DTW and LCS.

The decision that SC will outperform ERP based on the higher average accuracy of SC over the training data was most erroneous on the *GestureMidAirD3* (49) dataset, and next to it, the difference between the expected and the actual gain was greater than 0.03 also on the *BirdChicken* (9), *Car* (11), *Earthquakes* (29) and *Worms* (126) datasets (Fig. 4.40). The most pronounced disadvantage arising from IT's overconfidence was observed with the *PickupGestureWiimoteZ* (85) dataset, but it was greater than 0.03 (expressed as the difference between the expected and the actual gain) also for those numbered 9, 26, 29, 97, 101 and 126. Similar to DTW and LCS, the error in favoring IT over SC did not reach 0.03.

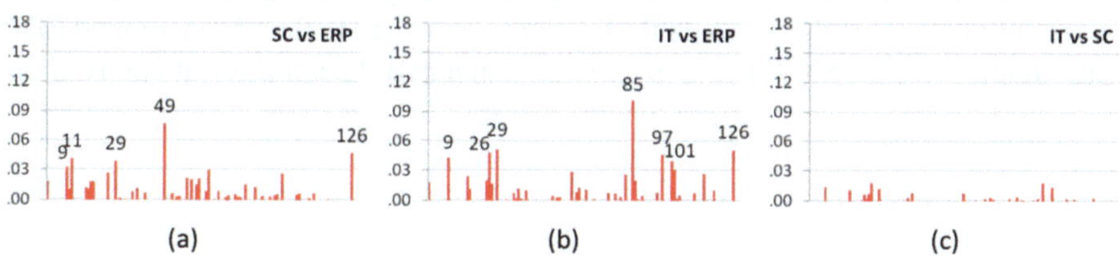

Fig. 4.40 ERP—differences between expected and actual accuracy gains of datasets belonging to the FP region: Sakoe-Chiba versus unconstrained ERP (**a**), Itakura versus unconstrained ERP (**b**), Itakura versus Sakoe-Chiba (**c**) (labels represent dataset identifiers)

4.5 Edit Distance on Real Sequence (EDR)

4.5.1 Changes in the NN Graph

The intensity of changes in the nearest neighbor graph generated by the application of global constraints was the lowest in the case of the EDR measure, which can be easily noticed by comparing Figs. 4.41 (EDR), 4.29 (ERP), 4.17 (LCS) and 4.5 (DTW). The average percentage of time series whose nearest neighbors were replaced was only about 41.5% (with a standard deviation of 28.4 percentage points (pp)) when their points were matched linearly (i.e. for $r = 0\%$), which represents the lowest value among the elastic measures analyzed. The rate of decrease in the amount of changes in the NN graph was noticeably slower when employing the Itakura parallelogram (IT) than the Sakoe-Chiba band (SC) in this case as well. Considering the average value across all 128 datasets, it was greater than 25% only for $r \leq 1\%$ when applying the SC constraint and it was above that threshold when using IT only up to $r \leq 3\%$. EDR also stands out from the other measures in that, when the warping window was constrained by SC, there was no change in the NN graph from $r = 25\%$ as compared to unconstrained EDR. The absence of changes was observed for LCS from $r = 40\%$, for ERP from $r = 75\%$ and for DTW only for $r = 95\%$.

The three datasets whose NN graphs were most affected by the SC constraint were: *PigCVP* (88), *BeetleFly* (8) and *FordB* (43). The warping window constraint proposed by Itakura generated the largest average changes in the nearest neighbor graphs of the *FordB* (43), *FordA* (42) and *PigArtPressure* (87) datasets. Compared to LCS and ERP, in addition to *Wine* (124), *OliveOil* (81) and *Meat* (70), neither SC nor IT changed the NN graph of the *ACSF1* (1) dataset for any considered value of the parameter *r*.

The average value of the percentages of nearest neighbors that altered their class as a result of applying the SC and IT constraints compared to the unconstrained EDR was slightly below 15% for $r = 0\%$, with a standard deviation of about 16.6 pp.

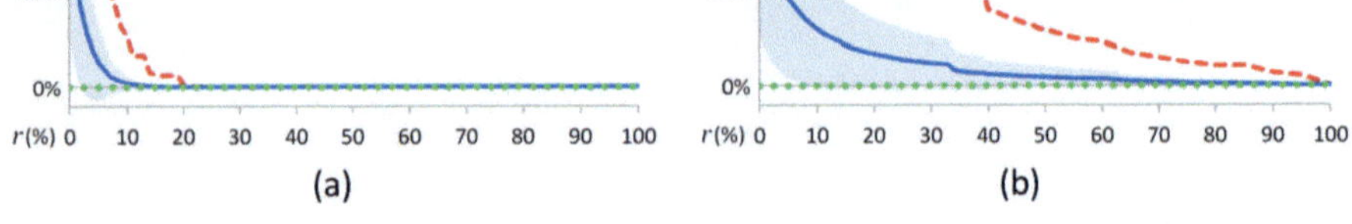

Fig. 4.41 EDR—proportion of time series that replaced their nearest neighbors as a result of applying the Sakoe-Chiba band (**a**) and the Itakura parallelogram (**b**)

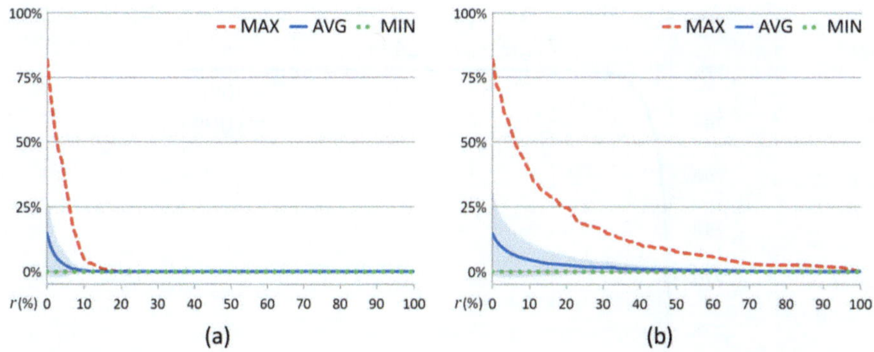

Fig. 4.42 EDR—proportion of time series for which a nearest neighbor class change was detected as a result of applying the Sakoe-Chiba band (**a**) and the Itakura parallelogram (**b**)

Both the mean value and the standard deviation were the lowest with EDR among the considered elastic measures. Moreover, in the case of SC, the average was greater than 10% exclusively at $r = 0\%$ and with IT only for $r \leq 2\%$ were changes greater than that magnitude detected. Additionally, when the Sakoe-Chiba constraint was applied to the warping window, no changes were observed for $r = 24\%$ and above, in either dataset (Fig. 4.42).

Class-level changes with both SC and IT were most noticeable in the nearest neighbor graphs of the *PigCVP* (88), *Phoneme* (84) and *PigArtPressure* (87) datasets. EDR also stood out in terms of the number of datasets in whose NN graphs not a single nearest neighbor changed its class under the influence of global constraints. While with DTW their number was 3, with LCS 6 and with ERP 8, here it amounted to 11. This feature was encountered in relation to the datasets numbered 1, 13, 14, 16, 46, 70, 73, 81, 92, 94 and 124.

4.5.2 Classification Accuracy—Individual Warping-Window Widths

At the level of individual values of the parameter r, the highest average classification accuracy obtained by applying SCV10x10 on whole datasets was approximately 77.06% (rounded to two decimal places) and it was achieved by the unconstrained variant of EDR (i.e. for $r = 100\%$). In Fig. 4.43, the maximum is marked at $r = 24\%$ for SC, but the same result was obtained for all other wider warping windows as well. These preliminary results suggest that applying global constraints may not be a suitable approach in an effort to improve the classification accuracy of EDR.

The potential inferiority of the constrained variants of EDR compared to the unconstrained one can be guessed from Fig. 4.44, which for each specific value of r depicts the number of datasets for which the classification accuracy of the

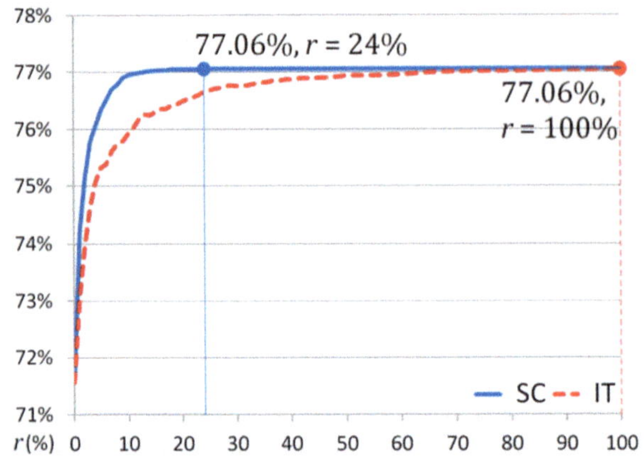

Fig. 4.43 EDR—average classification accuracy across all datasets for different warping-window widths

unconstrained EDR was lower, higher or equal to the accuracy of SC and IT. It can be seen that neither SC nor IT outperformed EDR for any value of r. In addition, it can also be noticed that at $r = 0\%$, among the analyzed elastic measures, the advantage of the unconstrained measure over the constrained ones was the largest in the case of EDR.

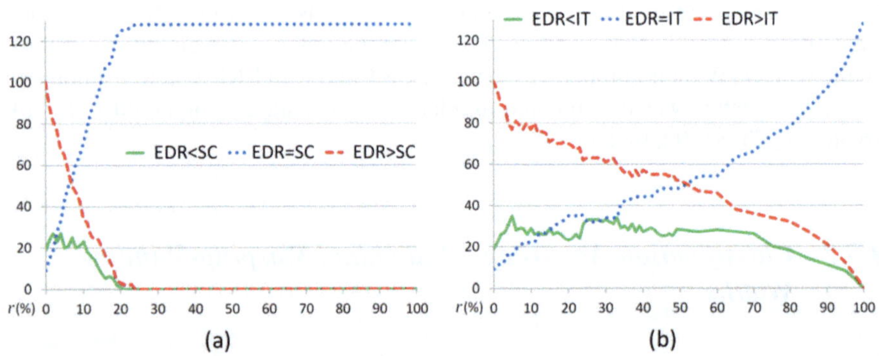

Fig. 4.44 EDR—number of datasets for which the accuracy generated by the unconstrained measure was lower (<), equal (=) and higher (>) than the result obtained by limiting the elasticity with the Sakoe-Chiba (**a**) and the Itakura (**b**) constraints

4.5.3 *Classification Accuracy—Optimal Warping-Window Widths*

The slight inferiority of SC and IT compared to the unconstrained EDR was also confirmed by their average classification accuracies across the 128 datasets of the UCR repository obtained by applying the SCV10x10 method where the optimal warping-window widths were chosen on the corresponding training subsets relying on the results of the SCV1x9 evaluation procedure (Table 4.10). It should be noted that the best (EDR) and worst (IT) results differed by only about 0.25 percentage points. The standard deviations of the different EDR variants were almost identical.

Considering all the 12,800 splits into testing and training subsets, the Itakura parallelogram selected on average approximately 4.6 times larger warping-window widths than the Sakoe-Chiba band and produced more than 5 times larger standard deviation.

Looking at individual datasets, the best result was most often achieved by the unconstrained EDR (89/128), followed by IT (36/128) and finally SC (31/128). The very close mean ranks (MR) of the constrained variants also suggest a high level of similarity in their performance.

Similar to the other elastic measures, the concentration of the number of splits on testing/training subsets for which the constrained variants chose the optimal value of r close to 0% was prominent (Fig. 4.45a). It is also visible that, while IT opted even for values close to $r = 100\%$, SC never relied on warping windows wider than 29% of the length of the time series.

A clear presentation of the difference between SC and IT regarding the preferred values of the parameter r is given in Fig. 4.45b. Namely, in 91% of the splits, the optimal width of the warping window was not greater than 13% of the length of the time series when using the SC constraint, but IT required widths up to $r = 75\%$ to reach a very close threshold. It should also be mentioned that in the case of the *ACSF1* (1), *FreezerRegularTrain* (44), *FreezerSmallTrain* (45), *Meat* (70), *OliveOil* (81) and *Wine* (124) datasets, the optimal value of r was 0% in each of the 100 test/train splits with both SC and IT. In addition to the above, SC demonstrated this characteristic with three other datasets: *MelbournePedestrian* (72), *ProximalPhalanxOutlineAgeGroup* (92) and *Strawberry* (109). Moreover, SC chose $r = 1\%$

Table 4.10 EDR—classification accuracy and optimal warping-window width (across 12,800 splits): average value (AV) and standard deviation (SD); the number of datasets for which a given variant produced the best result (WIN) and the mean rank (MR)

	EDR	SC	r_{SC}	IT	r_{IT}
AV (%)	77.06	76.82	5.22	76.80	23.99
SD (pp)	21.01	21.01	5.14	21.02	28.91
WIN	89	31		36	
MR	1.54	2.22		2.24	

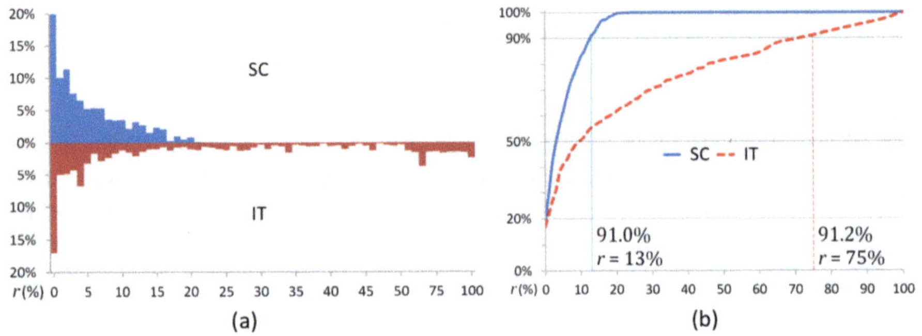

Fig. 4.45 EDR—frequencies (in percentage relative to the 12,800 splits) of different warping-window widths selected by the SCV1x9 algorithm over the training sets when evaluating classification accuracy using SCV10x10 (**a**); cumulative display of frequencies of optimal warping-window widths (**b**)

as the optimal option for classifying the *FordA* (42) time series in each SCV10x10 iteration.

Unlike the other three elasticity measures analyzed in this book, the average value of the parameter *r* generated by applying the Sakoe-Chiba constraint was less than 20% of the length of the time series in the case of each UCR dataset (Fig. 4.46a). Analogous to the other measures, IT produced an average *r*-value greater than 90% for *PigArtPressure* (87) and *PigCVP* (88) with EDR as well and in addition for *ShapesAll* (103), too (Fig. 4.46b).

Comparing Figs. 4.47 and 4.23, it is clear that unconstrained EDR has overtaken SC and IT even to a greater extent than LCS. EDR's win-loss ratio over SC was roughly 3.3:1 (90:27) and over IT it was slightly more moderate, close to 2.7:1 (86:32). IT fell slightly behind in the competition against SC, as SC proved to be better in a ratio of roughly 1.1:1 (57:50). As the datasets are positioned close to the diagonal, it appears that the dissimilarity between EDR, SC and IT is likely not to be significant as also indicated by the very close average differences in classification accuracies shown in Fig. 4.48.

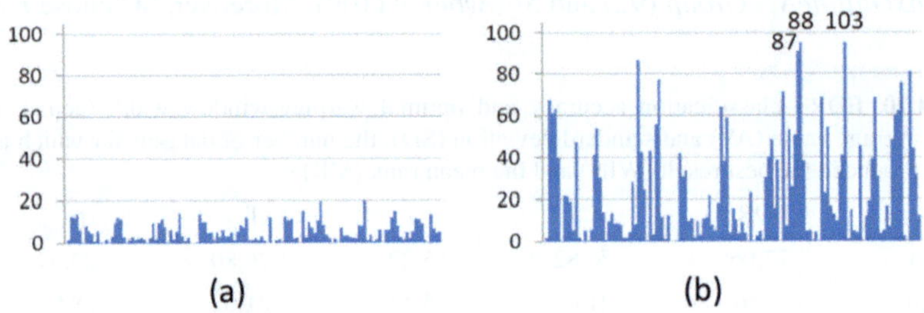

Fig. 4.46 EDR—average warping-window widths by datasets: Sakoe-Chiba (**a**), Itakura (**b**) (labels represent dataset identifiers)

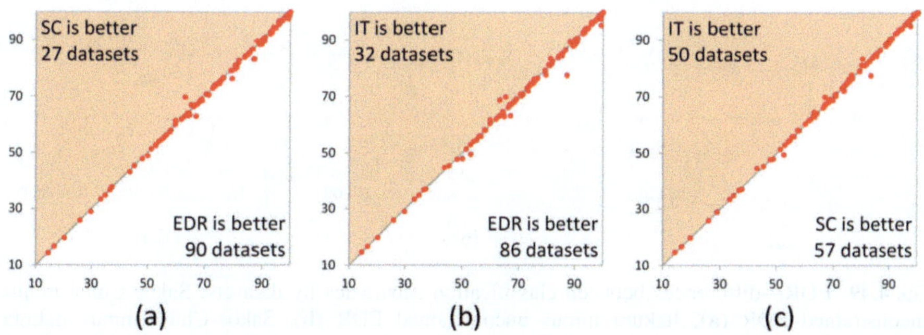

Fig. 4.47 EDR—comparison of similarity measure variants based on the number of datasets where they achieved higher classification accuracy: Sakoe-Chiba versus unconstrained EDR (**a**), Itakura versus unconstrained EDR (**b**), Itakura versus Sakoe-Chiba (**c**)

Fig. 4.48 EDR—average difference in classification accuracy of different EDR variants over datasets where the performance of a given variant was better (green) and worse (red) than the other one

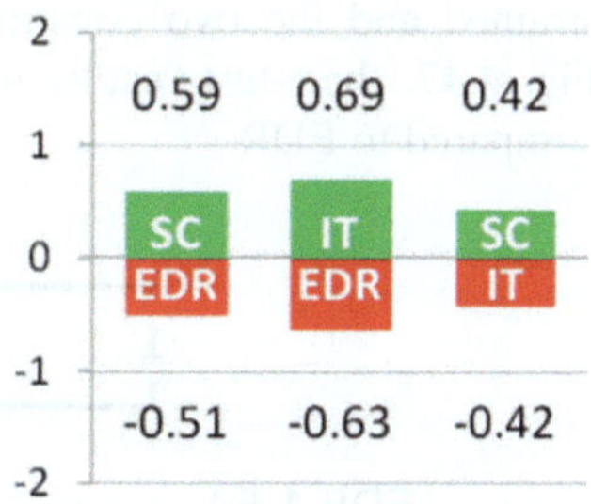

The similarity of the performance of various EDR variants is even more obviously visible in Fig. 4.49, which visualizes the differences in their classification accuracies at the level of individual datasets. The only dataset on which limiting the width of the warping window with the Sakoe-Chiba band was noticeably beneficial was *FordA* (42). The application of the Itakura parallelogram, in addition to *FordA* (42), gave a perceptible positive contribution also in the case of the *Earthquakes* (29) dataset. The advantage of the unconstrained EDR over SC was greater than 3 percentage points in the following three datasets: *PickupGestureWiimoteZ* (85), *BeetleFly* (8), *Rock* (96). Compared to EDR, IT showed the biggest disadvantage in the case of the *BeetleFly* (8) dataset (by almost 10 percentage points), but fell behind by more than 3 percentage points in *BirdChicken* (9), *PickupGestureWiimoteZ* (85) and *Herring* (59). An advantage of more than 3 percentage points of SC over IT was detected in the *BeetleFly* (8) and *Herring* (59) datasets. On the other hand, IT generated more than 3 percentage points better classification accuracy than SC only on the *DodgerLoopGame* (27) dataset.

The classification accuracies of EDR, SC and IT differ significantly, as confirmed by the result of the Friedman test: $\chi^2(2) = 44.66$, $p < .001$. In contrast to DTW where both constrained variants were ranked significantly better than the unconstrained one (Fig. 4.14), in the case of EDR the situation was reversed (Fig. 4.50): Nemenyi's post-hoc test confirmed the advantage of EDR over SC and IT. The

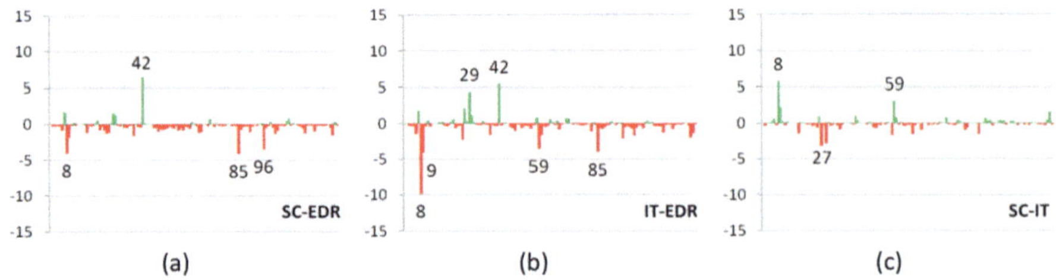

Fig. 4.49 EDR—differences between classification accuracies by datasets: Sakoe-Chiba minus unconstrained EDR (**a**), Itakura minus unconstrained EDR (**b**), Sakoe-Chiba minus Itakura (**c**) (labels represent dataset identifiers)

Wilcoxon signed-rank test (adjusted according to the Holm-Bonferroni scheme) identified a statistically significant difference between the performance of the unconstrained and the two constrained variants as well (Table 4.11). Consistent with Fig. 4.47, the *t*-test results in Table 4.12 reinforced the inferiority of SC and IT compared to EDR.

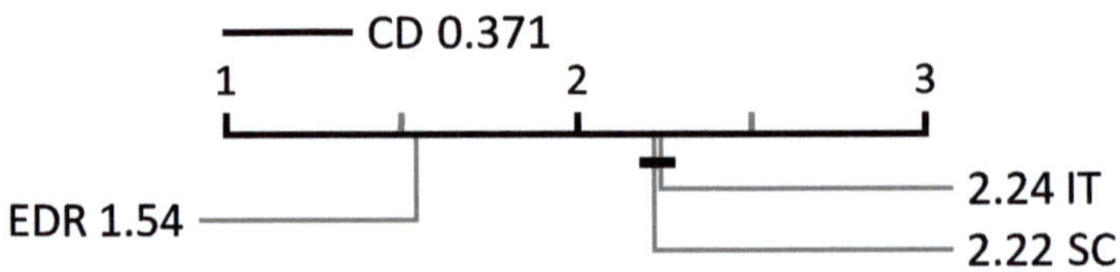

Fig. 4.50 EDR—critical difference diagram comparing average accuracy ranks based on Nemenyi's test: EDR variants connected by a solid bar did not differ statistically significantly

Table 4.11 EDR—Holm-Bonferroni-adjusted *p*-values of the pairwise Wilcoxon signed-rank test on classification accuracy

A		77.06	76.82	76.80
		EDR	SC	IT
77.06	EDR		< .001	< .001
76.82	SC			.841
76.80	IT			

Table 4.12 EDR—statistically significant wins (W) and losses (L) reported by the corrected resampled *t*-test on classification accuracy (across 12,800 splits)

	EDR	IT	SC
W	81	27	24
L	25	46	61
W-L	56	−19	−37

Unlike DTW, where the success rate of estimating the actual classification accuracy gain based on the gain on the training subset was much higher than the failure rate (Fig. 4.15), here the choice between the constrained and unconstrained measures was mostly unsuccessful (Fig. 4.51). Favoring SC over EDR turned out to be the wrong decision in somewhat above 66% (85/128) of the cases and was successful in only close to 23% (29/128) of the datasets. The frequency of wrong decisions was the same when choosing between IT and EDR and about 26% (33/128) of the choices were correct. When choosing between SC and IT, in nearly 58% (74/128) of the datasets, the better variant was successfully determined, but the choice was unsuccessful in nearly 27% (34/128) of the cases.

As can be seen from Fig. 4.52, the advantage of SC over EDR was the most overestimated in the *PickupGestureWiimoteZ* (85 dataset) and the largest bias in preferring IT over EDR appeared in the *BeetleFly* (8) dataset. In addition, both global constraints generated a difference between expected and actual gain greater than 0.03 for *DodgerLoopDay* (26), *GestureMidAirD3* (49), *Rock* (96) and *Worms* (126). When it comes to SC, the difference was greater than 0.03 also with the *Lightning2* (67) dataset and for IT, the list included *BirdChicken* (9), *Herring* (59) and *WormsTwoClass* (127) as well. EDR was the only elastic measure that produced an error greater than 0.03 in favoring IT over SC in a dataset (*BeetleFly* (8)).

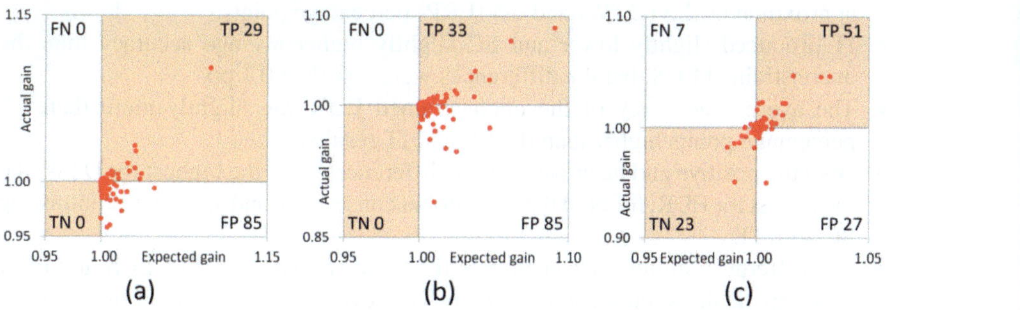

Fig. 4.51 EDR—comparing the actual and expected gain in classification accuracy: Sakoe-Chiba versus unconstrained EDR (**a**), Itakura versus unconstrained EDR (**b**), Itakura versus Sakoe-Chiba (**c**)

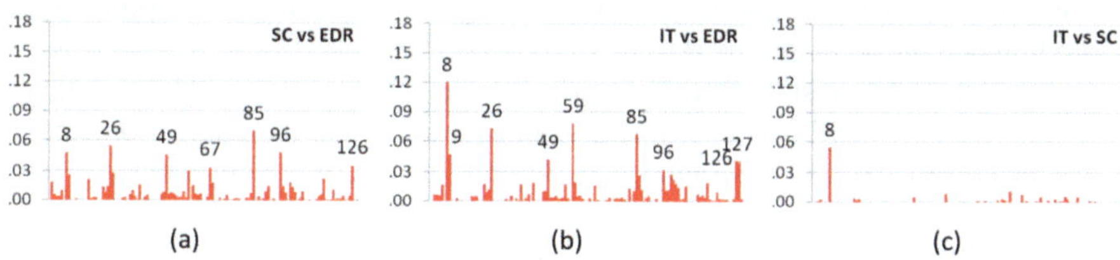

Fig. 4.52 EDR–differences between expected and actual accuracy gains of datasets belonging to the FP region: Sakoe-Chiba versus unconstrained EDR (**a**), Itakura versus unconstrained EDR (**b**), Itakura versus Sakoe-Chiba (**c**) (labels represent dataset identifiers)

4.6 Highlights

- The Itakura parallelogram (IT) generated noticeably more intense changes in the NN graph than the Sakoe-Chiba band (SC).
- The intensity of changes in the NN graph was the strongest in the case of DTW and the weakest in the case of EDR.
- By prohibiting any elasticity ($r = 0\%$) the average classification accuracy was considerably lower compared to the unconstrained measures ($r = 100\%$). The smallest difference was observed with DTW (3.3 percentage points (pp)) and the largest with LCS (6 pp).
- When using the SC constraint, the optimal width of the warping window in about 90% of the cases (out of a total of 12,800 splits into testing and training subsets) was not greater than 16% (DTW), 15% (LCS), 13% (ERP and EDR) of the length of the time series.
- When using the IT constraint, in order to cover approximately 90% of the cases (out of a total of 12,800 splits into testing and training subsets) it was necessary to take into account significantly wider warping windows, even up to 80% (DTW and LCS), 60% (ERP) and 75% (EDR) of the length of the time series.
- With DTW and ERP, both types of constraints generated slightly higher average classification accuracies than the unconstrained variant. The differences were approximately 2.5 (DTW) and 0.5 (ERP) percentage points.
- IT produced slightly lower and SC slightly higher average accuracy than the unconstrained LCS, but the differences were less than 0.1 pp.
- The average accuracy of the unconstrained EDR was slightly more than 0.2 percentage points higher than the SC and IT results.
- Both the positive and negative average differences were the highest for DTW and the lowest for EDR, for both types of global constraints, and also when contrasting SC with IT.
- The difference in the performance of the three variants of DTW, LCS and EDR was statistically significant at the $\alpha = 0.01$ level according to the Friedman test and borderline at the $\alpha = 0.05$ level in the case of ERP.
- According to Nemenyi's post-hoc test of the results of DTW and EDR, the performance of SC and IT was not significantly different from each other, but both were significantly different from the performance of the unconstrained variant. The Holm-Bonferroni-corrected Wilcoxon signed-rank test also confirmed these findings (at the level of $\alpha = 0.01$).
- Among the LCS results, the Nemenyi test detected a statistically significant difference between IT and the unconstrained variant and the Holm-Bonferroni-corrected Wilcoxon signed-rank test between IT and SC (but only at the $\alpha = 0.05$ level).
- The Nemenyi test did not reveal a statistically significant difference in the performance of the three ERP variants, but according to the Holm-Bonferroni-corrected Wilcoxon signed-rank test, both SC (at the $\alpha = 0.01$ level) and IT (at the $\alpha = 0.05$

level) were significantly different from the unconstrained measure. Furthermore, at the $\alpha = 0.05$ level, SC and IT were also significantly different from each other.

- When choosing between the constrained and the corresponding unconstrained variant, the actual accuracy gain was the most aligned with the expected accuracy gain in the case of DTW (the success rate was about 77% and 66% with SC and IT, respectively) and the most mismatched in the case of EDR (with an approximate success rate of only 23% and 26% with SC and IT respectively).
- Choosing between SC and IT based on their classification accuracy on the training set was the most successful with LCS and DTW (in about 66% of the datasets) and the least effective with EDR (in approximately 58% of the cases).
- *Worms* (126) was the only dataset for which, in combination with all four elastic measures, the expected accuracy gain of SC over the unconstrained measure was greater than 1 and the actual gain was less than 1, with their difference being greater than 0.03.
- The LCS and EDR lists of FP datasets for which the difference between expected and actual gain of SC over the unconstrained measure was largest were nearly identical (8, 26, 49, 85, 67, 96, 126). The only difference was the *WormsTwoClass* (127) dataset, which was included in the LCS, but not in the EDR list.
- *BirdChicken* (9) and *PickupGestureWiimoteZ* (85) were the only datasets for which, in combination with all four elastic measures, the expected accuracy gain of IT over the unconstrained measure was greater than 1 and the actual gain was less than 1, with their difference being greater than 0.03.
- When favoring IT over SC, EDR was the only elastic measure that produced a difference between expected and actual gain greater than 0.03 in one of the 128 UCR datasets (*BeetleFly* (8)).

References

1. Tan, P.-N., Steinbach, M., Kumar, V., Karpatne, A.: Introduction to data mining, 2nd ed. Pearson Education (2019)
2. Ding, H., Trajcevski, G., Scheuermann, P., Wang, X., Keogh, E.: Querying and mining of time series data: experimental comparison of representations and distance measures. In: Proceedings of VLDB Endow., VLDB Endowment, pp. 1542–1552 (2008). https://doi.org/10.14778/1454159.1454226
3. Cardoso, M.G.M.S., Alexandra Martins, A.: The performance of a combined distance between time series, pp. 71–83 (2022). https://doi.org/10.1007/978-3-031-12766-3_6
4. Górecki, T., Piasecki, P.: A comprehensive comparison of distance measures for time series classification. In: Steland, A., Rafajłowicz, E., Okhrin, O. (eds.) Stochastic Models, Statistics and Their Application, pp. 409–428. Springer, Cham (2019). https://doi.org/10.1007/978-3-030-28665-1_31
5. Wang, X., Mueen, A., Ding, H., Trajcevski, G., Scheuermann, P., Keogh, E.: Experimental comparison of representation methods and distance measures for time series data. Data Min. Knowl. Discov. **26**, 275–309 (2013). https://doi.org/10.1007/s10618-012-0250-5

6. Serrà, J., Arcos, J.L.: An empirical evaluation of similarity measures for time series classification. Knowledge-Based Syst. **67**, 305–314 (2014). https://doi.org/10.1016/j.knosys.2014.04.035
7. Yoshida, S., Chakraborty, B.: A comparative study of similarity measures for time series classification, pp. 397–408 (2017). https://doi.org/10.1007/978-3-319-50953-2_27
8. Bagnall, A., Lines, J., Bostrom, A., Large, J., Keogh, E.: The great time series classification bake off: a review and experimental evaluation of recent algorithmic advances. Data Min. Knowl. Discov. **31**, 606–660 (2017). https://doi.org/10.1007/s10618-016-0483-9
9. Shifaz, A., Pelletier, C., Petitjean, F., Webb, G.I.: Elastic similarity and distance measures for multivariate time series. Knowl. Inf. Syst. **65**, 2665–2698 (2023). https://doi.org/10.1007/s10115-023-01835-4
10. Dau, H.A., Bagnall, A., Kamgar, K., Yeh, C.-C.M., Zhu, Y., Gharghabi, S., Ratanamahatana, C.A., Keogh, E.: The UCR time series archive. IEEE/CAA J. Autom. Sin. **6**, 1293–1305 (2019). https://doi.org/10.1109/JAS.2019.1911747
11. Sakoe, H., Chiba, S.: Dynamic programming algorithm optimization for spoken word recognition. IEEE Trans. Acoust. **26**, 43–49 (1978). https://doi.org/10.1109/TASSP.1978.1163055
12. Itakura, F.: Minimum prediction residual principle applied to speech recognition. IEEE Trans. Acoust. **23**, 67–72 (1975). https://doi.org/10.1109/TASSP.1975.1162641
13. Berndt, D., Clifford, J.: Using dynamic time warping to find patterns in time series. In: Usama, R.U., Fayyad, M. (ed.) Knowledge Discovering Databases Paper from 1994 AAAI Workshop, pp. 359–370. AAAI Press, Seattle, Washington (1994). http://dblp.uni-trier.de/rec/bib/conf/kdd/BerndtC94
14. Vlachos, M., Kollios, G., Gunopulos, D.: Discovering similar multidimensional trajectories. In: Proceedings of 18th International Conference on Data Engineering, pp. 673–684. IEEE Computer Society (2002). https://doi.org/10.1109/ICDE.2002.994784.
15. Chen, L., Ng, R.: On the marriage of Lp-norms and edit distance. In: Nascimento, M.A., Özsu, M.T., Kossmann, D., Miller, R.J., Blakeley, J.A., Schiefer, K.B. (eds.) Proceedings of 2004 VLDB Conference, pp. 792–803. Elsevier (2004). https://doi.org/10.1016/B978-012088469-8.50070-X
16. Chen, L., Özsu, M.T., Oria, V.: Robust and fast similarity search for moving object trajectories. In: Proceedings of 2005 ACM SIGMOD International Conference on Management Data—SIGMOD'05, pp. 491–502. ACM Press, New York, USA (2005). https://doi.org/10.1145/1066157.1066213
17. Anh Dau, H., Keogh, E., Kamgar, K., Michael Yeh, C.-C., Zhu, Y., Gharghabi, S., Ann Ratanamahatana, C., Chen, Y., Hu, B., Begum, N., Bagnall, A., Mueen, A., Batista, G., Hexagon, M.L.: The UCR time series classification archive (2019). https://www.cs.ucr.edu/~eamonn/time_series_data_2018/
18. Paparrizos, J.: 2018 UCR time-series archive: backward compatibility, missing values, and varying lengths (2019). https://github.com/johnpaparrizos/UCRArchiveFixes
19. Kurbalija, V., Radovanović, M., Geler, Z., Ivanović, M.: The influence of global constraints on DTW and LCS similarity measures for time-series databases. In: Dicheva, D., Markov, Z., Stefanova, E. (eds.) Third International Conference on Software, Service Semantics Technology. S3T 2011 SE-10, pp. 67–74. Springer, Berlin, Heidelberg (2011). https://doi.org/10.1007/978-3-642-23163-6_10
20. Kurbalija, V., Radovanović, M., Geler, Z., Ivanović, M.: The influence of global constraints on similarity measures for time-series databases. Knowledge-Based Syst. **56**, 49–67 (2014). https://doi.org/10.1016/j.knosys.2013.10.021
21. Geler, Z.: Role of similarity measures in time series analysis. PhD thesis, University of Novi Sad, Serbia (2015). https://nardus.mpn.gov.rs/handle/123456789/1703
22. Geler, Z., Kurbalija, V., Ivanovic, M., Radovanovic, M., Dai, W.: Dynamic time warping: Itakura vs Sakoe-Chiba. In: 2019 IEEE International Symposium Innovation Intelligence System Application, pp. 1–6. IEEE (2019). https://doi.org/10.1109/INISTA.2019.8778300

23. Geler, Z., Kurbalija, V., Ivanović, M., Radovanović, M.: Elastic distances for time-series classification: Itakura versus Sakoe-Chiba constraints. Knowl. Inf. Syst. **64**, 2797–2832 (2022). https://doi.org/10.1007/s10115-022-01725-1
24. Kurbalija, V., Radovanović, M., Geler, Z., Ivanović, M.: A framework for time-series analysis. In: Dicheva, D., Dochev, D. (eds.) Artificial Intelligence Methodology System Application. SE-5, pp. 42–51. Springer, Berlin, Heidelberg (2010). https://doi.org/10.1007/978-3-642-15431-7_5
25. Kurbalija, V., Ivanović, M., Geler, Z., Radovanović, M.: Two faces of the framework for analysis and prediction, part 1—education. Inf. Technol. Control. **47**, 249–261 (2018). https://doi.org/10.5755/j01.itc.47.2.18746
26. Kurbalija, V., Ivanović, M., Geler, Z., Radovanović, M.: Two faces of the framework for analysis and prediction, part 2—research. Inf. Technol. Control. **47**, 489–502 (2018). https://doi.org/10.5755/j01.itc.47.3.18747
27. Preparata, F.P., Shamos, M.I.: Computational geometry. Springer, New York, NY (1985). https://doi.org/10.1007/978-1-4612-1098-6
28. Keogh, E., Ratanamahatana, C.A.: Exact indexing of dynamic time warping. Knowl. Inf. Syst. **7**, 358–386 (2005). https://doi.org/10.1007/s10115-004-0154-9
29. Ratanamahatana, C.A., Keogh, E.: Everything you know about dynamic time warping is wrong. In: 3rd Workshop Mining Temporal Seqence Data, Conjunction with 10th ACM SIGKDD International Conference on Knowledge Discover Data Mining (KDD-2004), Seattle, WA (2004)
30. Ratanamahatana, C.A., Keogh, E.: Making time-series classification more accurate using learned constraints. In: Proceedings of 2004 SIAM International Conference on Data Mining, pp. 11–22. Society for Industrial and Applied Mathematics, Philadelphia, PA (2004). https://doi.org/10.1137/1.9781611972740.2
31. Ratanamahatana, C.A., Keogh, E.: Three myths about dynamic time warping data mining. In: Proceedings of 2005 SIAM International Conference on Data Mining, pp. 506–510. Society for Industrial and Applied Mathematics, Philadelphia, PA (2005). https://doi.org/10.1137/1.9781611972757.50
32. Witten, I.H., Frank, E., Hall, M.A., Pal, C.J.: Data mining: practical machine learning tools and techniques, 4th ed. Morgan Kaufmann (2017). https://doi.org/10.1016/C2015-0-02071-8
33. Demšar, J.: Statistical comparisons of classifiers over multiple data sets. J. Mach. Learn. Res. **7**, 1–30 (2006)
34. Friedman, M.: The use of ranks to avoid the assumption of normality implicit in the analysis of variance. J. Am. Stat. Assoc. **32**, 675 (1937). https://doi.org/10.2307/2279372
35. Friedman, M.: A comparison of alternative tests of significance for the problem of m rankings. Ann. Math. Stat. **11**, 86–92 (1940). https://doi.org/10.1214/aoms/1177731944
36. Nemenyi, P.B.: Distribution-free multiple comparisons. Princeton University, Princeton (1963) (PhD thesis)
37. Benavoli, A., Corani, G., Mangili, F.: Should we really use post-hoc tests based on mean-ranks? J. Mach. Learn. Res. **17**, 1–10 (2016). http://jmlr.org/papers/v17/benavoli16a.html
38. Wilcoxon, F.: Individual comparisons by ranking methods. Biometr. Bull. **1**, 80 (1945). https://doi.org/10.2307/3001968
39. Holm, S.: A simple sequentially rejective multiple test procedure. Scand. J. Stat. **6**, 65–70 (1979). http://www.jstor.org/stable/4615733
40. Bouckaert, R.R., Frank, E.: Evaluating the replicability of significance tests for comparing learning algorithms. In: Dai, H., Srikant, R., Zhang, C. (eds.) Advanced Knowledge Discover Data Mining, pp. 3–12. Springer, Berlin, Heidelberg. https://doi.org/10.1007/978-3-540-24775-3_3
41. Batista, G.E.A.P.A., Wang, X., Keogh, E.J.: A complexity-invariant distance measure for time series. In: Proceedings of 2011 SIAM International Conference on Data Mining, pp. 699–710. Society for Industrial and Applied Mathematics, Philadelphia, PA (2011). https://doi.org/10.1137/1.9781611972818.60

Chapter 5
The Influence of Weighting Schemes on Classification Accuracy

This chapter is devoted to a detailed analysis of the influence of different kNN weighting schemes [1–9] (they are presented in Chap. 3 and summarized in Appendix B) on classification accuracy compared to the performance of the simple 1NN method combined with basic time series distance measures [10–14] discussed in this book (see Chap. 2 for detailed descriptions and Appendix A for brief presentations). For the sake of a clearer presentation of the results, we will refer to the included weighting schemes and distance measures in the text, tables, figures and their captions with the abbreviations introduced in Chaps. 2 and 3.

The examinations included all the 128 datasets of the UCR repository [15] preprocessed using Paparizzos' script [16]. The number of considered nearest neighbors of the kNN classifier (i.e., the value of the parameter k) ranged from 1 to 30. Since the focus of this book is not on comparing the examined distance measures based on their performance, in accordance with the recommendation in [14], in the case of LCS and EDR, we set the matching threshold parameter ε to a quarter of the maximum standard deviation of the time series of the given dataset. Given that all the datasets were z-normalized using Paparizzos' script, this value was equal to 0.25. In order for the results to reflect solely the influence of the weighting schemes, in this chapter only the unconstrained variants of the elastic measure were considered.

The experiments reported in this chapter build upon methodologies established in our prior publications [17–19] and, apart from the statistical tests, they were carried out relying on the *Framework for Analysis and Prediction* [20], whose suitability has been confirmed through its application in teaching [21] and in research [22] related to the application of time series in various fields.

In order to obtain an initial assessment of the performance of the weighting schemes, in the first step we compared their average accuracies calculated over the 128 datasets by applying 10 times repeated tenfold stratified cross-validation (SCV10x10) [23] for each individual value of the parameter k from the interval 1–30, as illustrated in Fig. 5.1.

Z. Gellér et al., *Recent Advances in Time-Series Classification—Methodology and Applications*, Intelligent Systems Reference Library 264,
https://doi.org/10.1007/978-3-031-77527-7_5

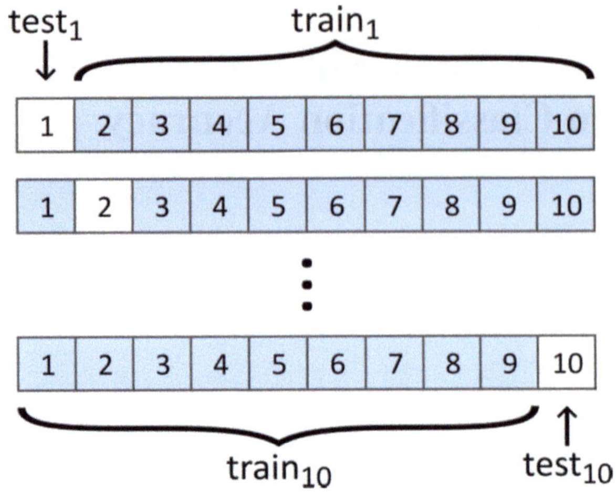

Fig. 5.1 Calculating classification accuracy for individual values of the parameter k applying tenfold cross-validation on the entire dataset

Each dataset was divided into 10 disjoint stratified folds of approximately equal size. One of the folds was used as the test set and the training set was obtained by merging the remaining 9 folds. After each of the 10 folds was used exactly once for testing, the elements of the dataset were shuffled and the entire procedure was repeated 9 more times. In this way, a total of 100 classification accuracy estimates were obtained for each dataset and their average value was calculated. Finally, the performance of a given kNN variant was calculated as the average of its 128 (average) accuracies.

The results of the preliminary experiments are shown in Figs. 5.3, 5.13, 5.23, 5.33, 5.43, 5.53 and 5.63 and their interpretation is given with regard to the accuracy of the 1NN classifier. The x-axis of the graphs in these figures represents the number of nearest neighbors and the y-axis the average accuracy of the given classifier. It should be noted here that the accuracy calculated with $k = 2$ always matches the result obtained with $k = 1$. This is because the weight of the second neighbor is always lower (so it cannot outvote the first) or equal to the weight of the nearest neighbor (when it cannot come to the fore due to the peculiarity of the implementation which always chooses the first among the classes with the most votes[1]). A more detailed discussion is given in Appendix D.

After the initial considerations described above, we conducted a more extensive examination of the performance of different variants of the kNN classifier based on nested cross-validation [23] in which the optimal number of nearest neighbors was selected on the training subset in each of the 100 iterations of dividing the given dataset into disjoint testing and training subsets. The applied evaluation technique is illustrated in Fig. 5.2.

In each of the 100 splits, one of the folds was used for testing and the union of the other 9 for tuning the parameter k using the SCV1x9 algorithm. For each value of k, we calculated the expected classification accuracy by applying ninefold stratified

[1] Random breaking of ties would make it impossible to replicate the results.

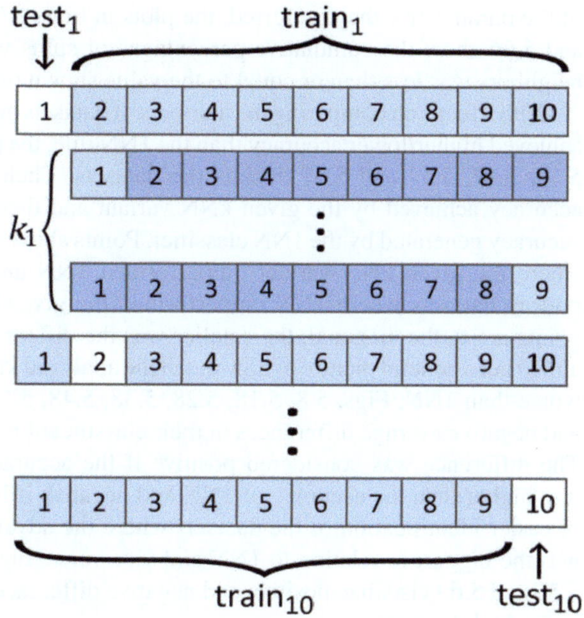

Fig. 5.2 Calculating classification accuracy using tenfold cross-validation where the optimal value of the parameter k is chosen using ninefold cross-validation on the training subset

cross-validation to the training subset and selected the smallest value that produced the best result. After that, the actual accuracy was calculated by classifying the time series of the testing subset based on the obtained optimal number of their nearest neighbors from the training subset.

Summary overviews of the performance of different kNN variants are given in Tables 5.1, 5.4, 5.7, 5.10, 5.13, 5.16 and 5.19 in the form of average classification accuracies with standard deviations, average values of the optimal numbers of nearest neighbors and their standard deviations, numbers of datasets on which the given classifiers achieved the best results and their mean ranks. The above descriptive statistics are calculated over 12,800[2] splits into testing and training subsets.

To investigate the extent to which a given weighting scheme relied on the first nearest neighbor, in Figs. 5.4, 5.14, 5.24, 5.34, 5.44, 5.54 and 5.64 we compare the percentages of splits in which the highest expected accuracy is achieved with $k = 1$ and with $k > 1$. The histograms in Figs. 5.5, 5.15, 5.25, 5.35, 5.45, 5.55 and 5.65 show the percentages of splits in which different values of the parameter k greater than 2 were chosen by the SCV1x9 algorithm as optimal over the training subsets. Since $k = 2$ could not be chosen as an optimal solution in any of the splits, it is not shown on these histograms. The reason is that the smallest value of k with the highest expected accuracy was always chosen and 2NN could never beat 1NN as explained in the description of experiments with individual k-values. To gain a clearer insight into the differences between the discussed kNN variants in terms of which values

[2] Each of the 128 datasets was split 100 times into testing and training subsets by the SCV10x10 procedure.

of the parameter k they preferred, the plots in Figs. 5.6, 5.16, 5.26, 5.36, 5.46, 5.56 and 5.66 show the cumulative percentages of splits where the optimal number of neighbors was less than or equal to the value shown on the x-axis.

With a focus on comparing the numbers of datasets on which a given kNN classifier achieved higher/lower accuracy than the 1NN rule, the points in Figs. 5.7, 5.17, 5.27, 5.37, 5.47, 5.57 and 5.67 depicts the datasets. Their y coordinate represents the accuracy achieved by the given kNN variant and their x coordinate represents the accuracy generated by the 1NN classifier. Points above the diagonal embody datasets where the given kNN variant outperformed 1NN and points below the diagonal portray datasets where 1NN outperformed the given kNN variant. The closer the points are to the diagonal, the smaller was the difference in performance. With the aim to get a general picture of how much the analyzed kNN variants performed better/ worse than 1NN, Figs. 5.8, 5.18, 5.28, 5.38, 5.48, 5.58 and 5.68 show the positive and negative average differences in their classification accuracies (see Appendix C). The difference was considered positive if the accuracy of a given kNN classifier was higher than the accuracy of 1NN and negative if it was lower. For the purposes of easier identification of the datasets where the advantage of a given kNN variant was the biggest in relation to 1NN and vice versa, Figs. 5.9, 5.19, 5.29, 5.39, 5.49, 5.59 and 5.69 visualize positive and negative differences in accuracies at the level of individual datasets.

The statistical analysis of the difference in the performance of the examined kNN classifiers was based on the Friedman test [24, 25] followed by the Nemenyi post-hoc test [26] as recommended in [27]. The corresponding critical difference diagrams are shown in Figs. 5.10, 5.20, 5.30, 5.40, 5.50, 5.60 and 5.70. In addition, complying with the guidance in [28], we also examined the statistical significance of the difference between the pairs of classifiers using the Wilcoxon signed-rank test [29] adjusted according to the Holm-Bonferroni correction method [30] (Tables 5.2, 5.5, 5.8, 5.11, 5.14, 5.17 and 5.20). Furthermore, we compared the numbers of statistically significant wins and losses obtained using the corrected resampled t-test [31] in Tables 5.3, 5.6, 5.9, 5.12, 5.15, 5.18 and 5.21. $\alpha = 0.01$ was chosen as the significance level of the statistical tests.

As recommended in [32], Figs. 5.11, 5.21, 5.31, 5.41, 5.51, 5.61 and 5.71 show *Texas Sharpshooter* (TS) scatter plots where the datasets are represented by 2D points. The x coordinate of the points represents the expected and the y coordinate the actual gain in accuracy. The gain of method A over method B is calculated as the quotient of their accuracies. If A achieved higher accuracy than B, the gain is greater than 1 and if A produced lower accuracy, it is less than 1. Whereby, the expected gain is calculated based on the accuracy over the training set and the actual gain over the test set.

Based on the coordinates of the points on the TS plot, the following regions are distinguished:

- $x > 1$ and $y > 1$—We expected A to be better than B and it really was better. This region is denoted as TP (*true positive*).

- $x < 1$ and $y < 1$—We expected A to be worse than B and it really was worse. This region is denoted as TN (*true negative*).
- $x > 1$ and $y < 1$—We expected A to be better than B, but it turned out to be worse. This region is denoted as FP (*false positive*) and contains datasets where, based on the results obtained on the corresponding training subset, we chose the classifier that proved to be less successful than its competitor on the testing subset.
- $x < 1$ and $y > 1$—We expected A to be worse than B, but it was actually better. This region is labeled FN (*false negative*) and contains datasets where we missed the opportunity to improve the classification accuracy compared to B (but at least we did not make it worse than the result produced by B as in the FP region).

To find out for which datasets of the FP region the expected and actual gains deviated the most, a visual representation of the differences is given in Figs. 5.12, 5.22, 5.32, 5.42, 5.52, 5.62 and 5.72.

When comparing expected and actual accuracy gains, the following points should be considered:

- The accuracy gains of the datasets were obtained by averaging the gains calculated on the 100 testing and training subsets generated by the SCV10x10 algorithm.
- Since $k = 1$ is included in the search for the optimal number of nearest neighbors, the classification accuracy of the kNN variants obtained over the training subset cannot be lower than the accuracy of the 1NN classifier. It follows that their expected gains (with respect to 1NN) are greater than 1, which means that the TN and FN regions are always empty.
- 1NN combined with DTW, ERP, L_2, L_1 and L_∞ misclassified each time series of the testing subset in, respectively, 3, 5, 2, 2 and 1 of the 100 splits of the *Beef* (7) dataset, therefore gains (both the expected and the actual) were calculated based on 97, 95, 98, 98 and 99 splits.
- 1NN combined with L_2, L_1 and L_∞ misclassified each time series of the testing subset in, respectively 1, 1 and 4 of the 100 splits of the *PigCVP* (88) dataset, therefore gains (both the expected and the actual) were calculated based on 99, 99 and 96 splits.
- 1NN combined with L_∞ misclassified each time series of the testing subset in 1 of the 100 splits of the *BeetleFly* (8), *PigAirwayPressure* (86) and *PigArtPressure* (87) datasets, therefore gains (both the expected and the actual) were calculated based on 99 splits.

In the rest of this chapter, we provide a comprehensive analysis of the impact of the discussed weighting schemes on kNN classifier's accuracy when using three distances based on L_p norms (Euclidean, Manhattan and Chebyshev) and the four basic unconstrained elastic similarity measures (DTW, LCS, ERP and EDR).

5.1 Euclidean Distance (L_2)

In Fig. 5.3, which shows the average classification accuracies calculated using the SCV10x10 algorithm for each value of the parameter k between 1 and 30 individually, the analyzed weighting schemes can be classified into three groups.

The first group was the most numerous and included kNN variants whose performance showed a decreasing trend as the number of nearest neighbors increased: kN, MC, RK, IN, UN, FB. As they all achieved the best average accuracy with $k = 1$ (75.90%), it seems that they will not be able to outperform the 1NN classifier. Between the minimum and the maximum, the difference was the largest when using the majority-voting rule (kN)—it was even slightly above 9 percentage points (pp). Except for UN and FB, all other representatives of the first group achieved their minimum at $k = 30$. Furthermore, the deterioration of accuracy was significantly slower with UN and FB than with the others: UN's minimum was 75.50% at $k = 29$ and FB reached its minimum of 75.71% at $k = 6$.

Unlike the first group, the performance of the classifiers of the second group (DN, DD, IS and ZA) went into decline only after a short period of improvement over the result of $k = 1$, suggesting that they could improve the accuracy of 1NN. All of them achieved the lowest average accuracy with $k = 30$, with the gap compared to $k = 1$ being the largest when using the IS (about 3.0 pp) and the smallest when applying the ZA weighting scheme (around 1.2 pp).

DU was the only scheme where the average accuracy was lowest for $k = 1$ and whose performance improved almost monotonically as the value of the parameter k increased. This suggests that DU might also be able to outperform 1NN to some

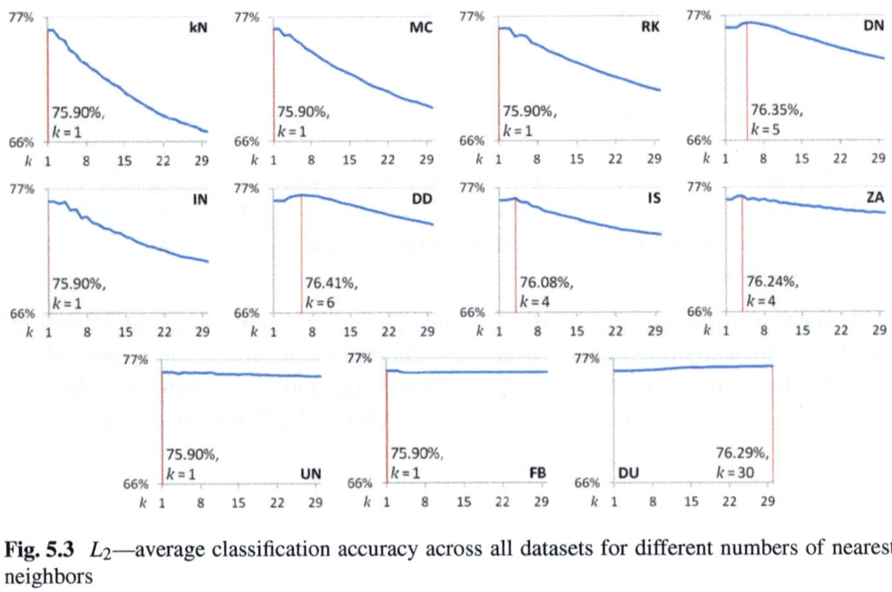

Fig. 5.3 L_2—average classification accuracy across all datasets for different numbers of nearest neighbors

extent. However, the difference between the maximum (76.29%) at $k = 30$ and the minimum did not even reach 0.4 pp.

The performance comparison results of various weighting schemes with the 1NN classifier (1N) are shown in Table 5.1. The evaluation was carried out using ten times repeated tenfold stratified cross-validation, where in each iteration of the SCV10x10 method, the number of nearest neighbors was chosen so as to maximize the classification accuracy computed with SCV1x9 over the training set.

Despite the preliminary results, in terms of average classification accuracy calculated with SCV10x10 over the 128 UCR datasets, 1N was outperformed by all 11 weighting schemes presented in this book. Still, the differences were very small and did not even reach 0.8 pp. An advantage greater than 0.7 was produced only by DD, ZA and DN, and in the case of MC, DU, FB and kN, it remained below 0.5 pp. Between the investigated NN classifiers, no noticeable difference was observed in terms of standard deviation either: it ranged in the interval of 21.79 ± 0.15 pp and it was the highest with 1N, as with all the other measures discussed in this chapter.

In line with the results based on the individual numbers of nearest neighbors, the average value of the parameter k was the highest for the DU (14.35), DD (8.99) and DN (8.65) schemes (similar to the other distance measures). The lowest average value was generated by the FB classifier followed by kN, and in both cases it was less than 5. Along with ZA, these two classifiers gave the lowest average k-values with the elastic similarity measures as well. In this respect, the Euclidean distance differed from the elastic measures since ZA was not among the schemes that produced the lowest averages of the optimal nearest neighbor numbers. The standard deviation was also lowest with kN and FB, while DU, DD and DN (along with UN) produced the highest ones.

When it comes to the number of datasets on which a given classifier achieved the best accuracy among all the analyzed ones, with a total of 36 wins, 1N left behind every kNN variant. In this ranking, the second place went to the ZA scheme with 30 wins, followed by DU with 26 wins. The number of wins was the lowest when using the IS (17), MC (17) and IN (13) schemes. Considering the mean rank, DD, DN and DU proved to be the best, but several kNN variants were outranked by 1N: UN, MC, IS, IN and kN.

Table 5.1 L_2—average value (AV) and standard deviation (SD) of classification accuracy (AV) and optimal number of nearest neighbors (k) across 12,800 splits; the number of datasets for which a given classifier produced the best result (WIN) and the mean rank (MR)

	DD	ZA	DN	IS	UN	RK	IN	MC	DU	FB	kN	1N
A_{AV} (%)	76.68	76.64	76.63	76.58	76.47	76.44	76.44	76.39	76.34	76.30	76.29	75.90
A_{SD} (pp)	21.72	21.71	21.74	21.64	21.82	21.74	21.69	21.75	21.91	21.92	21.77	21.94
k_{AV}	8.99	6.78	8.65	6.36	7.93	5.99	5.78	5.42	14.35	4.39	4.83	
k_{SD}	8.45	7.69	8.35	7.93	8.87	7.82	7.64	7.29	10.79	4.32	6.95	
WIN	22	30	22	17	21	21	13	17	26	21	20	36
MR	5.47	6.23	5.71	7.12	6.57	6.46	7.33	7.06	5.72	6.44	7.36	6.52

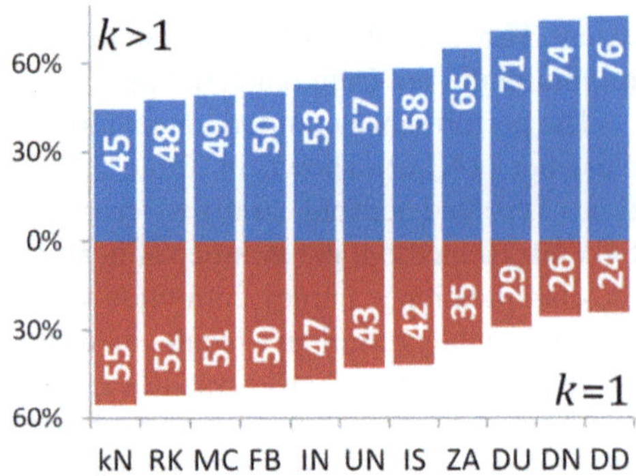

Fig. 5.4 L_2—percentages of splits in which SCV1x9 chose $k = 1$ and $k > 1$ as the optimal number of nearest neighbors on the training set when evaluating classification accuracy using SCV10x10

Figure 5.4 reveals that the dominance of the first nearest neighbor was the strongest with the kN, RK and MC weighting schemes where $k = 1$ was selected as the optimal solution in more than half of the 12,800 iterations by the SCV1x9 algorithm on the corresponding training subset. Besides them, $k = 1$ was preferred in more than 40% of the cases by FB, IN, UN and IS. In the initial analysis, all these schemes were outperformed by 1NN for $k > 2$, except IS. The lowest degree of dependence on the first nearest neighbor was shown by DU, DN and DD, which was also expected based on Fig. 5.3. They chose $k = 1$ in less than 30% of the splits.

The optimal value of k was 1 in each of the 100 splits of the *DiatomSizeReduction* (22) and *DodgerLoopWeekend* (28) datasets combined with each kNN variant. The number of datasets with this feature was the highest when using the kN (25), RK (23) and MC (21) and the lowest when relying on the DU (6), DN (5) and DD (3) schemes. While with FB the average value of the optimal numbers of nearest neighbors did not reach 16.1 with any dataset, with the other schemes there were several of them with averages higher than 20. Such datasets were the most numerous for DU (40), DD (15) and DN (15) and the least frequent for IN (9), IS (9), MC (7) and kN (7). The average of the optimal k-values was greater than 20 with each weighting scheme (except FB) on the following datasets: *FordA* (42), *FordB* (43), *MiddlePhalanxTW* (75), *Phoneme* (84), *ProximalPhalanxOutlineAgeGroup* (92).

Looking at the histograms in Fig. 5.5, it is striking that except for DU, all kNN variants favored values of the parameter k smaller than 10, which is in agreement with the preliminary findings shown in Fig. 5.3. It can also be seen that $k = 3$ did not appear as the optimal solution in any of the 12,800 splits when using the RK, DN, DD, UN, FB and DU schemes. Apart from LCS and EDR, this peculiarity will also appear in combination with other measures for all schemes listed here. At the same time, in more than 10% of cases, the classification accuracy over the training subset was the highest with $k = 3$ for kN and ZA. Excluding $k = 1$, DN, IN, DD, IS and FB relied to the greatest extent on $k = 4$ (in more than 10% of all the splits). The numbers of splits were most evenly distributed over different values of k with the DU and UN schemes where the standard deviation of the frequencies shown in Fig. 5.5 was the smallest: 1.7 and 1.8 pp. The distribution was the least uniform with FB, DN and DD, whose standard deviations were 3.3 pp, 2.9 pp and 2.8 pp, respectively.

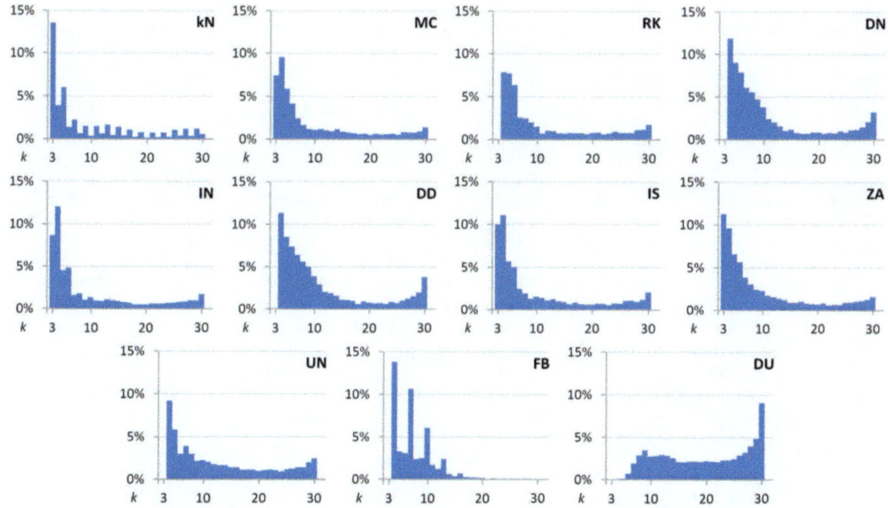

Fig. 5.5 L_2—frequencies (in percentage relative to the 12,800 splits) of different values of the parameter k ($\geq$ 3) chosen by the SCV1x9 algorithm over the training set when evaluating classification accuracy using SCV10x10

The cumulative plots in Fig. 5.6 further highlight the differences between the considered weighting schemes in terms of the extent to which they favored smaller or larger neighborhoods. FB opted for $k \leq 10$ in slightly more than 91% of cases. In about 90% of the splits, kN, MC and IN voted for $k \leq 15$, $k \leq 17$ and $k \leq 18$ respectively. In contrast to them, in order to capture a similar percentage of splits, DN and UN needed to consider k-values up to 24, DD up to 25 and DU up to 29.

In the largest number of datasets, 1N was outperformed by DD, DN and DU (Fig. 5.7), and at the same time, these same schemes were defeated by 1N in the smallest number of datasets, thus achieving the best win-loss ratios: 1.9:1 (DU) and 1.4 (DD and DN). Still, these ratios are smaller than in the case of elastic measures. Similar to DTW and ERP, at the level of individual datasets, the smallest difference in classification accuracy compared to 1N was generated by the DU scheme, which is manifested in the DU plot of Fig. 5.7 by very close positions of the points to the diagonal. This phenomenon can also be seen by the shorter lines on the DU plot of Fig. 5.9. Furthermore, the alikeness of the DU and 1N results is emphasized by the fact that both the positive and negative average differences were the smallest when using the DU weighting technique (Fig. 5.8).

It is also worth mentioning the fact that among the distance measures discussed in this chapter, L_2 and L_∞ produced the largest number (7 each) of weighting schemes that lost to 1N in terms of the number of wins and losses and the smallest number (4 each) that were overpowered 1N.

As the number of datasets on which kN, RK, IN and MC outperformed 1N was among the smallest, and at the same time, the number of those on which 1N outperformed them was among the largest, the win-loss ratio was the worst in their case

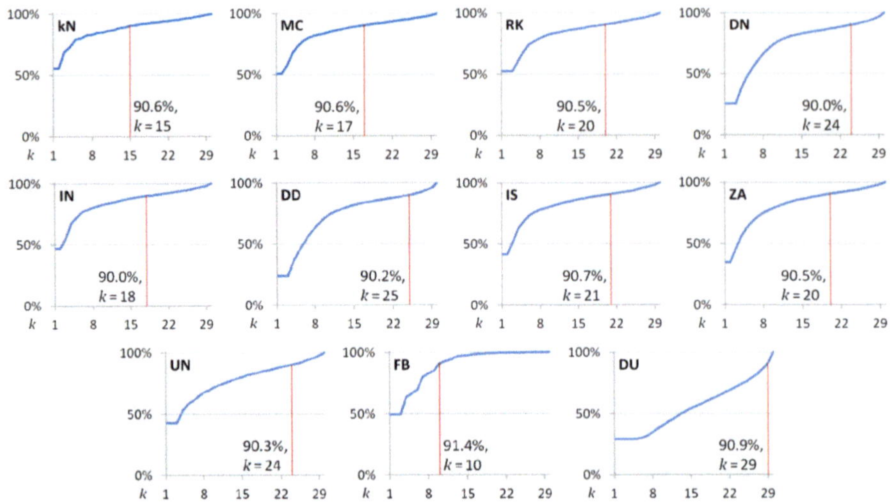

Fig. 5.6 L_2—cumulative display of the frequencies of optimal numbers of nearest neighbors chosen by the SCV1x9 algorithm over the training set when evaluating classification accuracy using SCV10x10

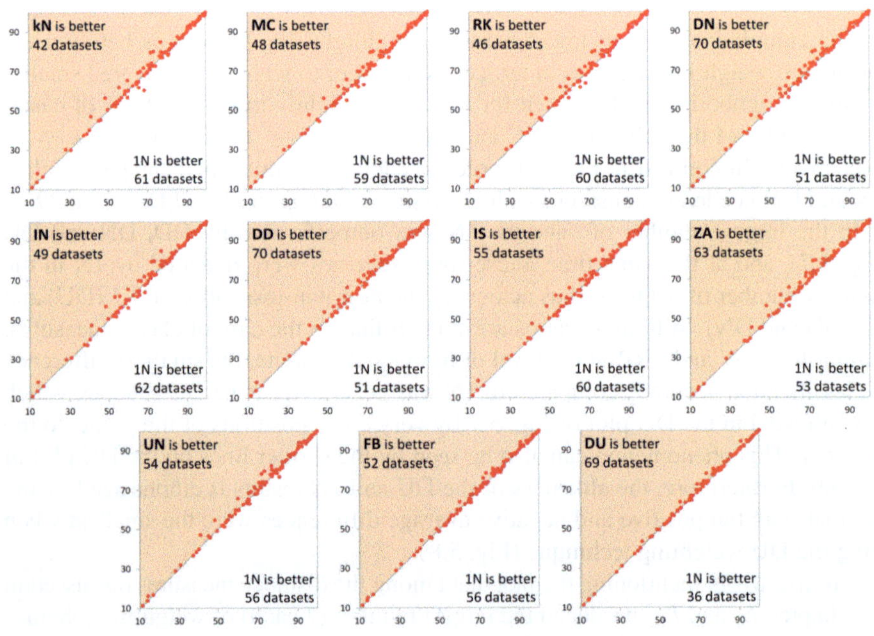

Fig. 5.7 L_2—comparison of the kNN variants with the 1N classifier based on the number of datasets where they achieved higher/lower classification accuracy

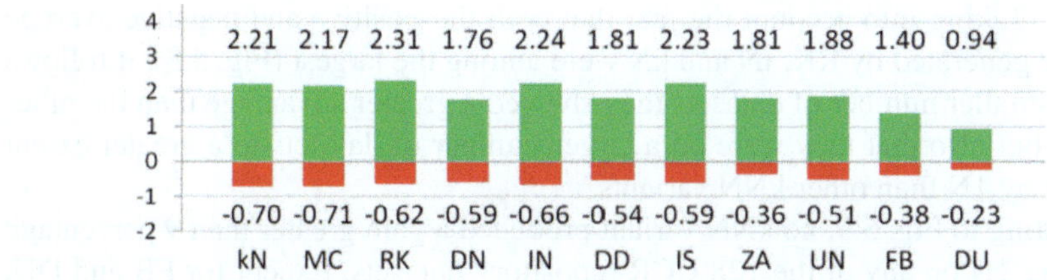

Fig. 5.8 L_2—average difference in classification accuracy of various kNN variants and 1N over the datasets where the performance of a given kNN variant was better (green) and worse (red) than 1N

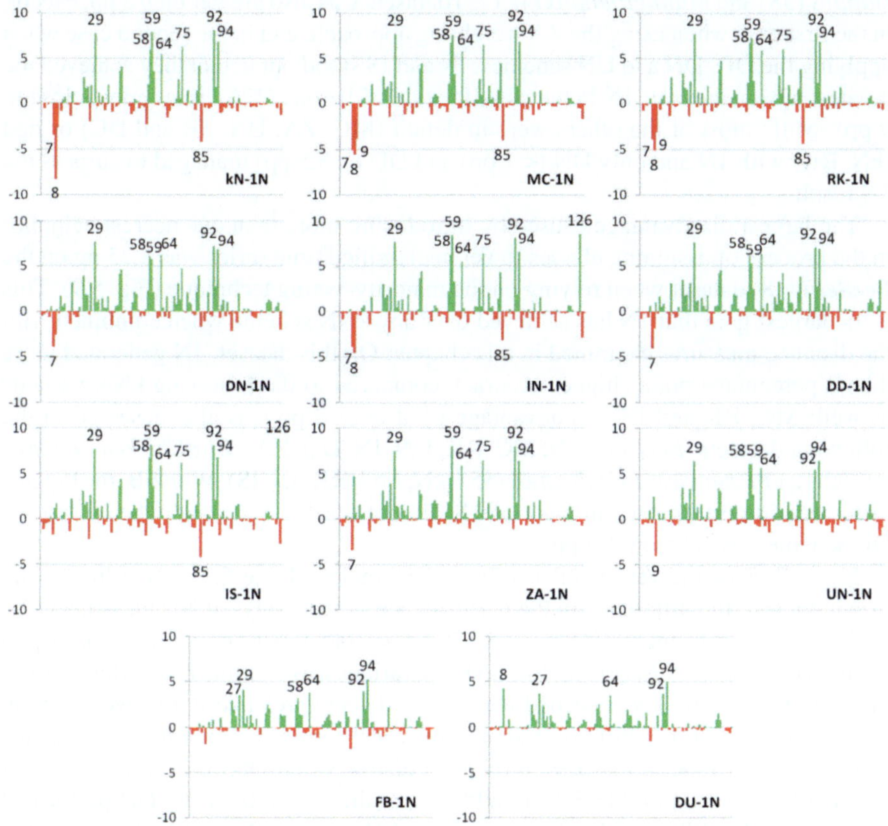

Fig. 5.9 L_2—differences between the classification accuracies of various kNN variants and 1N by datasets (labels represent dataset identifiers)

(Fig. 5.7). Taking into account the fact that both the positive and negative average difference generated by RK, IN and kN were among the largest (Fig. 5.8), it follows that on a smaller number of datasets they showed a greater advantage than the other schemes, but also that they were on a larger number of datasets to a greater extent surpassed by 1N than other kNN variants.

According to Fig. 5.9, no kNN variant produced a gain greater than 9 percentage points over 1N on any of the 128 UCR repository datasets. Except for FB and DU, all the other schemes outperformed 1N by more than 5 pp on the following datasets: *Earthquakes* (29), *Herring* (59), *InsectWingbeatSound* (64), *ProximalPhalanxOutlineAgeGroup* (92) and *ProximalPhalanxTW* (94). In addition, the advantage on the *Haptics* (58) and *MiddlePhalanxTW* (75) datasets was also greater than 5 pp, except, in the first case when using the ZA weighting approach, and in the second case when applying the DD, DN and UN schemes. IN and IS stood out in that they achieved the greatest advantage over 1N in the classification of *Worms* (126) time series (close to 9 pp), while most of the others were in deficit (MC, ZA, UN, FB and DU) or tied (kN, RK) with 1N and only DN (< 1 pp) and DD (< 3.5 pp) managed to surpass the 1N result.

The largest disadvantage caused by introducing more than one nearest neighbor in the process of assigning classes to yet unclassified time series was 8.25 pp on the *BeetleFly* (8) dataset when relying on the majority-voting technique (Fig. 5.9). This is the largest gain that 1N has achieved over any kNN scheme when combined with the distance measures examined in this chapter. On this dataset, 1N generated more than 3 percentage points higher accuracy compared to the following kNN variants as well: MC, RK and IN. An advantage of at least 3 pp was also observed in the following datasets: *Beef* (7) (kN, MC, RK, DN, IN, DD, ZA), *BirdChicken* (9) (MC, RK, UN), i *PickupGestureWiimoteZ* (85) (kN, MC, RK, IN, IS). With FB, the biggest difference in performance in favor of 1N did not reach 2.5 pp, and when using the DU scheme it was below 1.5 pp.

Regardless of the very similar average accuracies (Table 5.1), according to the Friedman test, the difference in the performance of the analyzed NN classifiers was statistically significant ($\chi^2(11) = 48.35, p < .001$). Nevertheless, according to Nemenyi's post-hoc test (Fig. 5.10), no kNN variant was significantly different from 1N and only IN and kN exceeded the critical distance relative to the average rank of DN. In contrast, the Wilcoxon signed-rank test reported a significant advantage of DD and DU (and at the $\alpha = 0.05$ level also ZA and DN) over 1N, as well as DD and DN over IN (Table 5.2). In addition, at the $\alpha = 0.05$ level, DD performed better than MC, FB and kN, ZA than kN and DN than FB and kN. Based on the number of statistically significant wins and losses shown in Table 5.3, DD, DN and ZA noticeably outperformed FB, IN, kN and 1N.

For approximately half of the considered weighting schemes (kN, MC, RK, IN and IS), the proportion of datasets where the assessment of the advantage of a given kNN variant over 1N based on the gain achieved over the training data was incorrect exceeded 45% (Fig. 5.11). The frequency of this type of overconfidence occurred in less than 40% of datasets only when relying on the DN, DD, ZA and DU algorithms. It should also be noted that these percentages were higher than those observed for

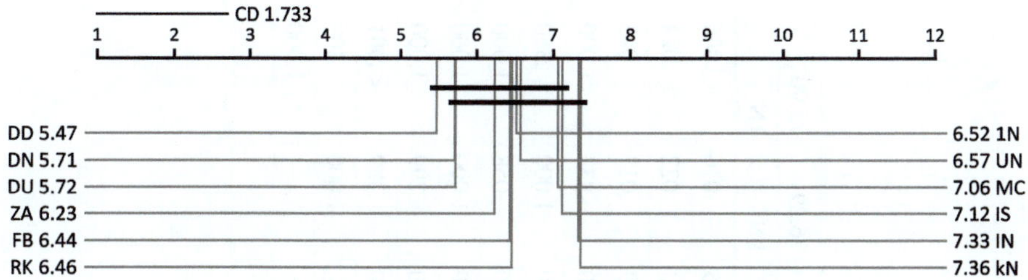

Fig. 5.10 L_2—critical difference diagram comparing average accuracy ranks based on Nemenyi's test: NN variants connected by a solid bar did not differ statistically significantly

the Manhattan distance and elastic measures, for each of the 11 weighting schemes. Moreover, except for DD, ZA and DU, they were also higher than the percentages obtained using the Chebyshev distance.

The number of datasets where both the expected and actual gains were greater than 1 (i.e., where, based on the training results, the given kNN variant was estimated to outperform 1N, which was confirmed on the test data) was greater than 50% of the total of the number of datasets for the DD, DN and DU schemes and less than 40% for the MC, RK and kN weighting approaches. In this respect, each kNN variant generated a better result in combination with the elastic measures (except for ZA whose lowest result was achieved with ERP).

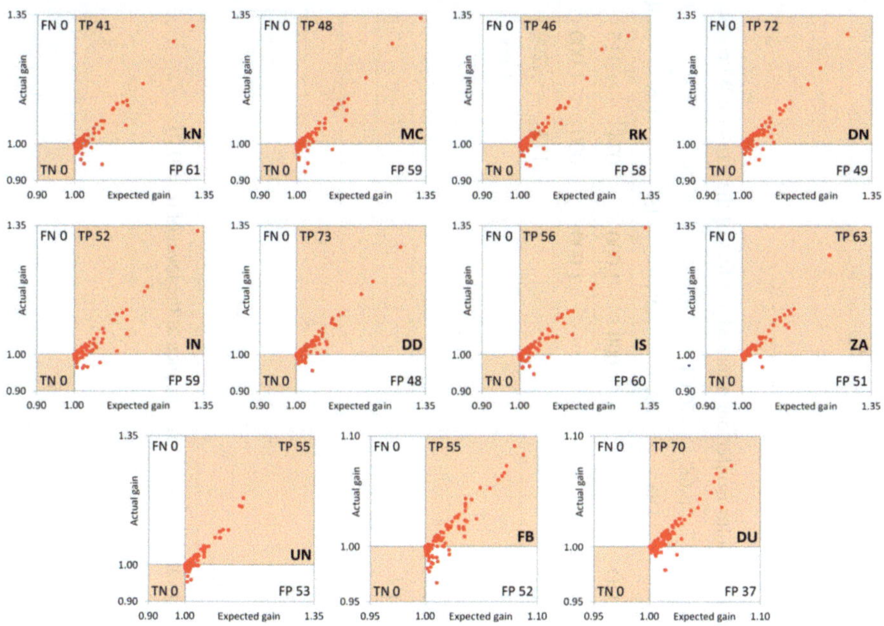

Fig. 5.11 L_2—comparing the actual and expected gain in classification accuracy: kNN variants versus 1N

Table 5.2 L_2—Holm-Bonferroni-adjusted *p*-values of the pairwise Wilcoxon signed-rank test on classification accuracy

A		76.68	76.64	76.63	76.58	76.47	76.44	76.44	76.39	76.34	76.30	76.29	75.90
		DD	ZA	DN	IS	UN	RK	IN	MC	DU	FB	kN	1N
76.68	DD		1.000	1.000	.074	.115	.360	**.002**	.022	1.000	.015	.016	**.006**
76.64	ZA			1.000	1.000	1.000	1.000	.106	.227	1.000	.493	.022	.024
76.63	DN				.173	.299	.494	**.003**	.057	1.000	.029	.017	.023
76.58	IS					1.000	1.000	1.000	1.000	1.000	1.000	.833	1.000
76.47	UN						1.000	1.000	1.000	1.000	1.000	1.000	.360
76.44	RK							.367	.409	1.000	1.000	.058	1.000
76.44	IN								1.000	1.000	1.000	1.000	1.000
76.39	MC									1.000	1.000	.299	1.000
76.34	DU										1.000	.489	< **.001**
76.30	FB											1.000	.409
76.29	kN												1.000
75.90	1N												

Note Bolded values indicate significance at the level of $\alpha = 0.01$

Table 5.3 L_2—statistically significant wins (W) and losses (L) reported by the corrected resampled *t*-test on classification accuracy (across 12,800 splits)

	DD	DN	ZA	DU	UN	IS	RK	MC	FB	IN	kN	1N
W	326	311	259	268	207	161	140	124	163	121	108	179
L	100	99	158	213	191	166	163	161	237	197	259	423
W–L	226	212	101	55	16	−5	−23	−37	−74	−76	−151	−244

Based on Fig. 5.12, the number of FP datasets with a difference between expected and actual accuracy gain greater than 0.03 was the highest with RK, DN and IS (9) and the lowest with ZA (1). The biggest differences were observed for *PickupGestureWiimoteZ* (85) (with RK, IS, UN, FB and DU), *Beef* (7) (with DN, IN, DD and ZA) and *BeetleFly* (8) (with kN and MC). Excluding ZA, FB and DU, *Beef* (7) was the only dataset for which the expected gain was more than 0.03 greater than the actual gain for each kNN variant. The average value of the differences (across all the FP datasets of the given kNN variant) was the highest for DN, MC and kN and the lowest for DU, FB and ZA.

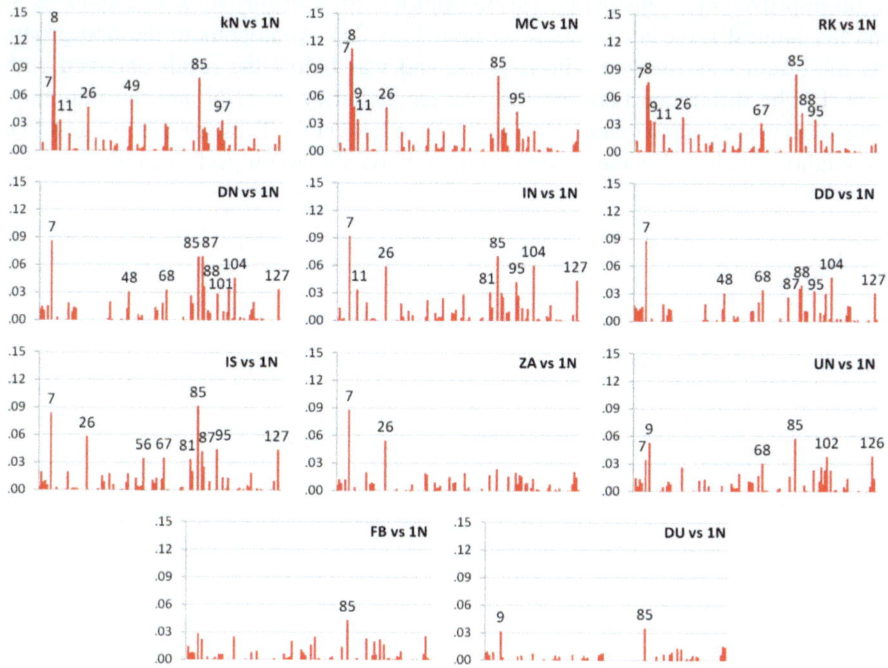

Fig. 5.12 L_2—differences between expected and actual accuracy gains of datasets belonging to the FP region: kNN variants versus 1N (labels represent dataset identifiers)

5.2 Manhattan Distance (L_1)

According to the preliminary analysis, which was based on the examination of the change in the classification accuracy of different kNN variants (calculated using the SCV10x10 evaluation method) when increasing the number of considered nearest neighbors from 1 to 30, the L_1 distance showed very similar properties to DTW (see Figs. 5.13 and 5.33). In both cases, the three recognizable groups were formed by the same classifiers—the only difference is that the UN and ZA schemes have swapped their groups.

In the first group, to which the kN, MC and RK weighting schemes belonged, the highest classification accuracy (76.97%) was achieved with $k = 1$ and by increasing the value of the parameter k, an almost monotonous deterioration of performance was observed until $k = 30$. The difference between the maximum and the minimum was the largest when relying on the majority-voting rule and amounted to about 7.8 percentage points (pp). These findings suggest that these schemes are unlikely to outperform the 1NN classifier.

The members of the second group (DN, IN, DD, IS and UN) were characterized by the peculiarity that the decline in performance occurred only after a period of slight improvement. This period was the shortest for IN (only for $k = 2$ and $k = 3$) and the longest for UN (from $k = 4$ until $k = 25$). Analogous to the first group, the minimum was reached with $k = 30$ and was below the result obtained with $k = 1$. The difference between the average accuracy of 1NN and the minimum ranged between approximately 1.4 pp (DD) and 4.4 (IN). These results suggest that the methods included in the second group may be able to outperform 1NN.

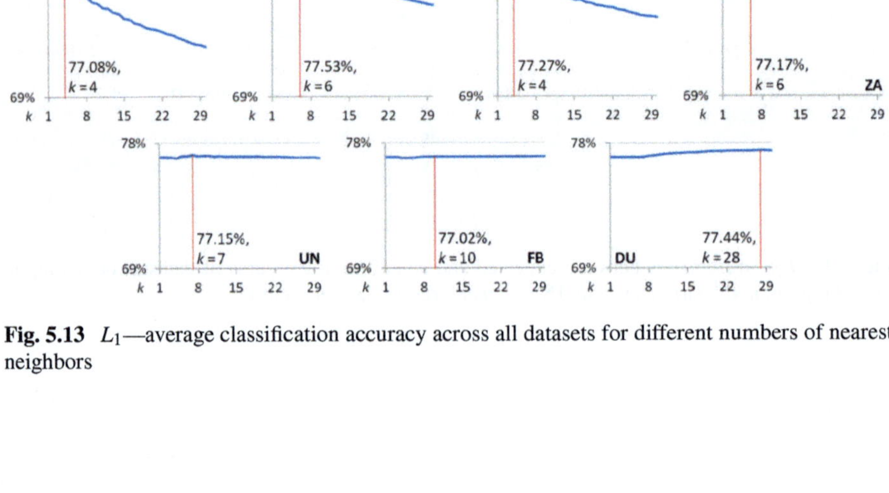

Fig. 5.13 L_1—average classification accuracy across all datasets for different numbers of nearest neighbors

When it comes to the ZA and DU schemes, they outperformed the average accuracy obtained with $k = 1$ for all k-values greater than 1. These two kNN variants, together with the FB weighting method, formed the third group. FB differed from ZA and DU in that for $4 \leq k \leq 6$ it generated slightly worse results than 1NN, but for $k > 6$ its average accuracy was also slightly better than that obtained with $k = 1$. It should also be noted that the advantage of the members of the third group over the result of 1NN was below 0.5 pp. However, it seems that to some extent they might also be able to surpass 1NN. Additionally, as with the other distance measures analyzed, DU's results suggest that it will prefer considerably higher k-values than other weighting schemes.

The averages and standard deviations of the classification accuracies (A) and optimal k-values detected within the evaluation of different NN classifiers using 10 times repeated tenfold stratified cross-validation (where the optimal numbers of nearest neighbors were chosen over the corresponding training subsets relying on the SCV1x9 procedure) are shown in Table 5.4 along with the numbers of UCR datasets on which the highest average accuracy was achieved by the given classifier (WIN) and the mean ranks (MR).

Although all kNN variants produced a higher average classification accuracy than 1NN, the difference did not even reach 1 pp. Analogous to the elastic measures, DD, IS, DN and IN proved to be the best in this regard. The smallest advantage over 1NN was achieved by FB, kN and ZA. As with all other similarity measures considered in this chapter, 1NN generated the largest standard deviation. Consistent with L_2 and L_∞, the smallest deviation was observed when using the IS weighting scheme. However, the differences between NN variants were marginal.

The expectation that the DU scheme would rely on larger numbers of nearest neighbors was confirmed by an average of optimal k-values (15.13) that was demonstrably higher than those of other kNN methods. It was immediately followed by DD (9.67) and DN (9.24), which will also be the case with the other measures (except DTW). On the other hand, kN (5.35), MC (6.10) and RK (6.59) produced higher average numbers of optimal nearest neighbors than FB (4.70) and ZA (3.70), which is inconsistent with preliminary predictions. Moreover, FB and ZA generated the smallest standard deviations (4.27 and 5.70) and DU the largest (10.62).

Table 5.4 L_1—average value (AV) and standard deviation (SD) of classification accuracy (AV) and optimal number of nearest neighbors (k) across 12,800 splits; the number of datasets for which a given classifier produced the best result (WIN) and the mean rank (MR)

	DD	IS	DN	IN	UN	RK	MC	DU	FB	kN	ZA	1N
A_{AV} (%)	77.88	77.85	77.83	77.65	77.60	77.53	77.52	77.44	77.42	77.42	77.33	76.97
A_{SD} (pp)	21.46	21.41	21.47	21.45	21.62	21.62	21.59	21.80	21.73	21.66	21.82	21.84
k_{AV}	9.67	6.94	9.24	6.39	9.06	6.59	6.10	15.13	4.70	5.35	3.70	
k_{SD}	8.36	7.76	8.23	7.53	9.02	7.94	7.46	10.62	4.27	7.14	5.70	
WIN	31	20	22	9	23	15	15	26	15	15	27	23
MR	4.61	6.48	4.98	7.32	6.51	7.17	7.19	5.84	6.68	7.61	6.41	7.19

1NN achieved the best accuracy on 23 out of 128 UCR datasets, the same as the UN weighting scheme. It was outperformed only by the DU, ZA and DD methods, which scored 26, 27 and 31 wins, respectively. In terms of mean rank, DD, DN and DU proved to be the most successful, but 1N overtook MC, IN and kN.

In line with the findings of the preliminary analysis, members of the first group (kN, RK, MC) preferred the first nearest neighbor in more than 40% of the 12,800 splits into test and training subsets (Fig. 5.14). What might not have been expected is that the FB scheme opted for $k = 1$ to a similar extent. Even more surprising was the behavior of ZA, which surpassed all of them and was reduced to 1NN in around 63% of cases (which was also the highest proportion detected among the considered kNN variants in combination with all seven distance measures). As with the other distance measures, $k = 1$ was the least preferred by the DU, DN and DD weighting schemes—in less than 30% of the splits.

Several datasets showed the property that the optimal number of nearest neighbors was 1 in each of the 100 iterations of the SCV10x10 algorithm. Their number was the highest when using the ZA (30), kN (21), RK (19) and MC (17) schemes and the lowest when classifying with the DN (4), DD (4) and DU (5) methods. Except for ZA, the other results faithfully reflected the indications of the introductory analysis at the level of individual values of the parameter k shown in Fig. 5.13. It is worth noting that *DodgerLoopWeekend* (28), *Fungi* (46) and *PigAirwayPressure* (86) exhibited the above-mentioned feature with all 11 variants of kNN. Furthermore, while with FB the average of the optimal k-values was not greater than 15 with any dataset, with the other schemes there were several of them with averages greater than 20. The largest number of such datasets was observed in the case of the DU (45) and the smallest when it came to the ZA (4) scheme. Disregarding ZA and FB, this property was invariant to the applied weighting scheme when *DistalPhalanxTW* (25), *FordA* (42), *FordB* (43), *Phoneme* (84), *ProximalPhalanxOutlineAgeGroup* (92) were concerned.

The greater degree of preference for larger values of the parameter k by the DU scheme compared to the other methods is also clearly visible in the histograms shown in Fig. 5.15, which visualize the frequencies (expressed as percentages in relation to a total of 12,800 splits into testing and training subsets) of different numbers of nearest neighbors between 3 and 30. In more than 10% of cases, kN, IN, IS and ZA opted for $k = 3$ as the optimal solution. In also more than 10% of divisions, $k = 4$

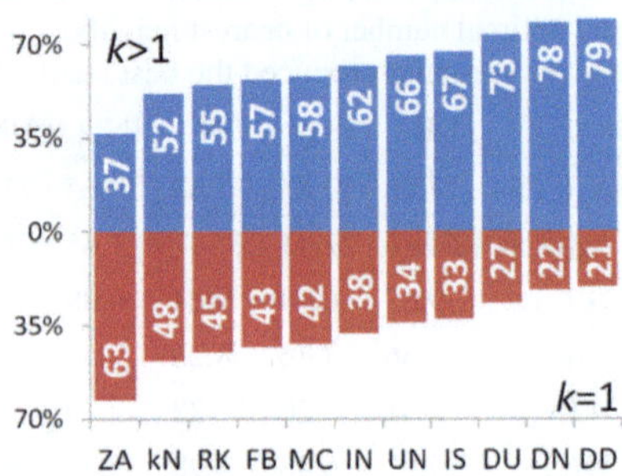

Fig. 5.14 L_1—percentages of splits in which SCV1x9 chose $k = 1$ and $k > 1$ as the optimal number of nearest neighbors on the training set when evaluating classification accuracy using SCV10x10

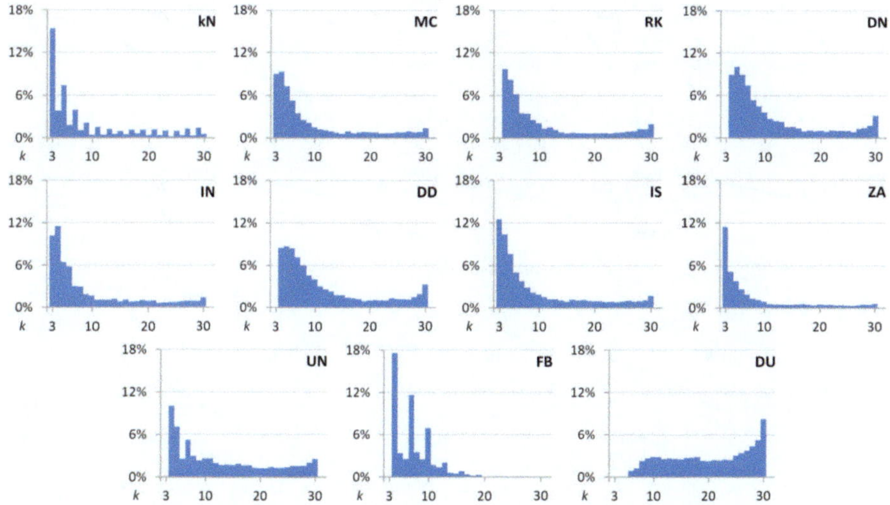

Fig. 5.15 L_1—frequencies (in percentage relative to the 12,800 splits) of different values of the parameter k ($\geq$ 3) chosen by the SCV1x9 algorithm over the training set when evaluating classification accuracy using SCV10x10

was chosen by IN, IS, UN and FB schemes and DN relied on $k = 5$. Furthermore, $k = 7$ also proved to be the best choice for FB in more than 10% of iterations. It is also worth noting that RK, DN, DD, UN, FB and DU never relied on $k = 3$ and DU did not even rely on $k = 4$. The frequencies were the most balanced when using the DU, UN and ZA schemes (with standard deviations around 1.6, 2.0 and 2.3 pp, respectively) and the least balanced when applying the FB, kN and IS weighting methods (where the deviation was approximately 4.0, 3.0 and 3.0 pp, respectively).

Based on the cumulative plots in Fig. 5.16, the optimal number of nearest neighbors was not greater than 9 in more than 90% of the splits when classifying with ZA. In order to cover slightly more than 91% of cases, it was necessary to consider up to 10 nearest neighbors when weighting with the FB scheme. The other classifiers covered a similar proportion of splits taking into account noticeably higher k-values, which, when it comes to kN, MC and RK, deviates from the initial assumptions reached by considering individual values of k (Fig. 5.13). The need to include up to 24, 25, 25 and 29 nearest neighbors to achieve a similar threshold further confirms that DN, DD, UN and especially DU, relied more often also on higher values of the parameter k.

L_1 was the only distance measure where the best win/loss ratio over 1N did not come from the DU (2.4:1) scheme (Fig. 5.17): it was overtaken by both DD (2.7:1) and ZA (2.7:1). The fourth best result was achieved by the DN (2.2:1) method. When interpreting these findings, it should be taken into account that DD and DN were among the best in terms of average accuracy and DU and ZA were among those who outperformed 1N to the smallest extent (see Table 5.4). The high degree of similarity of the performance of DU and ZA with 1N can be seen by the proximity

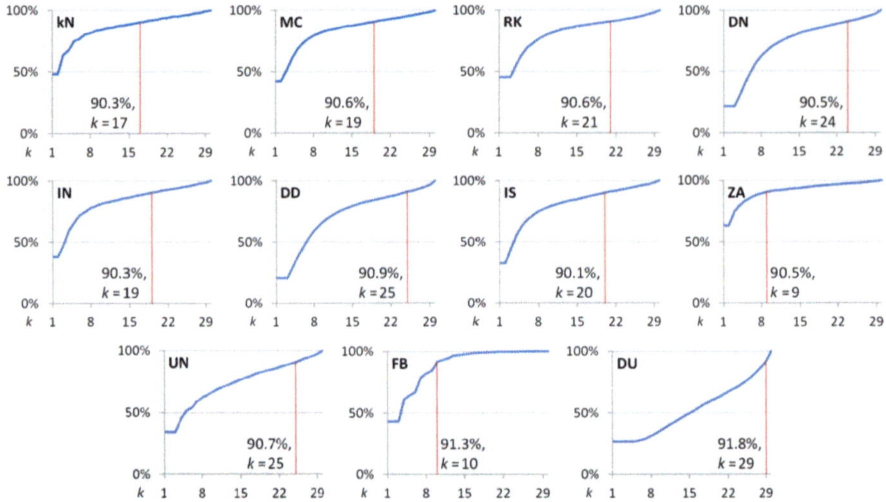

Fig. 5.16 L_1—cumulative display of the frequencies of optimal numbers of nearest neighbors chosen by the SCV1x9 algorithm over the training set when evaluating classification accuracy using SCV10x10

of the points to the diagonal on their plots in Fig. 5.17, by the smallest (both positive and negative) mean differences in Fig. 5.18, as well as by the less frequent and/or shorter lines on their plots in Fig. 5.19.

In addition to the members of the first group (kN, RK, MC), 1N also surpassed the IN scheme in terms of the number of wins and losses. Moreover, the number of datasets on which the members of the first group achieved better accuracy than 1N was among the smallest and the number of those datasets on which they produced lower accuracy was among the largest. In addition, similar to L_2, kN, RK and MC generated among the largest positive and also negative average differences. It follows that, compared to the other schemes, they achieved greater advantages over 1N on a smaller number of datasets, but also lost to a larger extent on a larger number of them.

With the exception of MC, all other weighting schemes achieved the smallest positive difference in relation to the accuracy of 1NN when relying on the Manhattan distance (taking into account the elastic measures as well), thus achieving the smallest average of approximately 1.7.

According to Fig. 5.19, the FB and DU schemes did not achieve an advantage over 1N of more than 5 percentage points on any dataset. When using the UN and ZA schemes, there were only 3 (numbered 29, 92 and 94) and 4 (labeled 73, 75, 92 and 94) such datasets, respectively. All other kNN variants produced between 5 and 10 pp higher accuracy than 1N on *Earthquakes* (29), *Haptics* (58), *MiddlePhalanxOutlineAgeGroup* (73), *MiddlePhalanxTW* (75), *ProximalPhalanxOutlineAgeGroup* (92) and *ProximalPhalanxTW* (94).

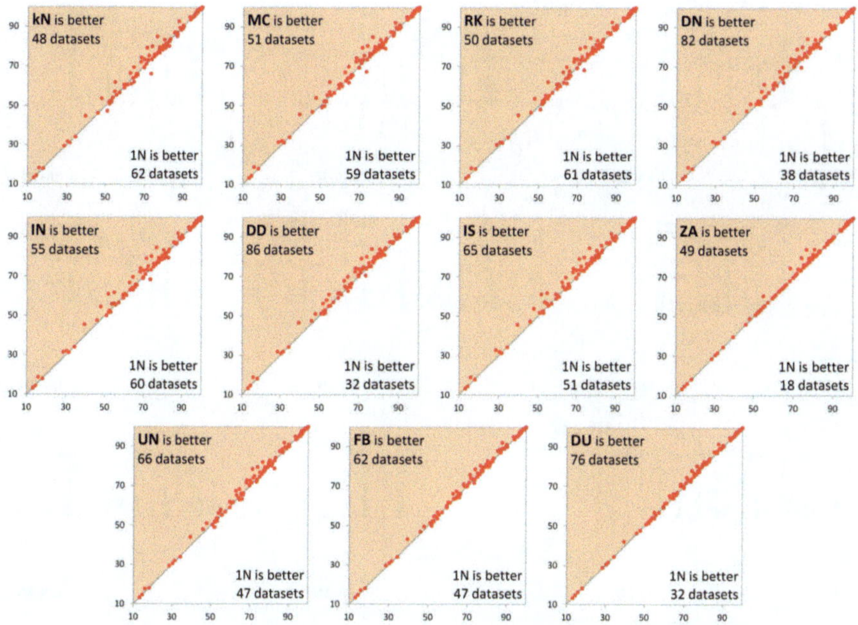

Fig. 5.17 L_1—comparison of the kNN variants with the 1N classifier based on the number of datasets where they achieved higher/lower classification accuracy

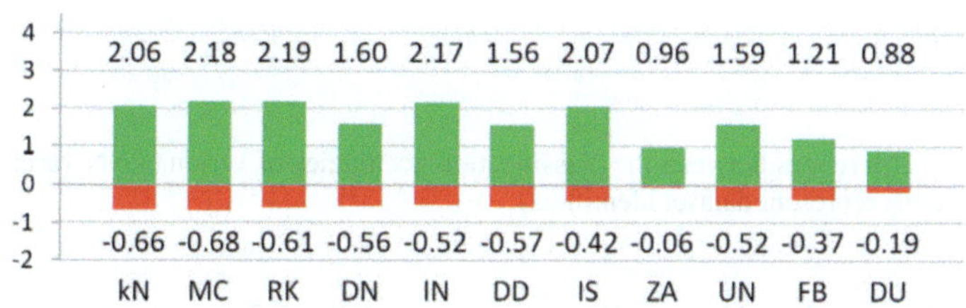

Fig. 5.18 L_1—average difference in classification accuracy of various kNN variants and 1N over the datasets where the performance of a given kNN variant was better (green) and worse (red) than 1N

The advantage of 1N over ZA did not reach even 0.3 pp on any dataset. The biggest difference compared to the DU scheme in favor of 1N was realized on the *BeetleFly* (8) dataset and it amounted to 1 pp. In the case of the IS and FB methods, the gap was limited by 2 pp. When it comes to the remaining kNN variants, 1N outperformed them by more than 3 pp only on *Beef* (7) (kN), *BeetleFly* (8) (kN, MC, RK, DN, IN, DD) and *BirdChicken* (9) (MC, DN, IN, UN).

Although the difference in average classification accuracy between 1N and the analyzed kNN classifiers did not even reach 1 pp, the difference in their performance was, according to the Friedman test ($\chi^2(11) = 103.19, p < .001$), statistically

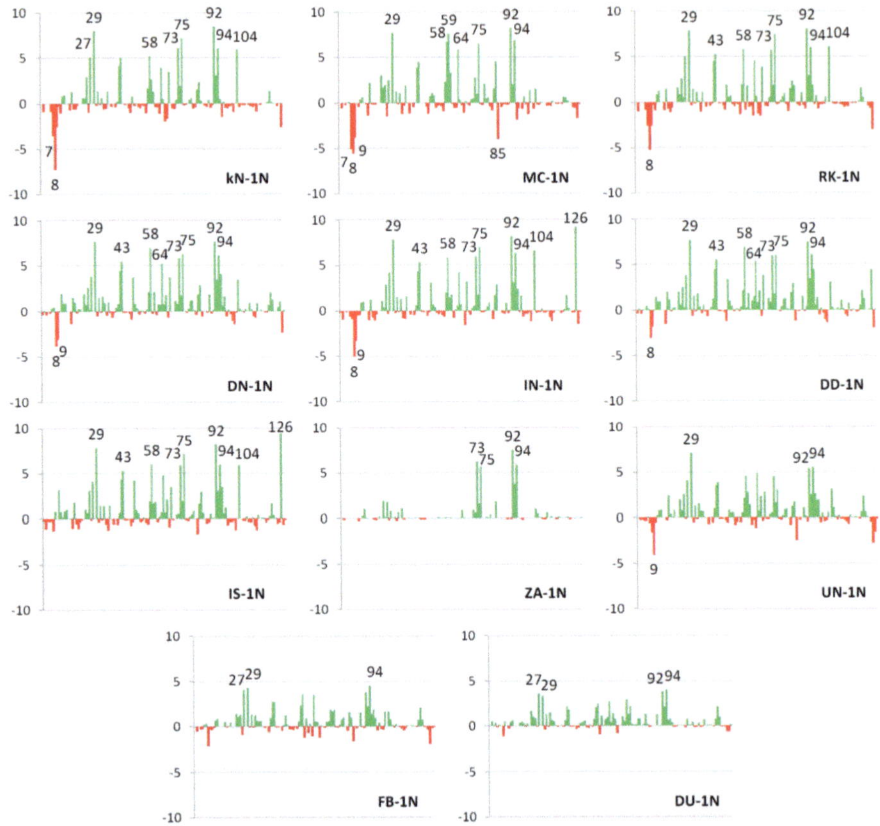

Fig. 5.19 L_1—differences between the classification accuracies of various kNN variants and 1N by datasets (labels represent dataset identifiers)

significant. As stated by Nemenyi's post-hoc test (Fig. 5.20), only the results of the DD and DN schemes were significantly different from those of 1N. In addition to them, the Wilcoxon signed-rank test (corrected using the Holm-Bonferroni scheme) also assessed the difference in relation to UN, DU and ZA as significant and at the $\alpha = 0.05$ level, also to US and FB (Table 5.5). In addition, the advantage of the DD and DN schemes over the majority-voting approach and the simple 1NN rule was confirmed by the numbers of statistically significant wins and losses obtained using the corrected resampled t-test (Table 5.6).

When evaluating the advantage of the considered kNN variants over 1N based on their expected gain, false positives were most frequent when classifying with the kN (45%), RK (43%), MC (41%) and IN (41%) schemes (Fig. 5.21). These are the same classifiers that 1N dominated in the number of wins and losses (see Fig. 5.17). In the FP region, datasets were the least numerous in the case of the DU (23%), FB (22%) and ZA (14%) methods. On the other hand, the estimation of the actual gain was the

Table 5.5 L_1—Holm-Bonferroni-adjusted p-values of the pairwise Wilcoxon signed-rank test on classification accuracy

A		77.88	77.85	77.83	77.65	77.60	77.53	77.52	77.44	77.42	77.42	77.33	76.97
		DD	IS	DN	IN	UN	RK	MC	DU	FB	kN	ZA	1N
77.88	DD		.088	1.000	< **.001**	.012	< **.001**	< **.001**	.031	< **.001**	< **.001**	< **.001**	< **.001**
77.85	IS			.343	.043	1.000	1.000	1.000	1.000	1.000	**.005**	1.000	.019
77.83	DN				< **.001**	.012	< **.001**	< **.001**	.152	< **.001**	< **.001**	**.010**	< **.001**
77.65	IN					1.000	1.000	1.000	1.000	1.000	.338	1.000	1.000
77.60	UN						1.000	1.000	1.000	1.000	.896	1.000	**.010**
77.53	RK							1.000	1.000	1.000	.343	1.000	1.000
77.52	MC								1.000	1.000	.343	1.000	1.000
77.44	DU									1.000	.151	.726	< **.001**
77.42	FB										1.000	1.000	.019
77.42	kN											1.000	1.000
77.33	ZA												**.001**
76.97	1N												

Note Bolded values indicate significance at the level of $\alpha = 0.01$

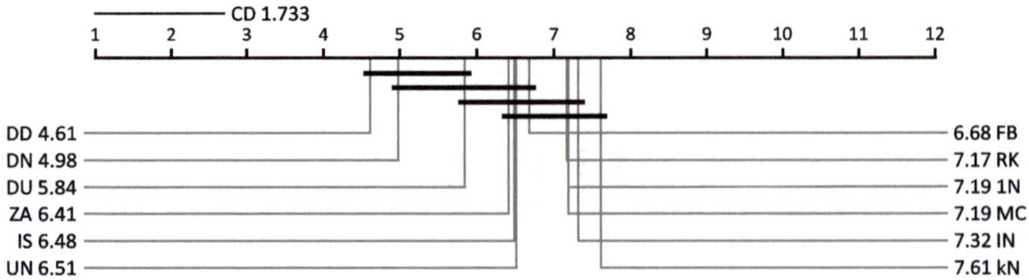

Fig. 5.20 L_1—critical difference diagram comparing average accuracy ranks based on Nemenyi's test: NN variants connected by a solid bar did not differ statistically significantly

Table 5.6 L_1—statistically significant wins (W) and losses (L) reported by the corrected resampled t-test on classification accuracy (across 12,800 splits)

	DD	DN	IS	DU	UN	IN	RK	MC	FB	ZA	kN	1N
W	356	352	227	262	232	154	152	145	184	195	106	101
L	60	57	135	222	196	164	206	202	241	278	280	425
W–L	296	295	92	40	36	−10	−54	−57	−57	−83	−174	−324

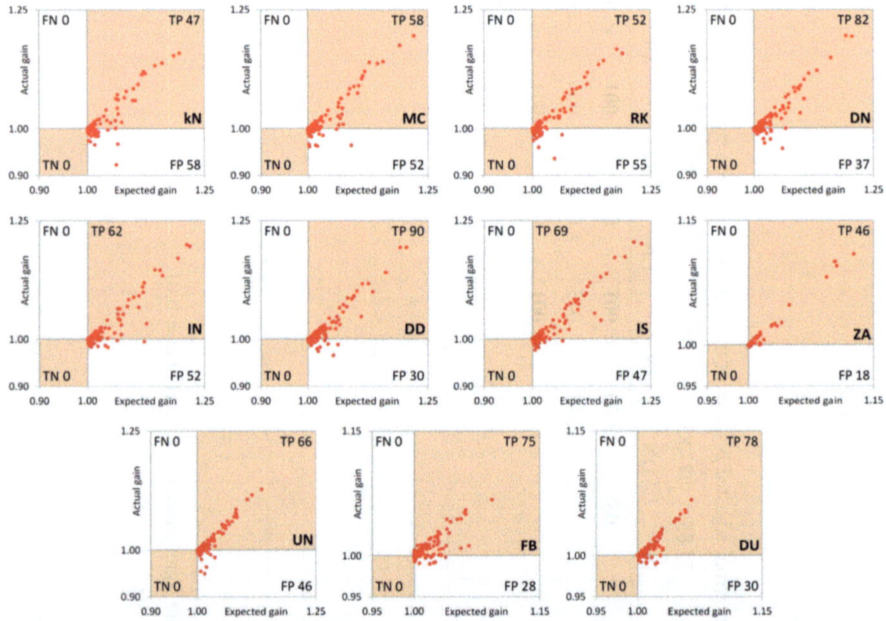

Fig. 5.21 L_1—comparing the actual and expected gain in classification accuracy: kNN variants versus 1N

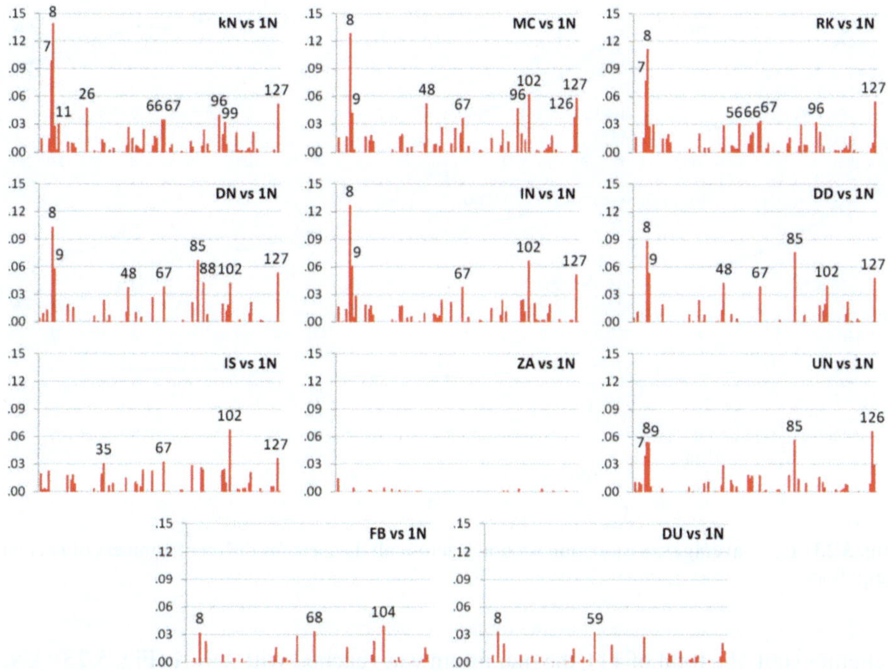

Fig. 5.22 L_1—differences between expected and actual accuracy gains of datasets belonging to the FP region: kNN variants versus 1N (labels represent dataset identifiers)

most successful for DD (70%), DN (64%) and DU (61%) and the least successful for RK (41%), kN (37%) and ZA (36%).

The difference between the expected and actual gain was less than 0.03 for all FP datasets when classifying time series using the ZA weighting scheme (Fig. 5.22). The DU method produced two (*BeetleFly* (8) and *Herring* (59)) and the FB variant generated three (*BeetleFly* (8), *Lightning7* (68) and *SmallKitchenAppliances* (104)) FP datasets where the estimation error was greater than 0.03. With the other kNN classifiers, their number ranged between 4 (IS) and 9 (kN). *BeetleFly* (8), *Lightning2* (67) and *WormsTwoClass* (127) exhibited this feature with more than half of the considered weighting schemes (kN, MC, RK, DN, IN and DD). The average error of gain estimation (calculated over all datasets in the FP region of a given kNN variant) was the highest for kN, MC and DN and the lowest for ZA, FB and DU.

5.3 Chebyshev Distance (L_∞)

Examining the change in classification accuracy (calculated using the SCV10x10 evaluation method) as a function of the number of considered nearest neighbors revealed that, when relying on the Chebyshev distance, in the case of 8 weighting

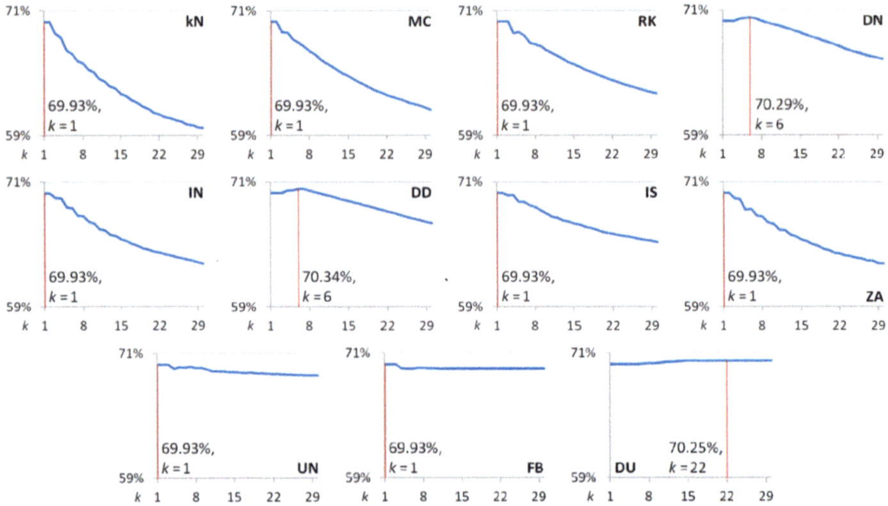

Fig. 5.23 L_∞—average classification accuracy across all datasets for different numbers of nearest neighbors

schemes (out of a total of 11), the maximum was reached with $k = 1$ (Fig. 5.23): kN, MC, RK, IN, IS, ZA, UN and FB. Based on these initial observations, it appears that 1NN will be very difficult to beat and that DN, DD and DU may be the only viable candidates for the task.

Analogously to the other analyzed distance measures, the deterioration of performance by increasing the value of the parameter k was the most pronounced when using the majority-voting rule and at $k = 30$ it reached about 10.2 percentage points (pp). This difference was also the largest detected lag in relation to $k = 1$ within the preliminary analyzes in this chapter. Among the members of the first group of classifiers, which consisted of kNN variants that for $k > 1$ produced worse results than those obtained with $k = 1$, the advantage of 1NN was the smallest (< 0.5 pp) when relying on the FB weighting scheme.

By expanding the neighborhood to between 3 and 8 (DN) and between 3 ad 9 (DD) nearest neighbors, DN and DD managed to slightly surpass the accuracy of 1NN (by less than 0.5 pp). Both schemes reached their maximum at $k = 6$. In both cases, a short period of growth was followed by a decline in performance, with a minimum at $k = 30$ that was below the result achieved with $k = 1$.

The only scheme that for all k-values greater than 1 produced better accuracy than that achieved with $k = 1$ was DU. Although the difference between the maximum (achieved at $k = 22$) and the minimum was not particularly large (it was slightly above 0.3 pp), these findings are encouraging regarding the ability of DU to outperform 1NN to some small extent. In addition, they indicate that it will prefer larger values of the parameter k than the other weighting schemes.

The results of a detailed analysis of the performance of different kNN variants compared to the simple 1NN rule (denoted as "1N") are given in Table 5.7. The presented average values (AV) and standard deviations (SD) of accuracies (A) and optimal nearest neighbor numbers (k) were obtained based on 12,800 splits of the 128 UCR datasets into test and training subsets within 10 times repeated stratified tenfold cross-validation. In each of the 128,000 splits, the best value of the parameter k was chosen over the training subset so that the classification accuracy calculated by the SCV1x9 method was maximal.

Regardless the quite discouraging initial evaluations, the average accuracy of all kNN variants was somewhat better than that achieved by the 1NN classifier. However, except for DD and DN, the advantage of all other weighting schemes was below 1.0 pp (and even in the case of DD and DN it was not much higher than one percentage point). It is also worth noting that 1N was outperformed to the least extent by UN, DU and FB, which does not reflect their initial results. The standard deviation fluctuated within a range of ±0.21 around 23.48 percentage points and was again the highest when classifying with 1N.

In accordance with the results of the initial experiments, the preference for a larger number of nearest neighbors was most noticeable with the DU scheme: the average value of k in its case was close to 15. As with the other measures (except DTW), it was directly followed by DD (8.98) and DN (8.59). The lowest average of the optimal numbers of nearest neighbors was generated by FB (4.29), which might not have been expected given its results over individual k-values. It was followed by the majority-voting rule (5.02). The order of kNN variants in terms of standard deviation did not differ noticeably from the order in terms of average value—it was the highest when using the DU and the lowest when relying on the FB scheme. Chebyshev's distance did not differ from the other measures in that UN generated a slightly higher SD than DD and DN.

With 42 wins, only DD surpassed 1N (32) in terms of the number of datasets on which the best result came from a given method, which further reinforces the indication that 1N is a strong contender when combined with the L_∞ distance. DD and 1N were followed by DU and kN with 28 and 22 wins, respectively. The lowest numbers of wins were observed for IS (15), IN (14) and ZA (12). 1N proved to be competitive in terms of mean rank as well: only DD, DN and DU surpassed it and

Table 5.7 L_∞—average value (AV) and standard deviation (SD) of classification accuracy (AV) and optimal number of nearest neighbors (k) across 12,800 splits; the number of datasets for which a given classifier produced the best result (WIN) and the mean rank (MR)

	DD	DN	IS	ZA	IN	RK	kN	MC	UN	DU	FB	1N
A_{AV} (%)	71.01	70.93	70.91	70.90	70.87	70.79	70.76	70.75	70.71	70.47	70.47	69.93
A_{SD} (pp)	23.43	23.45	23.27	23.33	23.34	23.47	23.45	23.43	23.56	23.68	23.64	23.70
k_{AV}	8.98	8.59	6.36	6.06	5.79	6.17	5.02	5.55	8.07	14.84	4.29	
k_{SD}	8.55	8.44	8.10	7.99	7.73	8.30	7.27	7.56	9.37	10.90	4.29	
WIN	42	17	15	12	14	19	22	17	16	28	20	32
MR	5.57	5.74	6.75	6.72	6.85	6.54	6.57	6.90	7.07	6.18	6.68	6.44

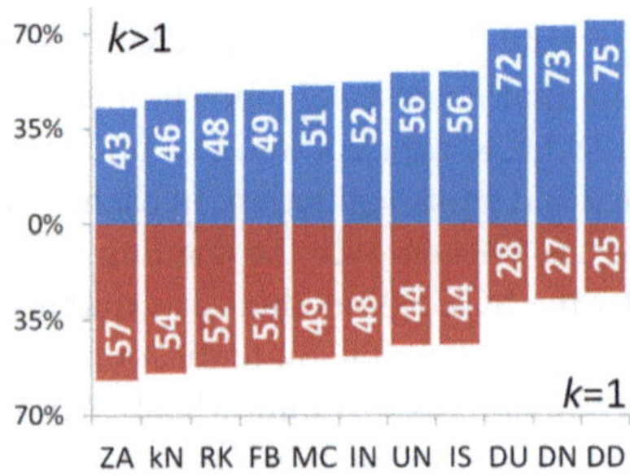

Fig. 5.24 L_∞—percentages of splits in which SCV1x9 chose $k = 1$ and $k > 1$ as the optimal number of nearest neighbors on the training set when evaluating classification accuracy using SCV10x10

all members of the first group (identified within the initial analysis) turned out to be weaker than 1N.

Averaged across all 11 weighting schemes, the proportion of those splits of the SCV10x10 method in which the kNN variants were reduced to 1NN (based on the decision of the SCV1x9 algorithm used to select the optimal number of nearest neighbors) was the highest when comparing time series using the Chebyshev distance. As with all other distance measures discussed in this chapter, $k = 1$ was the least relied upon by the DU, DN and DD schemes (in less than 30% of the cases, see Fig. 5.24). The dominance of the first nearest neighbor was most intense when applying the ZA, kN, RK and FB variants of the kNN classifier, in their case it was present in more than 50% of the splits.

The high degree of dependence of kNN variants on $k = 1$ when using the Chebyshev distance was also reflected in the fact that the average number of datasets for which the optimal k-value was 1 in each of the 100 iterations of the SCV10x10 procedure was the highest among the analyzed measures precisely for L_∞. Among the weighting schemes that we included in the first group in the preliminary analysis, their number ranged between 13 (IS) and 29 (kN). The datasets that we labeled with the numbers 12, 14, 15, 37, 39, 103, 115, 117 and 124 exhibited the mentioned feature in combination with all eight schemes of the first group. On the other hand, the Chebyshev distance was the only one for which there was a scheme (DD) for which such first-nearest-neighbor dominance did not manifest itself for any of the 128 UCR datasets. When it comes to DN and DU, $k = 1$ was the optimal choice in all 100 iterations of the evaluation algorithm for the following datasets: *FreezerSmallTrain* (45) and *TwoPatterns* (117) for DN and *Coffee* (16), *DiatomSizeReduction* (22), *Fungi* (46), *PigAirwayPressure* (86), *PigCVP* (88) and *Trace* (115) for DU.

Except for the FB scheme where the average value of the parameter k was less than 14 for all datasets, for the other kNN variants there were several of them with averages greater than 20. Their number was the lowest in combination with kN and MC (7 in both cases) and the highest with DU (39). In the following datasets, the average number of optimal nearest neighbors was greater than 20 when using all the investigated weighting methods (except FB): *Earthquakes* (29), *FordA* (42), *FordB* (43), *MiddlePhalanxOutlineAgeGroup* (73), *ProximalPhalanxOutlineAgeGroup* (92) and *RefrigerationDevices* (95).

The strong reliance on smaller k-values by all weighting schemes, except for DU, is clearly visible in the histograms shown in Fig. 5.25. This is well illustrated by the fact that in more than 10% of cases, the kN and IS variants found the consideration of the three nearest neighbors to be the most fruitful. Like them, in more than 10% of the 12,800 splits of the datasets into test and training subsets, DN, DD, IS and FB estimated that the optimal result would be achieved with $k = 4$. Unlike them, the most frequent values of the parameter k when using the DU weighting method were 29 and 30. DU, UN and RK generated the most uniform frequency distributions of different k-values (with standard deviations of 1.7, 1.8 and 1.9, respectively) and the differences were the greatest when applying the FB, DN and DD methods of calculating weights (with deviations of 3.1, 2.9 and 2.8, respectively).

Figure 5.26 with accumulative frequency plots of different optimal numbers of nearest neighbors indicates that FB covered more than 90% of the splits very quickly, already with $k = 10$. In contrast to FB, DU needed to include all k-values up to 29 to achieve a similar degree of coverage. The other weighting schemes crossed the threshold of 90% of cases with values of the parameter k between these two extreme values: kN and MC at the earliest (with $k = 17$ and $k = 18$) and DN, DD and UN at the latest (with $k = 24$, $k = 25$ and $k = 26$).

The numbers of wins and losses of the discussed weighting schemes over the 1N classifier shown in Fig. 5.27 confirm the high degree of competitiveness of the simple 1NN rule in combination with the Chebyshev distance. Namely, out of 11 weighting techniques, 7 of them were overcome by 1N and only 4 schemes (DN, DD, FB and DU) proved to be better than it in this regard. The same number of better and worse kNN variants was detected when using the Euclidean distance as well. However,

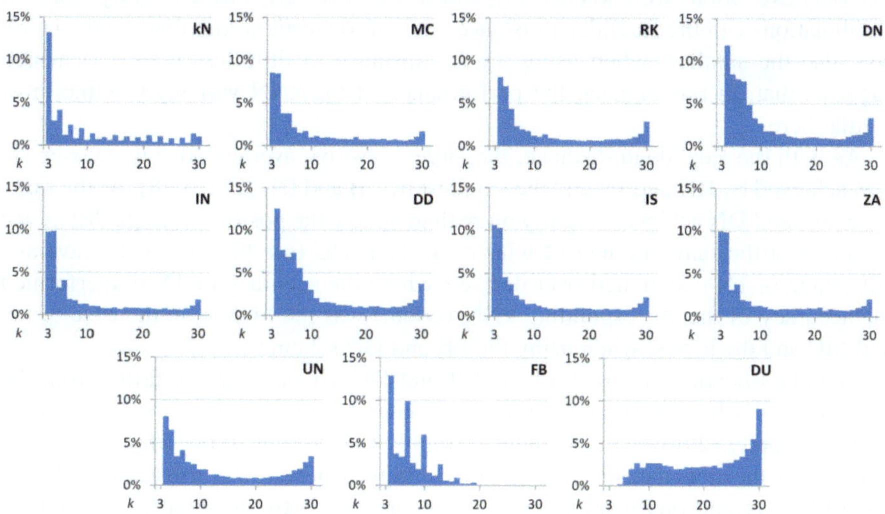

Fig. 5.25 L_∞—frequencies (in percentage relative to the 12,800 splits) of different values of the parameter k (≥ 3) chosen by the SCV1x9 algorithm over the training set when evaluating classification accuracy using SCV10x10

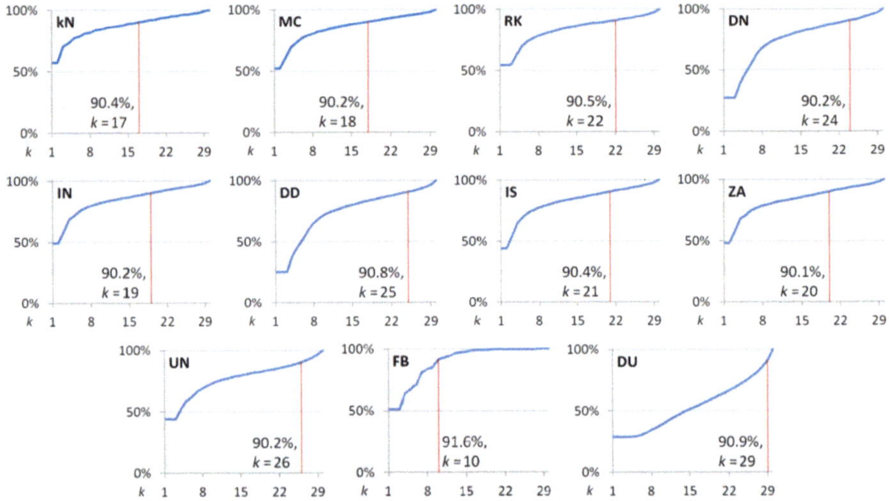

Fig. 5.26 L_∞—cumulative display of the frequencies of optimal numbers of nearest neighbors chosen by the SCV1x9 algorithm over the training set when evaluating classification accuracy using SCV10x10

L_∞ produced a lower average win-loss ratio (just under 1.0, averaged across 11 weighting schemes) not only than L_2, but also all other distance measures discussed in this chapter. In addition, even at the level of individual weighting schemes, the Chebyshev distance gave the lowest win-loss ratios among the 7 examined measures (the only exceptions were kN and FB, whose win–loss ratios were slightly worse in combination with the Euclidean distance). It is also worth noting that the deviation was also the smallest when using the L_∞ distance (a little less than 0.3), which suggests that, in this respect, the performance of the kNN variants was the most similar here.

As with the Euclidean distance, the largest positive average advantage over 1N was achieved by RK and IN and the smallest by FB and DU (Fig. 5.28). At the same time, DD and DN achieved slightly more than double the positive average difference with almost the same number of wins over 1N as DU (see Fig. 5.27). The average advantage of 1N (calculated over datasets where the accuracy of 1N outperformed the accuracy of the corresponding kNN variant) was the highest in the case of kN and MC and the lowest when using the FB and DU schemes.

The classification accuracy of the FB and DU scheme did not differ from the accuracy of 1N by more than 6.2 pp on any dataset, and when it comes to UN, only on *Earthquakes* (29) was the advantage slightly higher than 10 pp, while it was less than 8 pp on the other datasets (Fig. 5.29). The biggest benefit was also achieved by the other kNN variants on *Earthquakes* (29) and it ranged between about 12.06 pp (DN) and 13.64 pp (RK). In addition, there were also several other datasets where 1NN was overpowered by more than 8 pp, but only *ProximalPhalanxOutlineAgeGroup* (92) was common to all those methods.

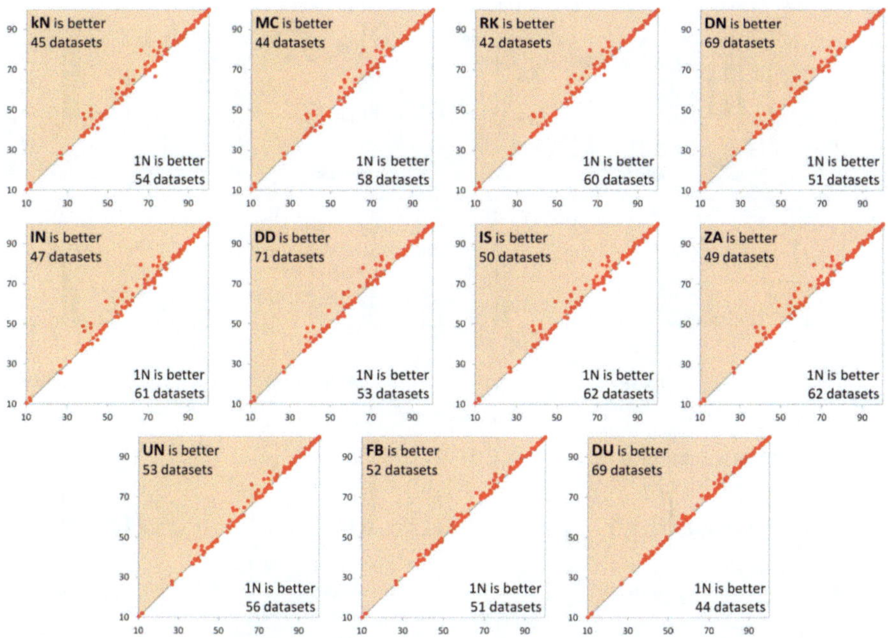

Fig. 5.27 L_∞—comparison of the kNN variants with the 1N classifier based on the number of datasets where they achieved higher/lower classification accuracy

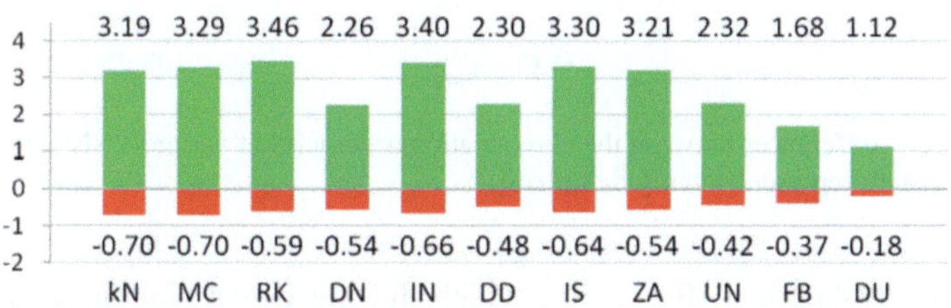

Fig. 5.28 L_∞—average difference in classification accuracy of various kNN variants and 1N over the datasets where the performance of a given kNN variant was better (green) and worse (red) than 1N

The difference in performance in favor of 1N did not reach 3 percentage points on any dataset when it came to DD, UN, FB and DU. All other weighting schemes were outperformed by more than 3 pp on *BeetleFly* (8). In addition, the threshold of 3 pp was crossed also in the case of *BirdChicken* (9) (only for IS) and *Lightning7* (68) (kN, MC, IN).

The Chebyshev distance was the only measure where, according to the Friedman test ($\chi^2(11) = 24.70$, $p = .010$), the differences in the performance of the NN classifiers were not statistically significant at the $\alpha = 0.01$ level, but at the $\alpha = 0.05$ level. The similarity of the results of the NN variants was also confirmed by

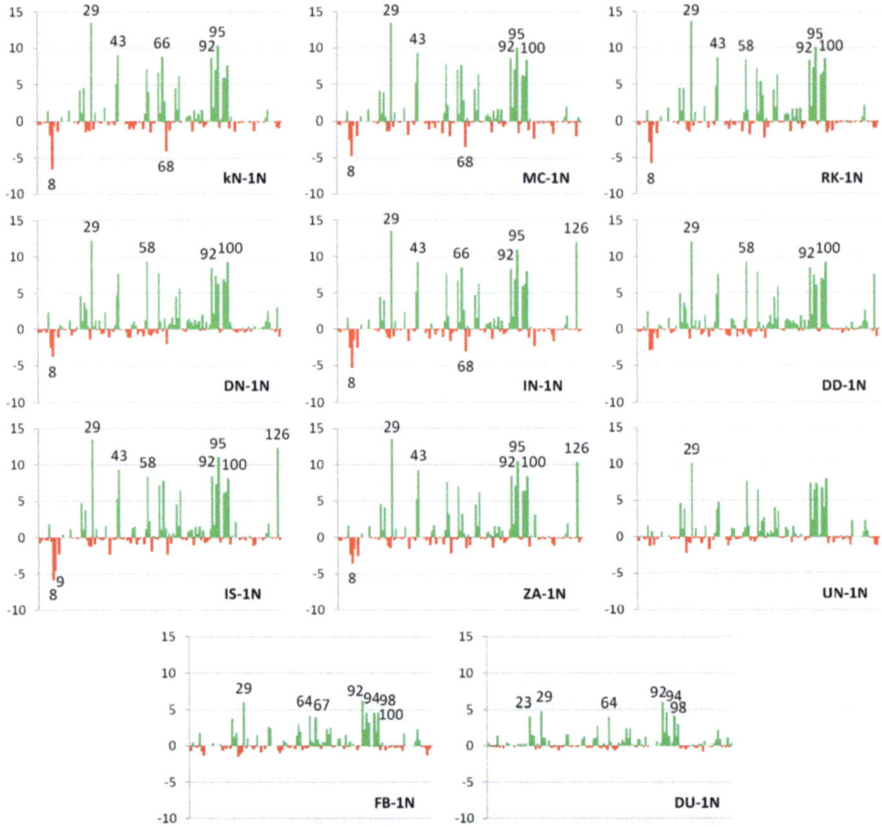

Fig. 5.29 L_∞—differences between the classification accuracies of various kNN variants and 1N by datasets (labels represent dataset identifiers)

Nemenyi's post-hoc test (Fig. 5.30), which did not detect significant differences between them. Nonetheless, the Wilcoxon signed-rank test (adjusted according to the Holm-Bonferroni rule) revealed a significant difference in the average classification accuracies of 1N and the three weighting schemes that in the initial analysis proved to be the most promising for surpassing it (Table 5.8): DN, DD and DU. However, with the exception of DD and UN, as well as DN and MC, the other pairs did not differ significantly even with the significance level raised to $\alpha = 0.05$. Regarding the number of statistically significant wins and losses in Table 5.9, DD and DN confirmed as more successful than FB and 1N.

According to Fig. 5.31, the frequencies of false positive (FP) cases arising from the comparison of the expected and actual accuracy gains of the considered kNN variants over 1N differed only by approximately 10 percentage points: when relying

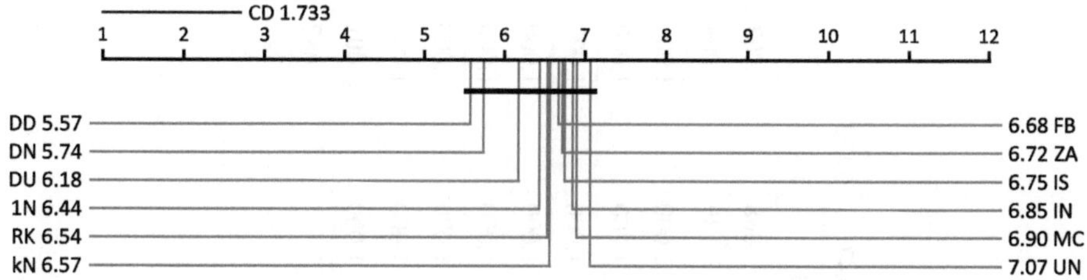

Fig. 5.30 L_∞—critical difference diagram comparing average accuracy ranks based on Nemenyi's test: NN variants connected by a solid bar did not differ statistically significantly

on the ZA scheme it was about 44% and when the DU method was used it was slightly above 34%. In fact, the standard deviation of the numbers of FP datasets was the lowest and the average value the second highest when applying the Chebyshev distance. In this regard, when it comes to DD, ZA and DU, their results were the worst in combination with L_∞ among all the measures discussed in this chapter.

The failure of the kNN variants to outperform the 1NN classifier was also reflected in the fact that the average number of true positive estimates was the smallest in combination with the Chebyshev distance (around 44%). DN (57%), DD (56%) and DU (54%) proved to be the most successful in this category as well, while kN (38%), RK (38%) and MC (37%) produced the weakest results.

Chebyshev's distance also stood out in that it generated the highest average number (about 7.2) of datasets in the FP region for which the difference between expected and actual gain was greater than 0.03 (Fig. 5.32). The lowest number of such cases was observed for DU (*Herring* (59) and *Lightning7* (68)) and FB (*BeetleFly* (8), *DodgerLoopDay* (26), i *Worms* (126)) and they were the most numerous in the FP region of the IS scheme (11). The error in assessing the superiority of all the examined kNN variants (except FB and DU) exceeded the threshold of 0.03 when classifying the *GestureMidAirD3* (49) time series, and if the UN scheme was excluded from consideration, the list included three more common datasets: *BeetleFly* (8), *Lightning7* (68) and *PigCVP* (88). Comparing the average values of the differences between the expected and actual gains in the FP region, DD, ZA and DN turned out to be the least successful, and UN, DU and FB were the most successful.

Table 5.8 L_∞—Holm-Bonferroni-adjusted *p*-values of the pairwise Wilcoxon signed-rank test on classification accuracy

A		71.01	70.93	70.91	70.90	70.87	70.79	70.76	70.75	70.71	70.47	70.47	69.93
		DD	DN	IS	ZA	IN	RK	kN	MC	UN	DU	FB	1N
71.01	DD		1.000	.430	.467	.135	.146	.354	.086	.025	.320	.068	**.003**
70.93	DN			1.000	1.000	.390	.320	.285	.049	.113	1.000	.139	**.006**
70.91	IS				1.000	1.000	1.000	1.000	1.000	1.000	1.000	1.000	1.000
70.90	ZA					1.000	1.000	1.000	1.000	1.000	1.000	1.000	1.000
70.87	IN						1.000	1.000	1.000	1.000	1.000	1.000	1.000
70.79	RK							1.000	1.000	1.000	1.000	1.000	1.000
70.76	kN								1.000	1.000	1.000	1.000	1.000
70.75	MC									1.000	1.000	1.000	1.000
70.71	UN										1.000	1.000	1.000
70.47	DU											1.000	< **.001**
70.47	FB												.682
69.93	1N												

Note Bolded values indicate significance at the level of $\alpha = 0.01$

Table 5.9 L_∞—statistically significant wins (W) and losses (L) reported by the corrected resampled t-test on classification accuracy (across 12,800 splits)

	DD	DN	DU	IS	ZA	IN	RK	UN	MC	kN	FB	1N
W	374	315	291	176	149	132	141	180	129	128	139	130
L	99	106	224	147	132	155	166	212	172	194	271	406
W–L	275	209	67	29	17	−23	−25	−32	−43	−66	−132	−276

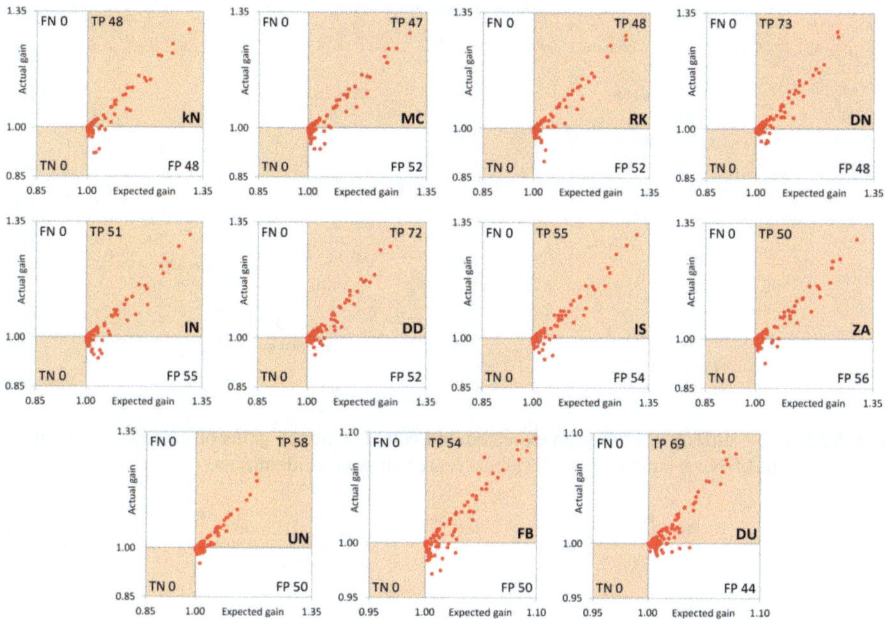

Fig. 5.31 L_∞—comparing the actual and expected gain in classification accuracy: kNN variants versus 1N

5.4 Dynamic Time Warping (DTW)

The average classification accuracies across the 128 datasets of the UCR repository for individual values of the parameter k (from 1 to 30) of the considered variants of the kNN classifier are given in Fig. 5.33. Very similar to the L_1 distance, three groups of the analyzed variants of kNN can be clearly distinguished based on the shape of the line that represents the change in average performance on the charts.

The first group included kN, MC and RK and in their case the maximum average accuracy (79.23%) was achieved with $k = 1$, which suggests that these classifiers may not be able to surpass the performance of 1NN. The negative impact of including a large number of nearest neighbors in the class selection process was most noticeable with kN, whose average efficiency monotonically decreased as the value of k

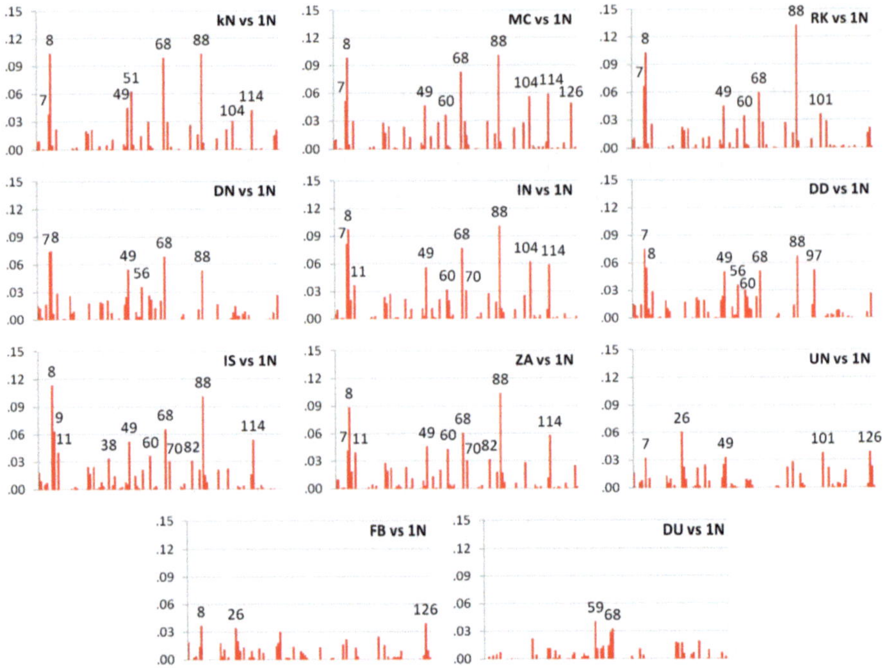

Fig. 5.32 L_∞—differences between expected and actual accuracy gains of datasets belonging to the FP region: kNN variants versus 1N (labels represent dataset identifiers)

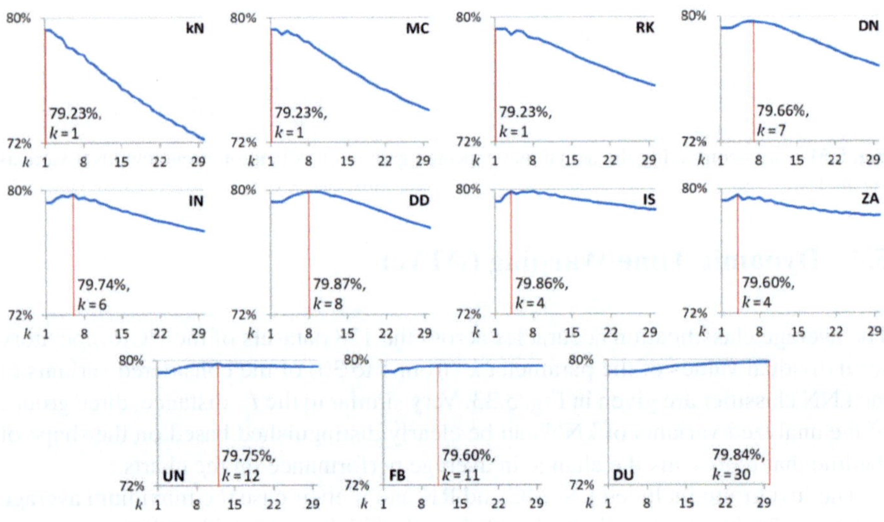

Fig. 5.33 DTW—average classification accuracy across all datasets for different numbers of nearest neighbors

increased, with a minimum of 72.24% at $k = 30$, which was almost 7 percentage points (pp) below the maximum.

In the second group, which consisted of DN, IN, DD, IS and ZA, for small values of the parameter k, a slight improvement in average accuracy was observed compared to 1N. The interval within which better results than 1N were detected was the longest when using IS (up to $k = 18$) and the shortest with ZA (up to $k = 10$). Except for ZA, this interval was followed by a monotonous decline in average performance. It is also worth mentioning that in the case of the ZA variant of kNN, starting from $k = 3$, an alternating increase and decrease of the average classification accuracy was observed (except between $k = 12$ and $k = 15$): every even value of the parameter k generated a better result than the previous odd value, and every odd value of the parameter k produced a worse result than the previous even value. These preliminary findings suggest that the classifiers from this group may outperform 1N to some extent.

The members of the third group were UN, FB and DU and their average accuracy for any observed value of the parameter k was not lower than the result of 1N. After reaching a maximum at $k = 12$ and $k = 11$, the average performance of both UN and FB began to decline, with the deterioration of FB being noticeably more moderate. However, unlike the classifiers of the second group, their results remained above that of 1NN (at least up to $k = 30$). DU stood out among all the considered variants of kNN due to the almost monotonic growth of its average classification accuracy up to $k = 30$. This initial review of the effectiveness of UN, FB and DU predicts that they too could outperform 1N to some degree. In addition, it appears that DU would prefer a larger number of nearest neighbors than the others.

Table 5.10 summarizes the performance analysis results of different variants of the kNN classifier based on the nested SCV10x10 evaluation method where the optimal number of nearest neighbors was chosen using the SCV1x9 procedure over the training subset in each of the 100 testing/training splits of each dataset.

Looking at the average classification accuracies, it can be seen that 1N was not only outperformed by the weighted versions but also by the majority voting kNN. However, the difference was not greater than 1.2 percentage points even compared to the best performer. The four most effective weighting schemes were IS, DD, DN and IS while DU, FB and kN classifiers were the weakest. The standard deviation

Table 5.10 DTW—average value (AV) and standard deviation (SD) of classification accuracy (AV) and optimal number of nearest neighbors (k) across 12,800 splits; the number of datasets for which a given classifier produced the best result (WIN) and the mean rank (MR)

	IS	DD	DN	IN	UN	RK	MC	ZA	DU	FB	kN	1N
A_{AV} (%)	80.39	80.35	80.28	80.19	80.16	79.98	79.94	79.94	79.93	79.92	79.86	79.23
A_{SD} (pp)	18.90	18.98	18.98	18.92	19.01	19.11	19.15	19.20	19.23	19.13	19.13	19.45
k_{AV}	7.66	9.23	8.54	6.92	9.60	6.62	5.82	5.71	16.04	5.01	5.13	
k_{SD}	7.81	8.05	7.83	7.57	9.37	7.61	6.91	6.65	11.04	4.37	6.69	
WIN	28	37	19	19	26	17	12	25	27	16	16	32
MR	5.80	4.96	5.47	6.97	6.05	7.02	7.63	6.56	5.95	6.45	7.65	7.48

was within a narrow range of 19.18 percentage points (± 0.27), where it was the largest for 1NN, DU, ZA and FB and the smallest for DD, DN, IS and IN.

Consistent with the expectations of the initial experiments with individual values of the parameter k, the average of the optimal numbers of nearest neighbors was by far the largest when using the DU weighting scheme (16.04). It was followed by UN (9.60), DD (9.23) and DN (8.54), which also supported the preliminary results. However, contrary to the incipient estimate that FB would rely on a larger neighborhood, it generated the smallest average value of k (5.01). In addition, the standard deviation was also the smallest with FB (4.37), and the largest with DU (11.04), UN (9.37), DD (8.05) and DN (7.83).

In terms of the number of datasets for which a given classifier achieved the highest accuracy, only DD (37) surpassed 1N (32)—similarly as in the case of the Chebyshev distance. In this respect, the least successful were kN (16), FB (16) and MC (12). According to the mean rank, the best ranked were DD, DN, IS and DU and the weakest were (beside 1N) those variants for which the result with $k = 1$ represented the maximum in the performance analysis of individual k values: RK, MC and kN.

Figure 5.34 confirms that among the analyzed classifiers, $k = 1$ was the most heavily relied upon by those three that formed the first group in the initial analysis (kN, MC, RK). Along with them, in more than 40% of the 12,800 splits into testing and training subset FB also estimated that the classification would be most successful relying solely on the first nearest neighbor, which we might not expect based on the preliminary analysis. The percentage of those SCV10x10 iterations in which $k = 1$ prevailed was less than 30% when using the IS, DN, DU and DD weighting schemes.

It is also worth noting that the optimal number of neighbors was 1 in each of the 100 SCV10x10 iterations for each considered variant of the kNN classifier for the following datasets: *ChlorineConcentration* (14) (except for DU), *DiatomSizeReduction* (22), *PigAirwayPressure* (86), *Plane* (90), *Trace* (115) and *TwoPatterns* (117). This feature was also present in some other datasets, but their list differed from classifier to classifier and their number varied between 20 (kN) and 6 (DD and DN). Excluding FB where the average value of the optimal numbers of nearest neighbors was not higher than 15, all other kNN variants generated averages higher than 20 with *DistalPhalanxTW* (25), *MiddlePhalanxOutlineAgeGroup* (73), *MiddlePhalanxTW* (75) and *ProximalPhalanxOutlineAgeGroup* (92). Such datasets were the

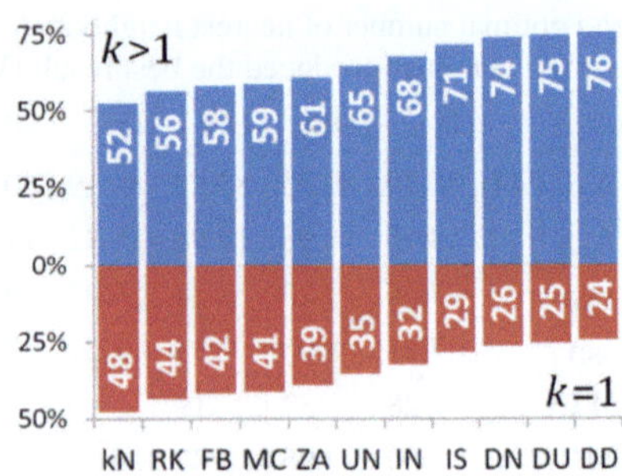

Fig. 5.34 DTW—percentages of splits in which SCV1x9 chose $k = 1$ and $k > 1$ as the optimal number of nearest neighbors on the training set when evaluating classification accuracy using SCV10x10

most numerous with the DU (54) and UN (19) schemes and the least frequent with kN, MC, RK, IN and ZA (7).

The histograms in Fig. 5.35 reveal that all kNN variants preferred smaller values of the parameter k, except DU which strongly relied on the largest values considered. kN, IS and ZA relied on $k = 3$ in more than 10% of the cases and MC and IN in more than 8% of the splits. On the other hand, the optimal number of nearest neighbors was never 3 when using any of the other weighting schemes. The highest standard deviation in percentage frequencies was detected for FB (about 3.56) followed by kN, IN and ZA (all three had an approximate deviation of 2.96), which emphasizes that these methods relied most strongly on small values of k. Different neighborhood sizes were most evenly chosen by the UN method, which can also be seen from the fact that it gave the smallest standard deviation, only about 1.38. It was followed by DU and DN, for which a deviation of approximately 2.24 percentage points was observed. These findings are also reflected in Fig. 5.36, which presents cumulative frequency plots of different k values evaluated by the SCV1x9 algorithm as optimal. In approximately 90% of the splits into testing and training subsets, FB opted for $k \leq 11$, kN for $k \leq 14$ and MC and ZA for $k \leq 15$. In order to cover a similar percentage of splits, it was necessary to take into account up to 21 nearest neighbors for DN and IS, 23 for DD, 26 for UN and all 30 of them for DU.

When it comes to wins and losses (Fig. 5.37), the greatest advantage compared to 1N was observed when applying the DU, ZA and DD weighting schemes in an approximate ratio of 3:1, 2.1:1 and 2:1, respectively. Given that DU and ZA were among the variants whose average accuracy differed the least from 1N's result (Table 5.10), it follows that they performed only a little better on a larger number of

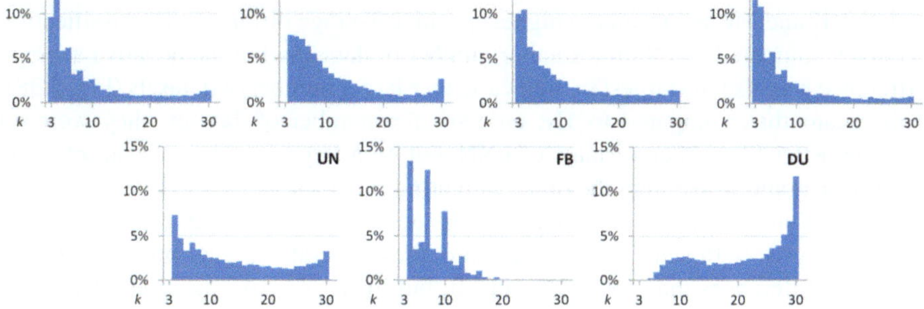

Fig. 5.35 DTW—frequencies (in percentage relative to the 12,800 splits) of different values of the parameter k (≥ 3) chosen by the SCV1x9 algorithm over the training set when evaluating classification accuracy using SCV10x10

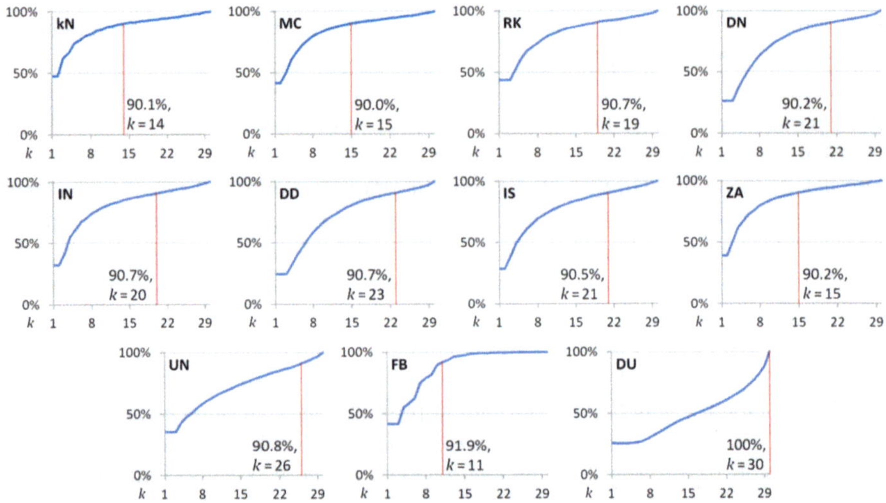

Fig. 5.36 DTW—cumulative display of the frequencies of optimal numbers of nearest neighbors chosen by the SCV1x9 algorithm over the training set when evaluating classification accuracy using SCV10x10

datasets and that the advantage of 1N was minor on the majority of datasets where DU and ZA proved to be worse. This can be seen, especially in the case of DU, in Fig. 5.37 by the proximity of the points to the diagonal and in Fig. 5.39 by the higher frequency and lower height of the (green) lines above zero and by the absence of the (red) lines below zero. Expressed in numbers, both positive and negative average accuracy differences were the lowest precisely for DU, ZA (and FB), as can be easily seen in Fig. 5.38.

In this category, 1N showed the greatest advantage compared to the members of the first group of classifiers, which were found in the preliminary analysis to have a lower result than 1N for each individual value of *k*: kN, RK and MC. However, the numbers of datasets on which 1N outperformed them were only slightly greater than the numbers on which it was inferior, so their win-loss ratios were very close.

kN, RK and MC achieved the highest positive average difference in classification accuracy compared to 1N on a smaller number of datasets, but the negative average difference was also the largest in their case, on a larger number of datasets (Fig. 5.38). This means that, compared to 1N, on a smaller number of datasets they were on average better than other variants of kNN, but on a larger number of datasets they were on average worse than the other weighting schemes.

DD, DN and IS may have proven to be the most promising alternatives to 1N in this analysis: with near 2:1 win-loss ratio, they generated higher positive average accuracy differences than DU, ZA and FB, but lower negative average differences than kN, RK and MC. It should also be emphasized that among the four elastic measures presented in this book, negative average differences were the largest for

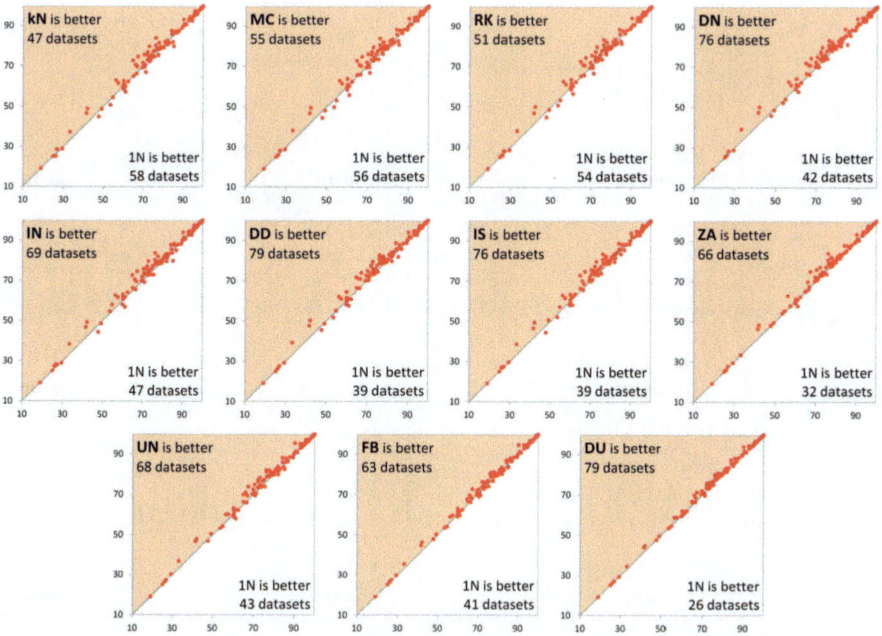

Fig. 5.37 DTW—comparison of the kNN variants with the 1N classifier based on the number of datasets where they achieved higher/lower classification accuracy

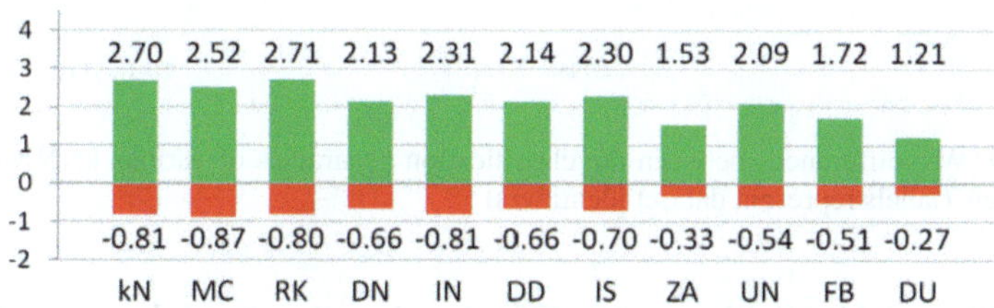

Fig. 5.38 DTW—average difference in classification accuracy of various kNN variants and 1N over the datasets where the performance of a given kNN variant was better (green) and worse (red) than 1N

most kNN variants in combination with DTW (the only exceptions were ZA, UN and DU).

At the level of individual datasets, there was no kNN variant whose accuracy exceeded that of 1N by more than 8.2 percentage points, with the difference being less than 4 percentage points in the case of DU (Fig. 5.39). Except for ZA, FB and DU, all other weighting schemes outperformed 1N by more than 5 pp on the following datasets: *Earthquakes* (29), *FordB* (43), *InsectWingbeatSound* (64), *MiddlePhalanxOutlineAgeGroup* (73) and *ProximalPhalanxOutlineAgeGroup* (92). If we also exclude IS and UN from consideration, the list will expand to include

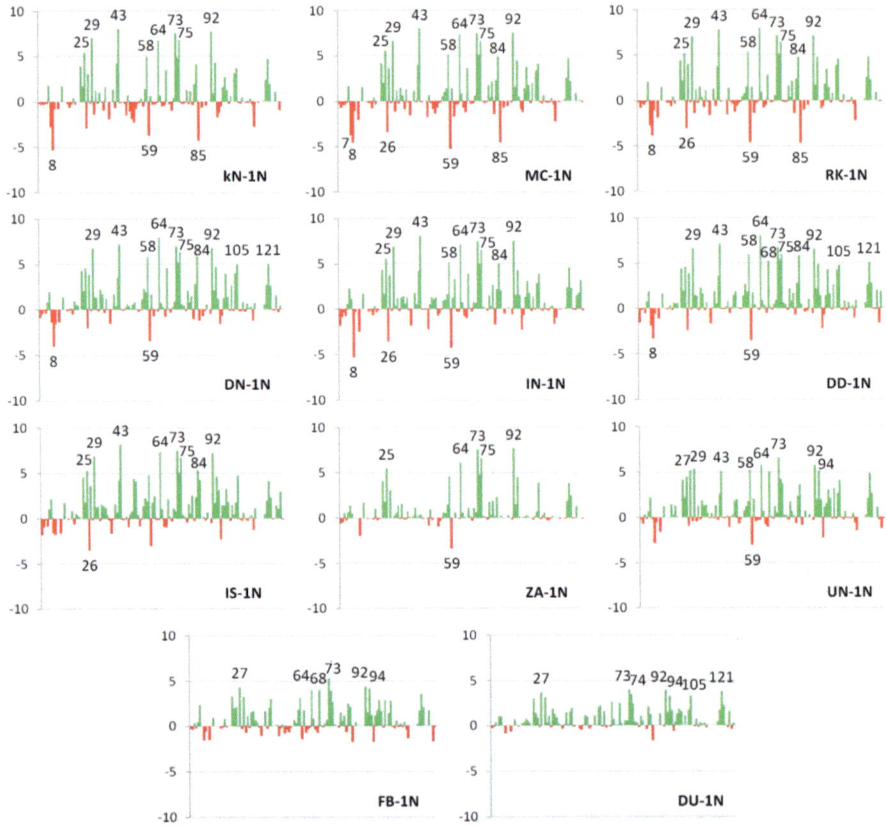

Fig. 5.39 DTW—differences between the classification accuracies of various kNN variants and 1N by datasets (labels represent dataset identifiers)

Haptics (58) and *MiddlePhalanxTW* (75). On the other hand, by more than 5 pp 1N outperformed only kN and IN on the *BeetleFly* (8) and MC on the *Herring* (59) dataset. When it comes to FB and DU, the advantage of 1N on none of the UCR datasets reached 2 percentage points.

According to the Friedman test, the difference in the performance of the evaluated NN classifiers was statistically significant ($\chi^2(11) = 92.31$, $p < .001$). A distance greater than the critical difference between the average rank of 1N and the kNN variants, Nemenyi's post-hoc test measured only in the case of the DD and DN classifiers (Fig. 5.40). At the same time, the Holm-Bonferroni-adjusted p-values of the Wilcoxon signed-rank test in Table 5.11 testify that the classification accuracy of 1N was statistically significantly different from all the considered kNN variants, except for those that we included in the first group in the introductory analysis: kN, MC and RK. Furthermore, the t-test results in Table 5.12 also support the inferiority of 1N compared to the kNN classifiers. From this point of view, DD, DN and IS proved to be the most and RK, MC and kN the least successful in overpowering 1N.

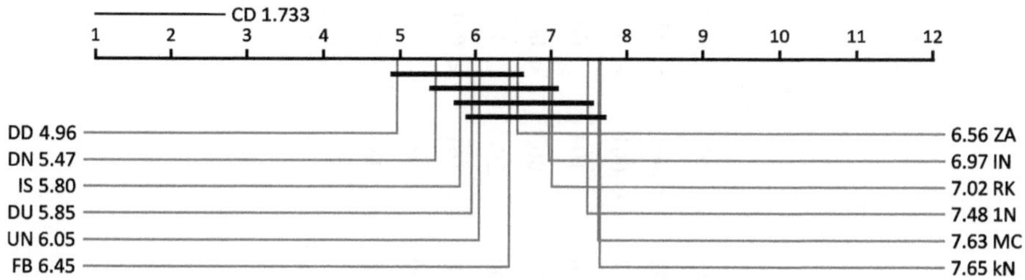

Fig. 5.40 DTW—critical difference diagram comparing average accuracy ranks based on Nemenyi's test: NN variants connected by a solid bar did not differ statistically significantly

To assess the extent to which expected and actual accuracy gains are aligned, in the final step of comparing the performance of different weighting schemes and the 1N classifier, we examined the Texas Sharpshooter plots shown in Fig. 5.41. The gain over the training set was misleading to the greatest extent in the case of kNN classifiers for which, within the initial analysis (based on individual values of the parameter k), we found that 1N surpassed them in terms of classification accuracy: kN, MC and RK (Fig. 5.33). We preferred them over 1N in approximately 45%, 42% and 40% of the datasets, respectively, based on their superiority over the training data, but they achieved lower accuracy than 1N on the test data. Such a misjudgment occurred most rarely with the DD, ZA and DU weighting schemes, in about 29%, 23% and 20% of the cases. The assessment of superiority compared to 1N based on the results of the training data was the most successful with DD (63%), DU (63%), DN (62%) and IS (59%) and the least successful with MC (43%). RK (43%) and kN (38%).

According to the number of FP datasets where the difference between the expected and actual gain was greater than 0.03, the MC scheme turned out to be the worst and DU the best: 15 such datasets were observed with MC and only one with DU (Fig. 5.42). The dataset over which the difference was the greatest varied from scheme to scheme: *Beef* (7) (MC, RK, DN, DD), *Herring* (59) (IN, ZA, UN), *PickupGestureWiimoteZ* (85) (FB, DU), *DodgerLoopDay* (26) (IS) and *GestureMidAirD3* (49) (kN). Furthermore, excluding ZA, FB and DU, all other kNN variants overestimated their gain over 1N by more than 0.03 in *BeetleFly* (8), *DodgerLoopDay* (26) and *Haptics* (58). Averaged across all FP datasets, the difference was the largest with MC, DD and kN, and the smallest with DU, ZA and FB.

Table 5.11 DTW—Holm-Bonferroni-adjusted *p*-values of the pairwise Wilcoxon signed-rank test on classification accuracy

A		80.39	80.35	80.28	80.19	80.16	79.98	79.94	79.94	79.93	79.92	79.86	79.23
		IS	DD	DN	IN	UN	RK	ZA	MC	DU	FB	kN	1N
80.39	IS		1.000	1.000	.012	1.000	.024	.035	**.002**	.317	.143	**.001**	< **.001**
80.35	DD			.161	**.003**	.143	< **.001**	.032	< **.001**	.035	**.001**	< **.001**	< **.001**
80.28	DN				.033	1.000	< **.001**	.161	< **.001**	.412	.017	< **.001**	< **.001**
80.19	IN					1.000	1.000	1.000	1.000	1.000	1.000	.104	**.001**
80.16	UN						.405	1.000	.072	1.000	.020	.063	< **.001**
79.98	RK							1.000	1.000	1.000	1.000	.296	.479
79.94	ZA								1.000	1.000	1.000	.161	< **.001**
79.94	MC									1.000	1.000	1.000	.612
79.93	DU										1.000	.363	< **.001**
79.92	FB											.612	**.001**
79.86	kN												1.000
79.23	1N												

Note Bolded values indicate significance at the level of $\alpha = 0.01$

Table 5.12 DTW—statistically significant wins (W) and losses (L) reported by the corrected resampled *t*-test on classification accuracy (across 12,800 splits)

	DD	DN	IS	UN	DU	IN	ZA	FB	RK	MC	kN	1N
W	407	350	340	277	292	224	238	187	158	153	141	128
L	105	101	165	190	260	203	249	270	242	245	316	549
W–L	302	249	175	87	32	21	−11	−83	−84	−92	−175	−421

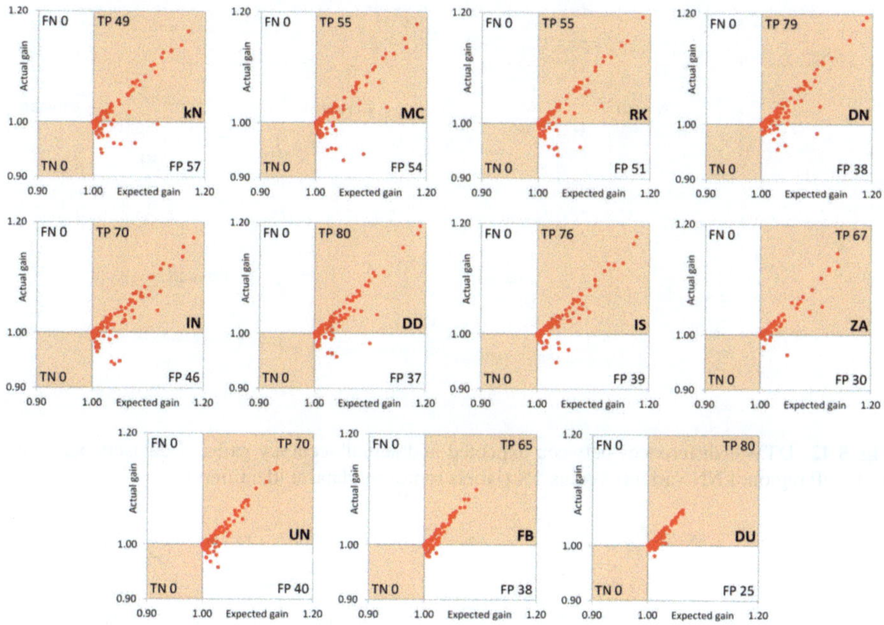

Fig. 5.41 DTW—comparing the actual and expected gain in classification accuracy: kNN variants versus 1N

5.5 Longest Common Subsequence (LCS)

The results of the preliminary analysis of the performance of different weighting schemes used in classification based on the NN rule are shown in Fig. 5.43 in the form of average accuracies for different numbers of nearest neighbors computed employing the SCV10x10 evaluation technique. Based on the shape of the plots and the position of the maximum values, three characteristic groups of kNN variants can be distinguished here as well.

The first group includes only kN whose performance decreased monotonically from $k = 1$ to $k = 30$, which suggests that it may not be able to surpass the performance of 1NN.

The second group consists of weighting techniques where, for smaller values of the parameter k, a slight increase in the average classification accuracy was observed

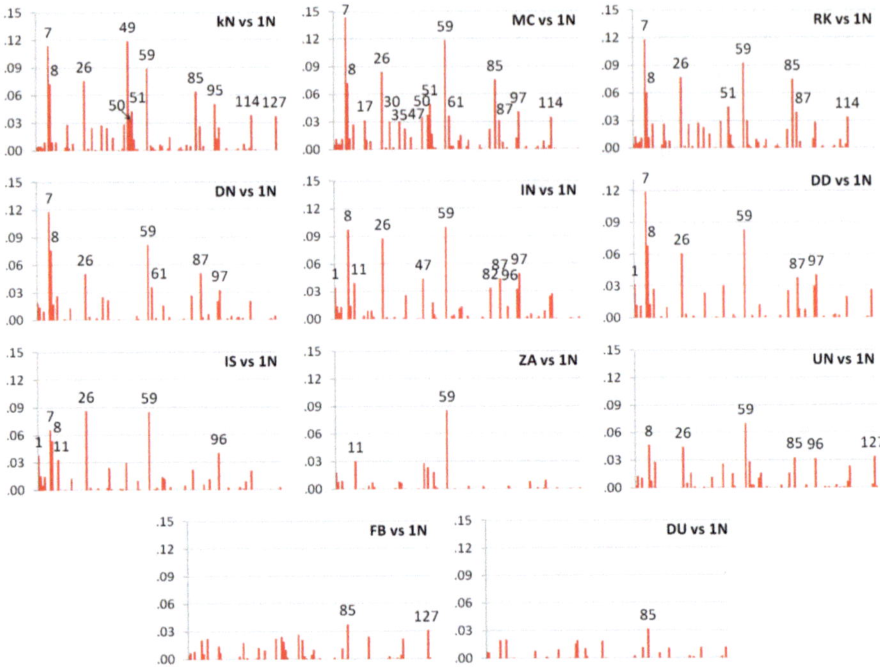

Fig. 5.42 DTW—differences between expected and actual accuracy gains of datasets belonging to the FP region: kNN variants versus 1N (labels represent dataset identifiers)

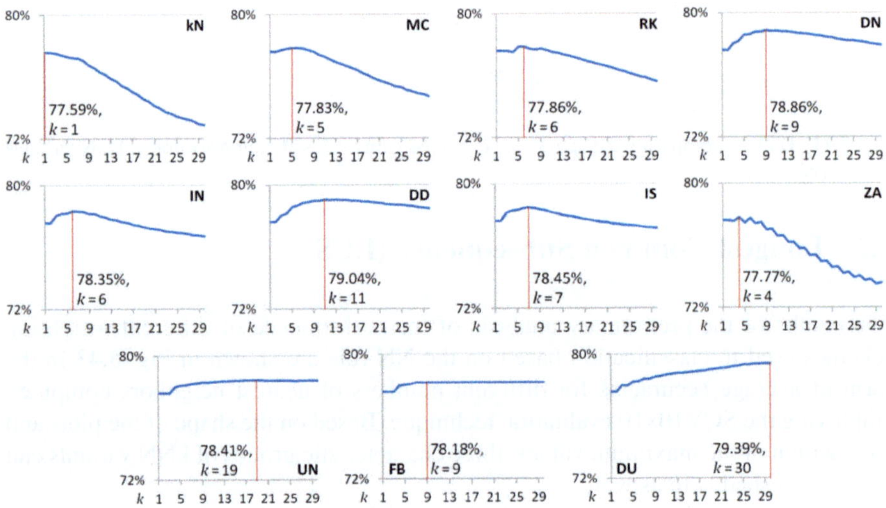

Fig. 5.43 LCS—average classification accuracy across all datasets for different numbers of nearest neighbors

compared to the result of 1NN: MC, RK, IN, IS and ZA. Based on this feature, it seems that these kNN variants might be able to outperform 1NN. After reaching the maximum, the performance of the members of the second group went into (almost) monotonous deterioration until $k = 30$, where their minimums were lower than the accuracy of 1NN. The alternation of growth and decline observed with DTW when using ZA was also present in combination with LCS: the classification accuracy with $2t$ number of nearest neighbors was always higher than that obtained with $2t - 1$ and the accuracy detected for $2t - 1$ was always lower than that observed for $2t - 2$, where $t = 2, \ldots, 15$. Thus, unlike the others, ZA did not reach its minimum at $k = 30$, but at $k = 29$.

The third group (DN, DD, UN, FB and DU) stands out for the peculiarity that, for any value of the parameter k greater than 1, their average classification accuracy was not lower than that obtained with 1NN. At the same time, it is also very easy to see the difference between the included kNN variants in that they showed different trends in the change of their results as the number of nearest neighbors increased. The curves describing the performance of DN and DD resemble those present in the members of the second group: a period of growth is followed by a period of decline. In the case of UN, and especially FB, after reaching the maximum, the curve becomes almost horizontal, i.e. the results were very close to the maximum. As with DTW, when using the DU weighting scheme, the increase in average accuracy was almost monotonic from $k = 1$ up to $k = 30$, with the advantage over the 1NN result being even more pronounced here. These characteristics of the members of the third group suggest that they could be successful in overpowering 1NN.

A summary of the various efficiency indicators of the analyzed variants of the kNN classifier obtained by applying ten times repeated tenfold stratified cross-validation where the optimal value of the parameter k was chosen over the corresponding training subset relying on the SCV1x9 evaluation method is given in Table 5.13.

In terms of average classification accuracy, 1N was outperformed by all the kNN variants discussed in this book. The biggest advantage was achieved by the DD and DN schemes, in the amount of about 2.2 and 2.1 percentage points (pp). They were followed by IS and IN with an approximate difference of 1.9 pp compared to 1N. On the other hand, 1NN had the smallest performance gap compared to RK, UN and FB.

Table 5.13 LCS—average value (AV) and standard deviation (SD) of classification accuracy (AV) and optimal number of nearest neighbors (k) across 12,800 splits; the number of datasets for which a given classifier produced the best result (WIN) and the mean rank (MR)

	DD	DN	IS	IN	MC	DU	ZA	kN	RK	UN	FB	1N
A_{AV} (%)	79.82	79.74	79.55	79.54	79.38	79.33	79.32	79.22	79.14	78.76	78.39	77.59
A_{SD} (pp)	19.38	19.43	19.46	19.44	19.54	19.69	19.59	19.61	19.89	20.47	20.69	20.76
k_{AV}	11.83	11.21	8.56	8.47	7.76	16.04	7.30	6.95	8.54	10.53	5.27	
k_{SD}	9.07	8.95	8.60	8.68	8.39	10.58	8.29	8.32	8.95	9.68	4.23	
WIN	38	24	19	18	15	27	12	14	14	13	15	25
MR	4.42	4.9	6.11	6.51	6.69	5.83	7.04	7.7	6.7	6.68	7.13	8.3

The largest standard deviations were present in the classification accuracies of the least successful classifiers (1N, FB, UN, RK) and the smallest in the most successful ones (IS, IN, DN, DD), but the differences were not large: all the deviations were in the interval of 20.07 ± 0.69.

Similar to DTW and EDR, the lowest average value of the parameter k came from the FB weighting scheme (5.27), which may not have been expected based on the preliminary results of the individual k values. In accordance with the initial results, the average of the optimal number of nearest neighbors was the highest when relying on the DU algorithm (16.04), followed by DD (11.83), DN (11.21) and UN (10.53), just as with the other elastic measures. The standard deviation was also the largest with these schemes and the smallest with FB.

With regard to the number of datasets on which the highest accuracy was achieved by a given variant of kNN, 1NN (25) was surpassed only by DN (38) and DU (27). In this category, RK (14), kN (14), UN (13) and ZA (12) proved to be the least powerful. When it comes to average ranks, DD, DN and DU were among the best, and FB, kN and 1N were among the worst classifiers.

As the optimal option, $k = 1$ was most often chosen by the majority voting NN approach (Fig. 5.44), as it could be guessed based on the initial analysis (Fig. 5.43). The first nearest neighbor had a very dominant role for the FB, RK and ZA weighting schemes as well, covering more than 30% of the 12,800 splits into testing and training subsets. Again, FB's strong reliance on the nearest neighbor disagrees with the preliminary results based on individual k-values. The smallest dependence on $k = 1$ was revealed for DU, DN and DD, where its share was less than 20%. Looking at the averages of the percentages of splits where the kNN variants chose $k = 1$ as the optimal number of neighbors, LCS produced the lowest value among the elastic measures.

At the level of individual datasets, *BME* (10), *Fungi* (46), *PigAirwayPressure* (86) and *TwoPatterns* (117) stood out in that all kNN variants in each of the 100 SCV10x10 iterations chose the first nearest neighbor as the best solution. However, it should be noted that this feature was also observed in several other datasets, but their list varied from scheme to scheme. The largest number of such datasets was observed with FB (13), kN (12), RK (12) and ZA (12) and the smallest with DN (4) and DD (4). In addition, excluding FB, the average of the optimal value of the

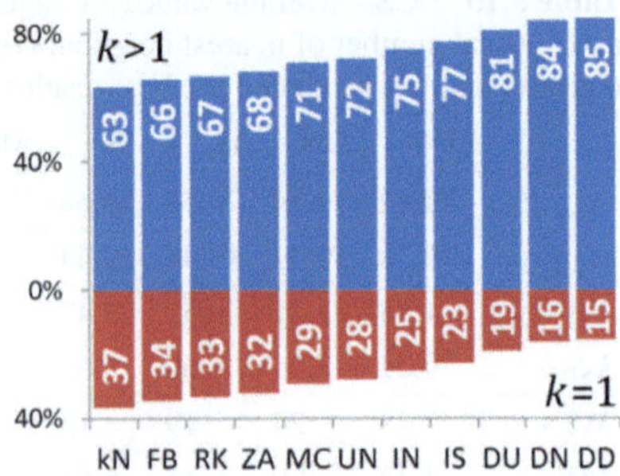

Fig. 5.44 LCS—percentages of splits in which SCV1x9 chose $k = 1$ and $k > 1$ as the optimal number of nearest neighbors on the training set when evaluating classification accuracy using SCV10x10

parameter k was greater than 20 for all kNN variants in the case of several datasets. Their number ranged between 49 (DU) and 15 (ZA) and datasets numbered 36, 43, 74, 75, 83, 92, 94, 109 exhibited this feature with each of the weighting schemes (except FB).

Figure 5.45 highlights that, except for DU, all other kNN variants preferred smaller numbers of nearest neighbors more than larger ones. MC, IN, IS and ZA opted for both $k = 3$ and $k = 4$ in slightly more than 10% of cases. Furthermore, kN chose $k = 3$ in a little more than 15% of the 12,800 splits and in nearly 18% of them, FB marked 4 as the optimal number of nearest neighbors while in almost 15% of the splits it opted for $k = 7$. The largest deviation in the frequencies (expressed in percentages) of different values of k was observed in the case of FB (4.4 pp), followed by ZA and kN (3.2 pp). The frequencies of different numbers of optimal nearest neighbors were most balanced with DD, DD and DU (the standard deviations were 1.7, 1.7 and 1.9 percentage points, respectively). However, given that LCS relied much less on the first nearest neighbor than DTW (compare Figs. 5.44 and 5.34), in order to capture more than 90% of the splits, except for FB, it was necessary to consider larger values of the parameter k (Fig. 5.46). This was especially pronounced with kN, MC, RK and ZA.

Comparing the numbers of wins and losses shown in Fig. 5.47, 1N was outmatched in the highest ratio by DU (4.3:1), DD (3.1:1) and DN (3.1:1). These ratios were the second best among the elastic measures analyzed (higher than them were observed only for EDR). In addition, analogous to EDR and unlike DTW and ERP, DU showed greater superiority over 1N in average classification accuracy as well, ranking in the middle of kNN variants based on their gain in average performance over 1N

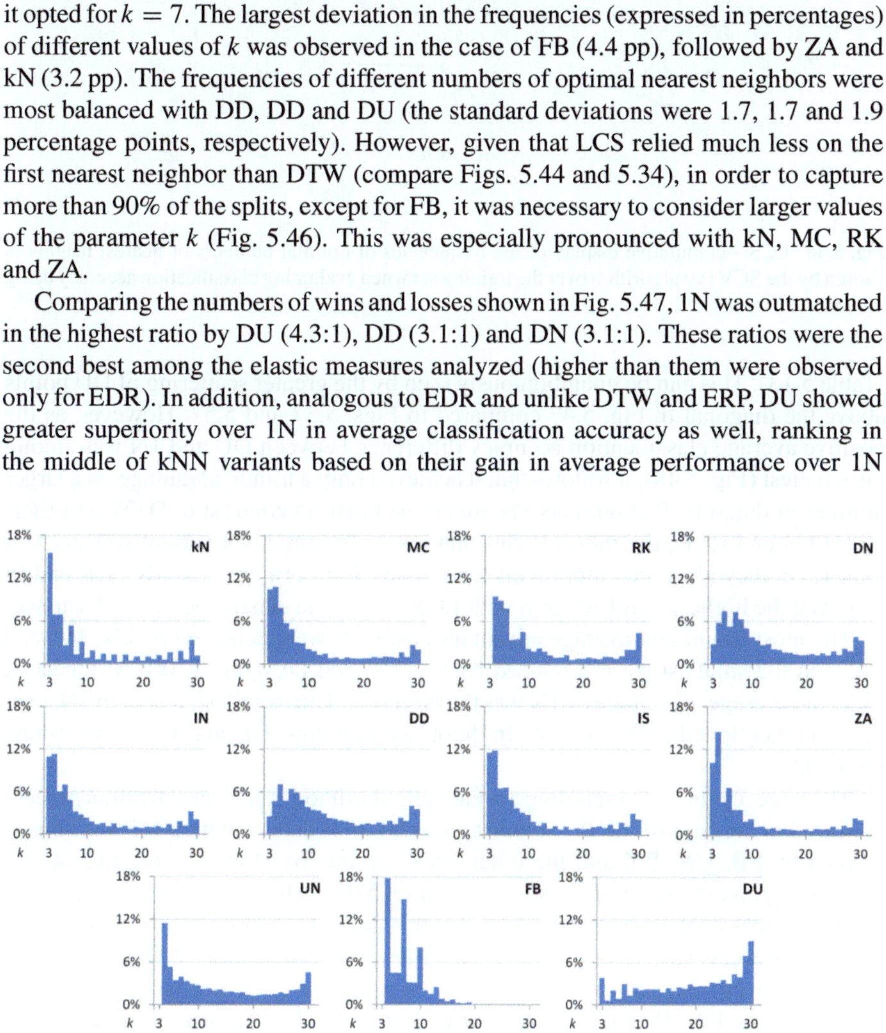

Fig. 5.45 LCS—frequencies (in percentage relative to the 12,800 splits) of different values of the parameter k ($\geq$ 3) chosen by the SCV1x9 algorithm over the training set when evaluating classification accuracy using SCV10x10

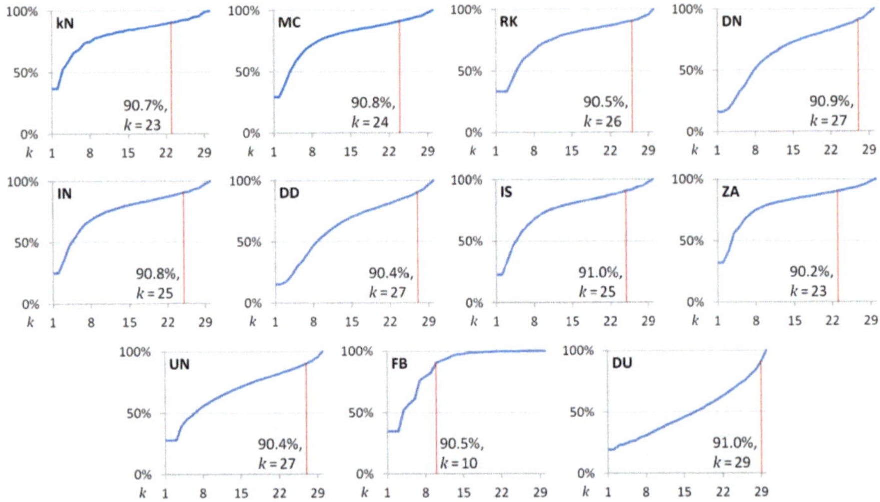

Fig. 5.46 LCS—cumulative display of the frequencies of optimal numbers of nearest neighbors chosen by the SCV1x9 algorithm over the training set when evaluating classification accuracy using SCV10x10

(Table 5.13). This can be unambiguously seen by the greater scattering of DU points above the diagonal in Fig. 5.47 compared to Figs. 5.37 and 5.57. However, as the positive average classification accuracy difference between DU and 1N was among the smallest (Fig. 5.48), it follows that it achieved only a minor advantage on a larger number of datasets. It should also be mentioned that, in contrast to DTW and ERP, with LCS and EDR, the ratio of wins and losses for any kNN variant was not less than 1, i.e. they all scored more wins than losses. Furthermore, kN, MC, RK and IS achieved the highest win-loss ratio with LCS among the elastic measures discussed.

The most prominent average advantages over 1N were achieved by kN, MC and ZA, and the smallest were produced by FB, UN and DU (Fig. 5.48). At the same time, the average advantage of 1N was the largest and the smallest, respectively, over the aforementioned kNN variants. In these two rankings, DD and DN were around the middle.

The largest gain from including a wider neighborhood in the classification process appeared in the case of the *ProximalPhalanxTW* (94) dataset where 1N was outperformed by FB, UN, RK and the other kNN variants by slightly more than 4, 10, 22 and 25 percentage points, respectively (Fig. 5.49). Moreover, the average of the optimal values of the parameter *k* was, for this dataset combined with LCS, greater than 27 for every weighting scheme except FB where it was about 9.5. In addition to LCS, it showed very similar properties when using the EDR similarity measure as well. kN, MC, DN, IN, DD, IS, ZA and DU proved to be very useful in classifying time series of the *Meat* (70), *MiddlePhalanxOutlineCorrect* (74), *OliveOil* (81) and *Strawberry* (109) datasets as well. The performance improvement compared to 1N in their case ranged between 13.5 and about 17.5 percentage points.

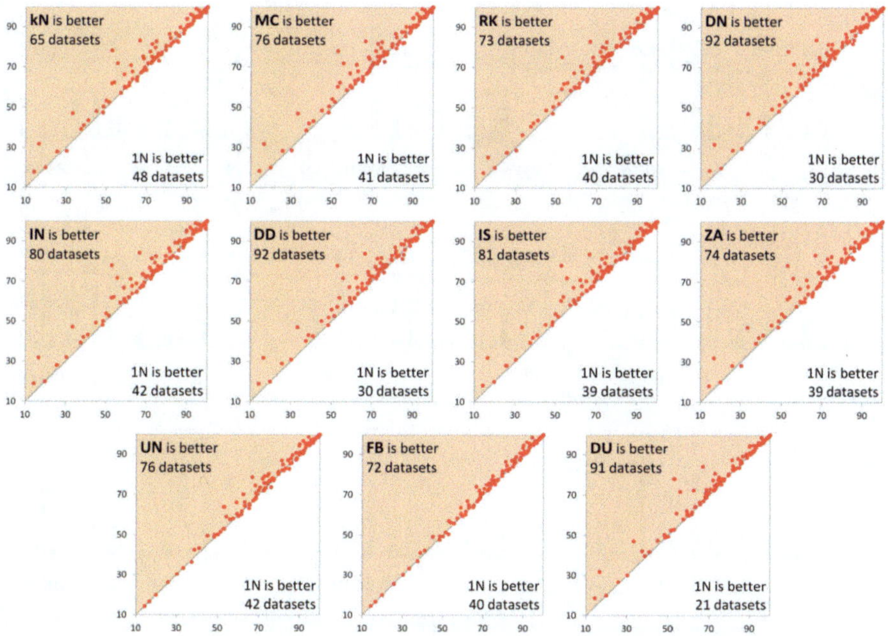

Fig. 5.47 LCS—comparison of the kNN variants with the 1N classifier based on the number of datasets where they achieved higher/lower classification accuracy

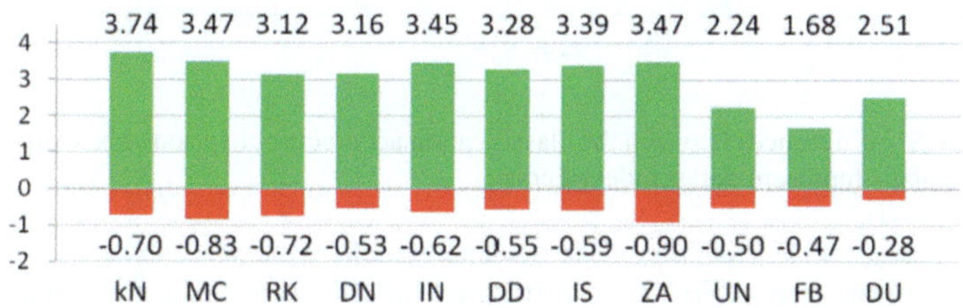

Fig. 5.48 LCS—average difference in classification accuracy of various kNN variants and 1N over the datasets where the performance of a given kNN variant was better (green) and worse (red) than 1N

1N did not surpass any KNN variant by more than 4 percentage points (Fig. 5.49), moreover, the amount of its biggest advantage over RK, UN and FB did not reach 2.5 pp and over DU it was even less than 1.5 pp. The difference in classification accuracy in favor of 1N was not less than 3 percentage points only for the following three datasets: *BeetleFly* (8) (kN, MC, IN, ZA), *InsectWingbeatSound* (64) (MC, DN, DD) and *WormsTwoClass* (127) (MC, ZA).

The result of the Friedman test confirmed the statistical significance of the difference in the accuracies of the considered NN classifiers ($\chi^2(11) = 137.62, p < .001$).

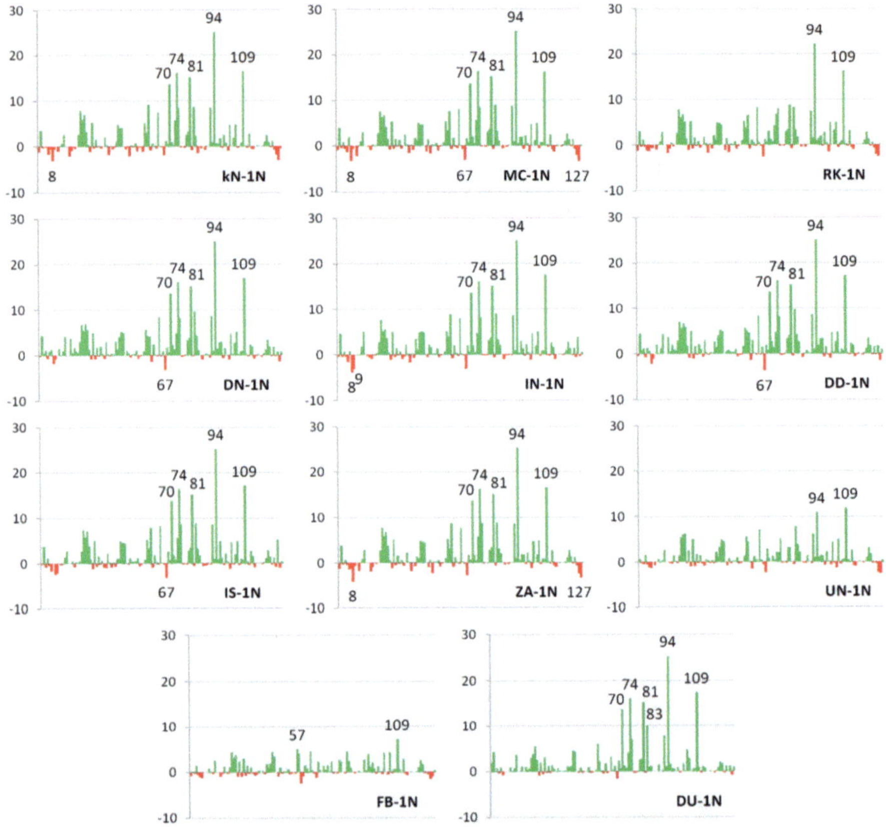

Fig. 5.49 LCS—differences between the classification accuracies of various kNN variants and 1N by datasets (labels represent dataset identifiers)

Nemenyi's post-hoc test (Fig. 5.50) attributed the difference to a statistically significant distinctness in performance between 1N and DD, DN, DU, IS and IN, between kN and DD, DN and DU, between DD, DN and FB, ZA, RK, MC and UN and between DD and IN.

According to the results of the Wilcoxon signed-rank test adjusted in agreement with the Holm-Bonferroni scheme (Table 5.14), the efficiency of 1N was significantly weaker than all kNN variants and the performance of DD was significantly better than all other NN classifiers, except DN. Furthermore, at the $\alpha = 0.05$ level, the performance of DN was not significantly better only compared to DU.

When it comes to the numbers of statistically significant wins and losses in Table 5.15, DD and DN noticeably surpassed the others and the lag of kN, FB and especially 1N was obvious.

The decision on the superiority of a given kNN variant over 1N established on their classification accuracy over the training data was most often wrong with kN, IN and RK (Fig. 5.51), in approximately 35%, 32% and 31% of the cases, respectively.

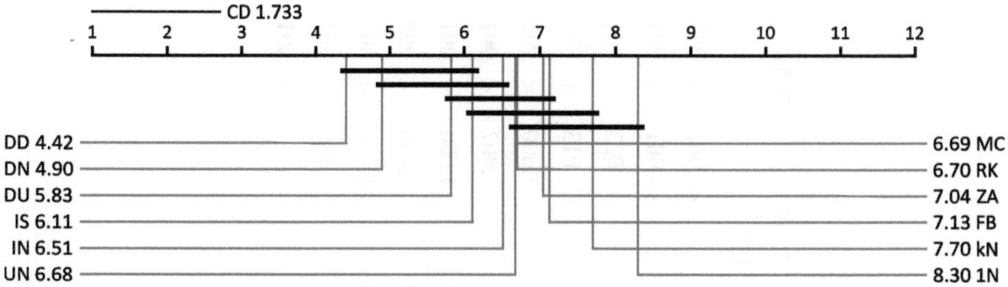

Fig. 5.50 LCS—critical difference diagram comparing average accuracy ranks based on Nemenyi's test: NN variants connected by a solid bar did not differ statistically significantly

Advantage overestimation was least frequent in the experiments performed with the DD, DN and DU weighting schemes where we misjudged in 23%, 21% and 16% of the datasets, respectively. The actual accuracy gain was the most consistent with the expected precisely in these three classifiers, in 74% (DN), 73% (DD) and 72% (DU) of the cases. The lowest degree of consistency was detected when relying on the RK (59%), ZA (59%), FB (58%) and kN (54%) methods.

Combined with the DU weighting scheme, when comparing the time series using the LCS measure, the difference between the expected and actual accuracy gain was not greater than 0.03 for any of the FP datasets (Fig. 5.52). Unlike the other distance measures, the number of datasets for which the difference exceeded 0.03 was the highest when using the ZA (9) and the lowest with the DN (2) weighting algorithm (excluding DU). The advantage of a given scheme was overestimated to the greatest extent in the case of the *GestureMidAirD3* (49) (kN, MC, RK, ZA), *Lightning2* (67) (DN, DD), *BeetleFly* (8) (IN, IS), *Herring* (59) (FB) and *Worms* (126) (UN) datasets. The largest average differences (calculated for all FP datasets of the given scheme) were observed for ZA, MC and RK and the smallest ones for DU, DN and UN.

5.6 Edit Distance with Real Penalty (ERP)

Figure 5.53 visualizes the change in the average accuracy of different variants of the kNN classifier as a function of the number of nearest neighbors in the interval from 1 to 30, calculated with ten times repeated tenfold stratified cross-validation. Analogously to DTW, ERP separates the different weighting approaches covered in this book into the same three characteristic categories, with the only difference being that here ZA moved from the second to the third group.

The first group is formed by the kNN variants whose accuracy showed a decreasing trend with the growth of the number of considered nearest neighbors: kN, RK, MC. As they all reached their maximum at $k = 1$, it seems that they may not be able to surpass 1NN in the following experiments either. The steepest decline was present with the majority voting kNN, where the difference between the minimum average

Table 5.14 LCS—Holm-Bonferroni-adjusted p-values of the pairwise Wilcoxon signed-rank test on classification accuracy

A		79.82	79.74	79.55	79.54	79.38	79.33	79.32	79.22	79.14	78.76	78.39	77.59
		DD	DN	IS	IN	MC	DU	ZA	kN	RK	UN	FB	1N
79.82	DD		.052	**.003**	**.002**	< **.001**	**.002**	< **.001**	< **.001**	< **.001**	< **.001**	< **.001**	< **.001**
79.74	DN			.029	**.010**	< **.001**	.141	< **.001**	< **.001**	< **.001**	< **.001**	< **.001**	< **.001**
79.55	IS				1.000	1.000	1.000	.055	< **.001**	1.000	1.000	.023	< **.001**
79.54	IN					1.000	1.000	.539	.042	1.000	1.000	.013	< **.001**
79.38	MC						1.000	1.000	.161	1.000	1.000	1.000	< **.001**
79.33	DU							1.000	.098	.967	1.000	**.007**	< **.001**
79.32	ZA								.440	1.000	1.000	1.000	< **.001**
79.22	kN									.440	1.000	1.000	**.008**
79.14	RK										1.000	1.000	< **.001**
78.76	UN											.085	< **.001**
78.39	FB												< **.001**
77.59	1N												

Note Bolded values indicate significance at the level of $\alpha = 0.01$

Table 5.15 LCS—statistically significant wins (W) and losses (L) reported by the corrected resampled *t*-test on classification accuracy (across 12,800 splits)

	DD	DN	DU	IS	IN	MC	RK	ZA	UN	kN	FB	1N
W	506	447	376	260	276	227	204	174	233	161	178	148
L	72	107	258	169	187	200	258	245	305	336	401	652
W–L	434	340	118	91	89	27	−54	−71	−72	−175	−223	−504

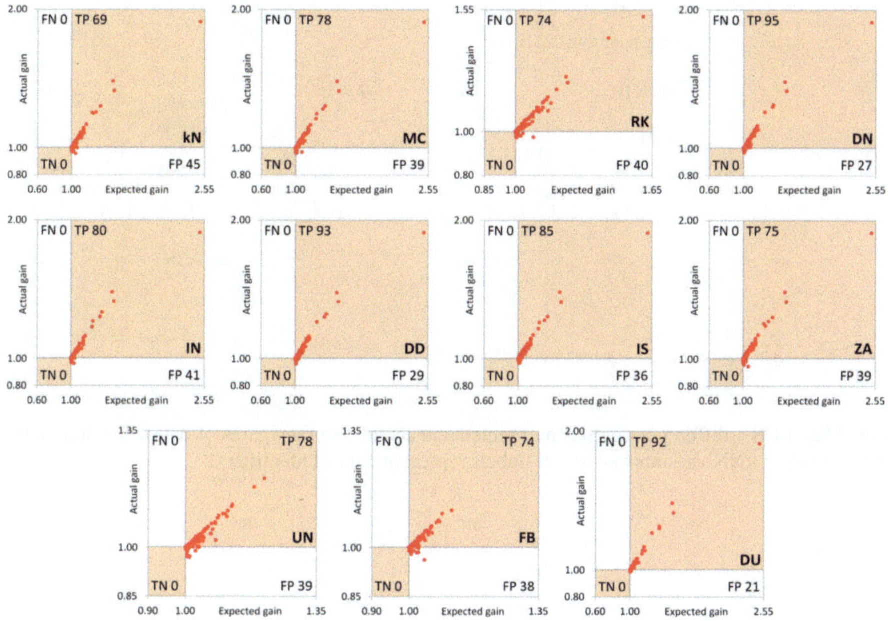

Fig. 5.51 LCS—comparing the actual and expected gain in classification accuracy: kNN variants versus 1N

accuracy at $k = 30$ and the maximum was even a bit greater than 8 percentage points (pp).

With the members of the second group (DN, IN, DD, IS), the initial phase in which the performance improved compared to $k = 1$, after reaching the maximum, turned into a monotonous deterioration. They all reached their minimum, which was below the accuracy of 1NN, with $k = 30$, whereby the intensity of the decline was the strongest in the case of IN, where the lag from the results with $k = 1$ reached nearly 4.4 pp. The period of growth was the shortest with IN (up to $k = 4$) and the longest with DD (up to $k = 8$). These preliminary observations suggest the conclusion that members of the second group may be capable of surpassing 1NN.

The performance of the third group of classifiers (ZA, UN, FB and DU) was not below the results of 1NN in the entire examined interval of *k*-values. However, there are noticeable differences in their behavior. After reaching the highest accuracy in

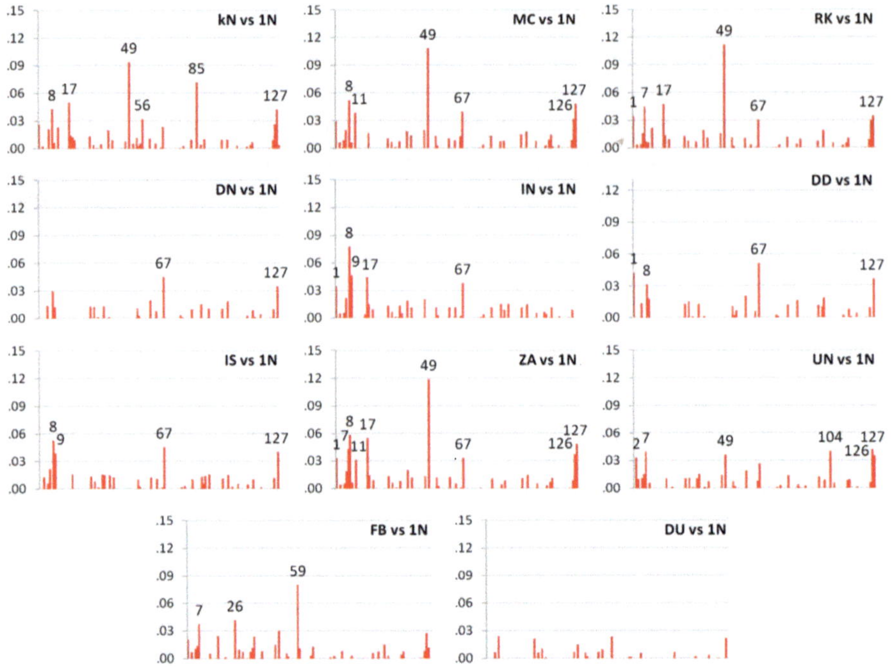

Fig. 5.52 LCS—differences between expected and actual accuracy gains of datasets belonging to the FP region: kNN variants versus 1N (labels represent dataset identifiers)

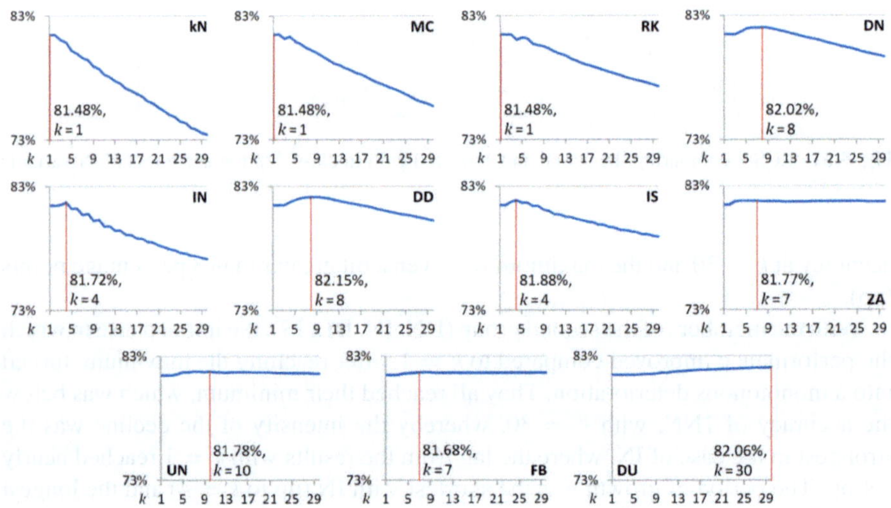

Fig. 5.53 ERP—average classification accuracy across all datasets for different numbers of nearest neighbors

$k = 7$, the deviation of ZA and FB from the result of 1NN stabilized and amounted to an almost constant value, with variations smaller than 0.02 pp. In contrast to them, after reaching the maximum at $k = 10$, the average classification accuracy of UN showed a decreasing trend, while with DU it continuously increased until $k = 30$. Based on the described properties, it seems that members of the third group could also achieve better performance than 1NN in the more detailed investigations that follow in the rest of this subsection and that UN combined with ERP will also prefer a wider area of nearest neighbors than other kNN variants.

Table 5.16 provides a comparative review of the efficiency of the analyzed NN classifiers obtained by applying the SCV10x10 algorithm in which, in each of the 100 splits of the datasets into a testing and a training subset, the best value of the parameter k was chosen relying on the SCV1x9 method so as to maximize the classification accuracy over the training subset.

The average accuracy of the procedure depending solely on the first nearest neighbor was outperformed by all the kNN variants, as with the other distance measures. However, the differences were very small, even the largest one, achieved by DD, was below one percentage point. Moreover, the advantage of each individual kNN variant was the lowest with ERP among the analyzed elastic measures. Again, DN and IS were also among the best-ranked schemes. In overcoming 1N, kN and ZA were the least successful, their advantage did not even reach 0.5 pp. The standard deviation varied around 17.66 pp in a 0.31 pp wide window. The largest deviations were generated by 1N, ZA, FB and DU, and the smallest by DN, IN, DD and IS. Similar to the intensity of the advantage over 1N, the deviation was the smallest among the considered the elastic measures when ERP was used to assess the similarity between time series.

The average value of the optimal numbers of nearest neighbors was the highest for DU (15.81), in accordance with the expectations derived from the results of the preliminary experiments with individual values of k. It was followed by DD (9.10), DN (8.71) and UN (8.70), which is also in line with the initial assessment. Since, when examining individual values of k, MC and RK were (together with kN) overpowered by 1NN, we expected that among the considered kNN variants they would rely on the smallest neighborhood, however, it turned out that they were overtaken by FB

Table 5.16 ERP—average value (AV) and standard deviation (SD) of classification accuracy (AV) and optimal number of nearest neighbors (k) across 12,800 splits; the number of datasets for which a given classifier produced the best result (WIN) and the mean rank (MR)

	DD	DN	IS	IN	UN	MC	RK	FB	DU	kN	ZA	1N
$\mathbf{A_{AV}}$ (%)	82.46	82.40	82.37	82.23	82.19	82.11	82.10	82.07	82.07	81.97	81.93	81.48
$\mathbf{A_{SD}}$ (pp)	17.36	17.43	17.35	17.43	17.59	17.62	17.60	17.72	17.72	17.70	17.85	17.97
$\boldsymbol{k}_{\mathbf{AV}}$	9.10	8.71	5.85	5.48	8.49	5.11	5.74	4.49	15.81	4.41	4.18	
$\boldsymbol{k}_{\mathbf{SD}}$	7.88	7.71	6.67	6.52	8.70	6.25	6.83	4.01	11.04	5.93	6.12	
WIN	29	22	19	17	20	20	23	23	39	22	34	39
MR	5.09	5.35	6.80	7.21	6.36	7.14	6.9	6.49	5.75	7.58	6.32	7.01

and ZA. The largest and smallest standard deviations were detected with the same weighting schemes where the average value of the parameter k was the largest and smallest.

ERP was the only elastic measure where no kNN variant achieved the best average accuracy on more datasets than 1N. Thus, the first place was shared by 1N and DD, with 39 wins and the third place was won by ZA with 34 wins. At the end of the list were MC, IS and IN with, respectively, 20, 19 and 17 wins. Despite that, according to the mean rank, DD, DN and DU turned out to be the best choices, and 1N, MC, IN and kN were the least successful.

Averaged over all the kNN variants, the degree of dependence on the first nearest neighbor was the most pronounced in the case of the ERP similarity measure (Fig. 5.54). $k = 1$ was chosen as the best option by ZA and kN in more than half of the 12,800 splits into testing and training subsets. For kN, a strong reliance on the first neighbor was expected since it generated weaker results than 1NN at the level of individual values of the parameter k (Fig. 5.53), but it seemed that this would not be the case for ZA. Besides them, the share of the first neighbor was higher than 40% also with RK, MC and FB. Again, for RK and MC, this was quite expected. In line with the preliminary expectations, DU, DN and DD relied the least on the first neighbor, in less than 30% of the cases.

When it comes to individual datasets, in each of the 100 iterations of the SCV10x10 procedure, SCV1x9 chose $k = 1$ as the best solution in combination with all the 11 weighting methods for *CBF* (12), *DiatomSizeReduction* (22), *Fungi* (46) and *TwoPatterns* (117). There were more datasets with the above-mentioned feature, but their list differed from method to method. Their number was the highest in kN (21), RK (19) and ZA (19), and the lowest in DN (5), DD (5) and DU (8), which is also in accordance with Fig. 5.54. On the other hand, there were also datasets for which the average of the optimal numbers of nearest neighbors was greater than 20. Excluding FB, where the average did not exceed 14 in any dataset, *MiddlePhalanxOutlineAgeGroup* (73), *MiddlePhalanxTW* (75), *ProximalPhalanxOutlineAgeGroup* (92), *ProximalPhalanxTW* (94) showed this peculiarity with all the other weighting schemes. The number of datasets with this specific feature was the highest with DU (53), UN (13) and DD (12), and lowest with kN (4) and MC (4).

Observing the frequencies of splits in which values of the parameter k greater than 2 were chosen as optimal (Fig. 5.55), it is immediately noticeable that DU

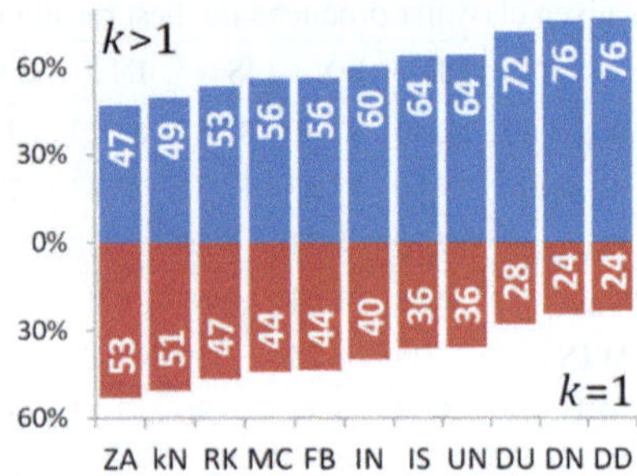

Fig. 5.54 ERP—percentages of splits in which SCV1x9 chose $k = 1$ and $k > 1$ as the optimal number of nearest neighbors on the training set when evaluating classification accuracy using SCV10x10

preferred values around 30, while the others more often opted for smaller numbers of neighbors. Specifically, $k = 4$ was chosen by FB in close to 19% and $k = 3$ in almost 18% and in somewhat more than 15% of the cases by ZA and kN, respectively. Both of these k-values occurred in more than 10% of the splits with both IS and IN and in addition, $k = 4$ was favored in more than 10% of iterations by MC, RK and UN as well. Analogously to DTW, $k = 3$ was not selected as the best solution in any of the divisions by any of the following schemes: RK, DN, DD, UN, FB and DU. The biggest differences in the frequencies of different numbers of nearest neighbors were recorded for FB and ZA with standard deviations of 4.0 and 3.5 percentage points. They were followed by IS, IN and kN with a deviation of approximately 3.2. The frequency was most evenly distributed with the DU and UN schemes, where the deviation was 2.0 and 2.1 pp.

The differences between the discussed kNN variants in the degree of preference for a smaller or larger neighborhood are visualized by the cumulative plots in Fig. 5.56. On the one hand, in slightly more than 90% of the splits, ZA and FB opted for $k \leq 10$ and kN for $k \leq 11$. On the other hand, in order to cover a similar percentage of splits, with UN it was necessary to increase the number of neighbors to 24 and with DU even to 29. Furthermore, comparing Figs. 5.36, 5.46, 5.56 and 5.66, it can be spotted that, among the examined elastic measures, these cumulative values were achieved with the smallest k-values with ERP.

In ranking the weighting schemes based on the numbers of datasets on which, compared to 1N, they achieved higher/lower classification accuracy, DU, ZA, DD and DN were among the best (Fig. 5.57). Their win-loss ratios, rounded to one decimal place, were 3.3:1, 1.9:1, 1.7:1 and 1.6:1, respectively. Similar to DTW, 1N

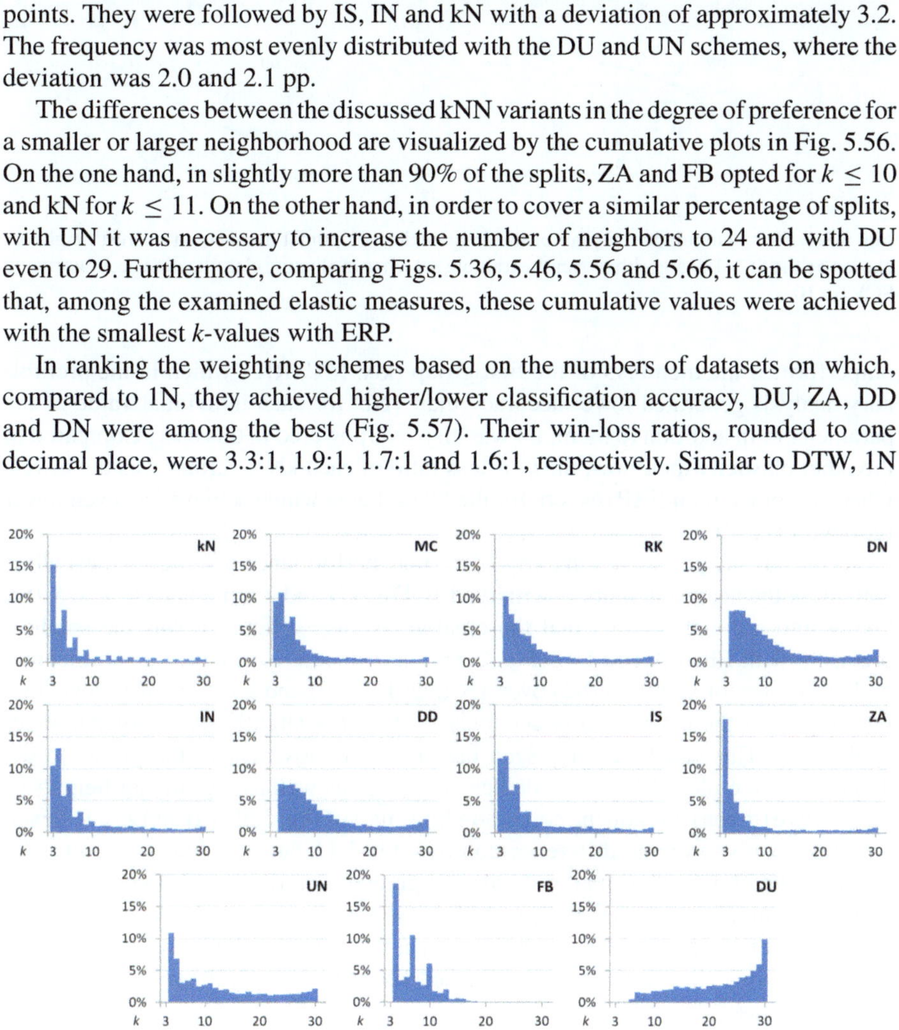

Fig. 5.55 ERP—frequencies (in percentage relative to the 12,800 splits) of different values of the parameter k (≥ 3) chosen by the SCV1x9 algorithm over the training set when evaluating classification accuracy using SCV10x10

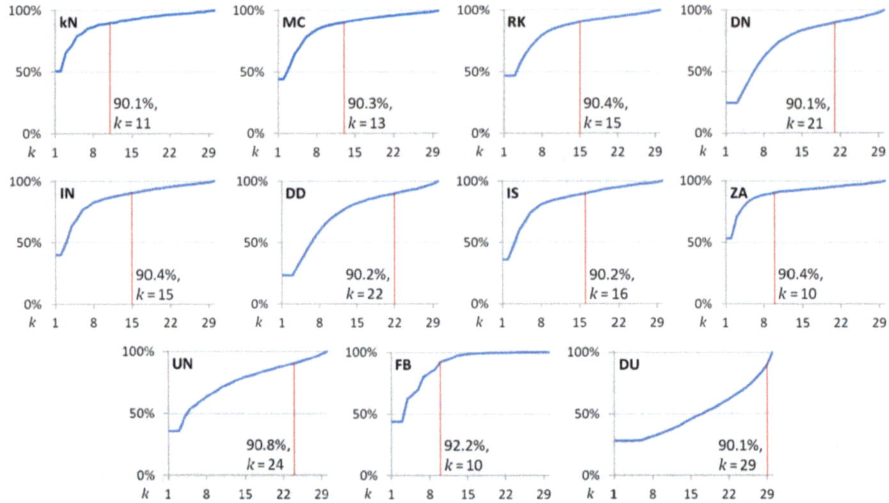

Fig. 5.56 ERP—cumulative display of the frequencies of optimal numbers of nearest neighbors chosen by the SCV1x9 algorithm over the training set when evaluating classification accuracy using SCV10x10

outperformed the members of the first group (kN, MC, RK), which in the preliminary analysis generated lower accuracy than 1NN for each individual value of the parameter *k*. In addition to them, IN was also overpowered in this case. Compared to the other elasticity measures, the win-loss ratio of each kNN variant was the lowest when combined with ERP (except for the DU scheme which achieved an even lower ratio with DTW).

As the advantage over 1N in terms of average performance was the smallest when weighting the nearest neighbors with the FB, DU, kN and ZA methods (Table 5.16), taking into account the fact that the positive average difference was the smallest precisely with FB, ZA and DU (Fig. 5.58), and since DU achieved a significantly higher number of victories (82) over 1N than FB (59) and ZA (53), it follows that DU improved the classification accuracy of 1N only slightly on a larger number of datasets. This can be clearly seen based on the closeness of the points to the diagonal on the plot of Fig. 5.57 dedicated to DU, as well as the smaller heights of the positive (green) lines on the corresponding chart of Fig. 5.59. FB and ZA achieved a higher positive average difference compared to DU, but with fewer datasets. ZA accomplished this with a smaller number of greater victories, and FB with a greater number of smaller victories, which can be easily seen in Fig. 5.59. In contrast, kN was characterized by the lowest number of wins with an even larger positive average difference, which means that it outperformed 1N more strongly than FB, ZA and DU on fewer datasets.

Comparing the positive average differences achieved using different elastic similarity measures it can be seen that they were lowest with ERP for each weighting scheme except for IN and IS which generated the lowest values with DTW.

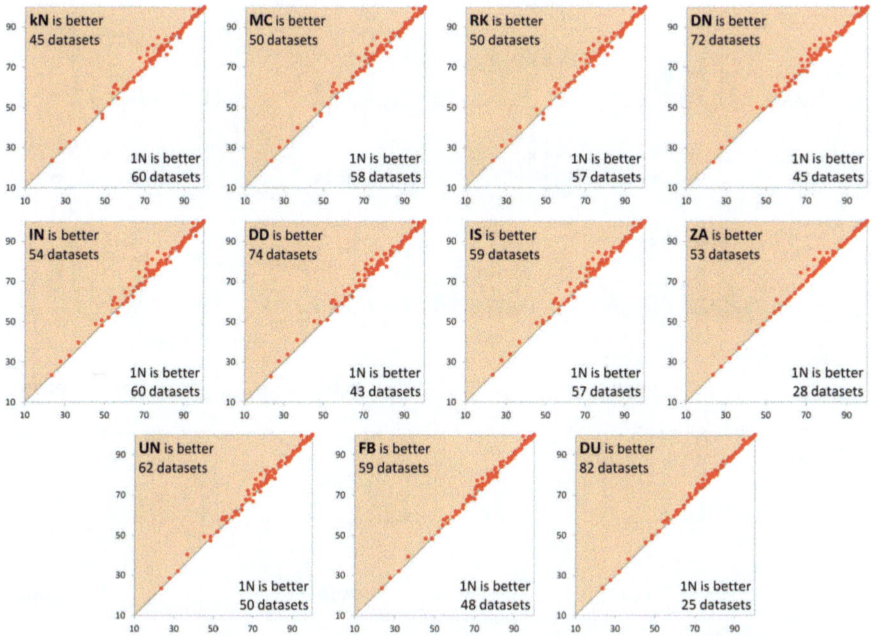

Fig. 5.57 ERP—comparison of the kNN variants with the 1N classifier based on the number of datasets where they achieved higher/lower classification accuracy

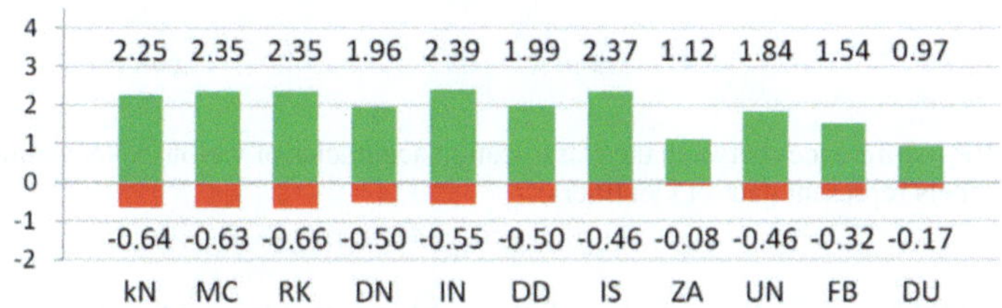

Fig. 5.58 ERP—average difference in classification accuracy of various kNN variants and 1N over the datasets where the performance of a given kNN variant was better (green) and worse (red) than 1N

The negative average difference was the smallest in DU and ZA (Fig. 5.58), where at the same time the number of wins of 1N was the smallest, which means that on the respective datasets, DU and ZA only slightly failed compared to 1N, as it is visually evident in Fig. 5.59. The biggest negative differences were detected when contrasting 1N with the members of the first group formed within the preliminary analysis: RK, kN and MC. Since the numbers of their losses were also among the highest, it follows that in the case of a larger number of dataset, they fell behind 1N to a greater extent.

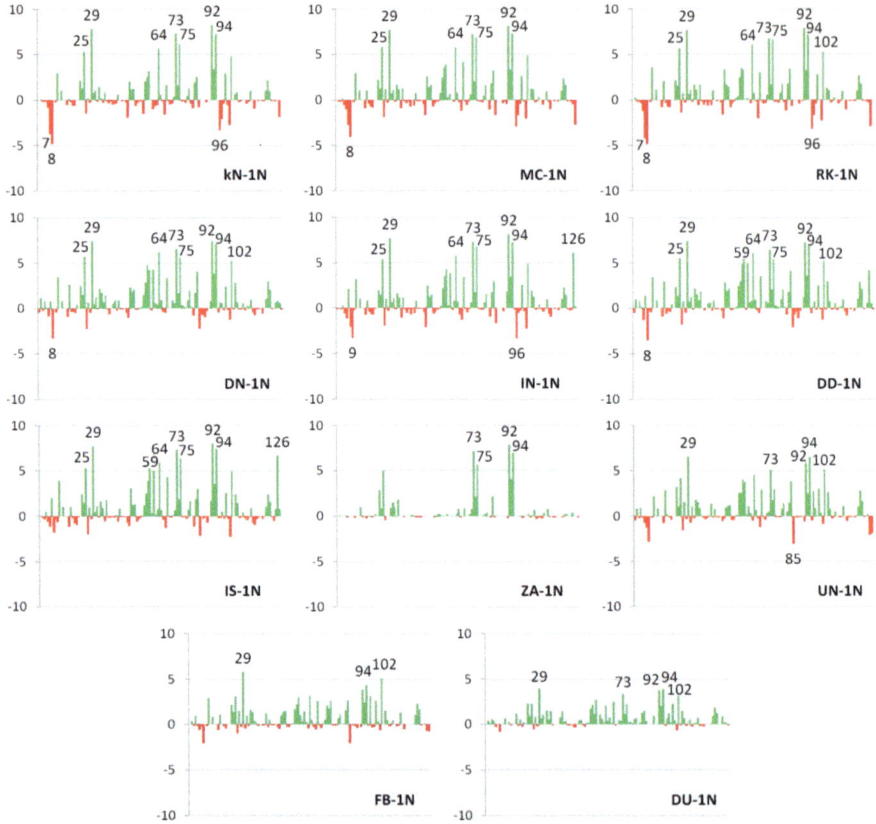

Fig. 5.59 ERP—differences between the classification accuracies of various kNN variants and 1N by datasets (labels represent dataset identifiers)

DD and DN were in the middle of the ranking in terms of both positive and negative average differences, with their number of wins being among the highest and their number of losses among the lowest.

The advantage of each of the considered variants of kNN was less than 8.3 percentage points over the accuracy of 1NN for each of the 128 UCR datasets, and in the case of DU it did not even reach 4 pp (Fig. 5.59). Very similar upper limits were observed for DTW as well. Except for FB and DU, all other weighting schemes outperformed 1NN by more than 5 pp over the following four datasets: *MiddlePhalanxOutlineAgeGroup* (73), *MiddlePhalanxTW* (75), *ProximalPhalanxOutlineAgeGroup* (92), *ProximalPhalanxTW* (94). If we disregard ZA and UN as well, this list will also include *DistalPhalanxTW* (25), *Earthquakes* (29) and *InsectWingbeatSound* (64).

By relying on 1N, the smallest benefit was achieved with ZA and DU, where the difference remained below 0.5 pp and 0.8 pp, respectively. They were followed by FB and IS, in whose case it was no higher than 2.2 pp. Only for the following datasets did the lag of the weighted kNNs exceed 3 pp compared to 1N: *Beef* (7) (kN, MC, RK), *BeetleFly* (8) (kN, MC, RK, DN, DD), *BirdChicken* (9) (IN), *Rock* (96) (kN, MC, RK, IN). It should be noted that all of them are among the smallest datasets of the UCR collection.

Despite the very close average accuracies of the NN classifiers, the Friedman test reported a statistically significant difference between them ($\chi^2(11) = 70.40$, $p < .001$). Nemenyi's post-hoc test, the results of which are presented in Fig. 5.60, testified that 1N was significantly different only from DD, and along with that, DD was also significantly different from RK, MC, IN and kN, furthermore, DN from MC, IN and kN and finally DU from kN. The Holm-Bonferroni-adjusted Wilcoxon signed-rank test detected a significant difference between 1N and DD, DN, UN, FB, DU, ZA and IS (at the $\alpha = 0.05$ level) (Table 5.17). In addition to 1N, according to this test, DD also differed significantly from IN, MC, RK, FB and kN. In the case of DN, the difference was, in addition to 1N, also significant in relation to MC, RK and kN, and IN and FB at the $\alpha = 0.05$ level. The numbers of statistically significant wins and losses in Table 5.18 confirmed, above all, the superiority of DD and DN over kN and 1N.

Among the four elastic measures, the prediction of the potential advantage of the kNN variants in relation to 1NN, based on the expected accuracy gain over the training data, was the most erroneous precisely when using the ERP technique for comparing time series. In more than 40% of the cases, the assessment that relying on a larger number of neighbors will yield higher accuracy on the test data turned out to be incorrect for almost half of the weighting schemes: kN (47%), MC (45%), IN (45%), RK (44%), IS (43%). The error rate was the lowest with ZA (23%) and DU (20%). The largest number of datasets where the actual accuracy gain was in line with the expected one was detected for DU (64%), DD (59%) and DN (57%), and the smallest number for the members of the first group (established within the initial experiment described at the beginning of this subsection): MC (41%), RK (40%), kN (35%). ERP proved to be the most inferior elastic measure in this regard as well.

The results of the analysis of the relationship between actual and expected gains in classification accuracy for the ERP measure are given in Figs 5.61 and 5.62. The largest difference between the expected and actual accuracy gains among the datasets in the FP region was the smallest when using the ZA, DU and FB schemes. In the

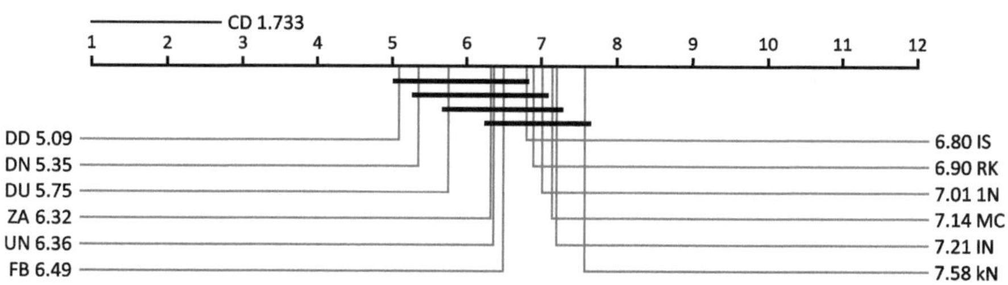

Fig. 5.60 ERP—critical difference diagram comparing average accuracy ranks based on Nemenyi's test: NN variants connected by a solid bar did not differ statistically significantly

Table 5.17 ERP—Holm-Bonferroni-adjusted p-values of the pairwise Wilcoxon signed-rank test on classification accuracy

A		82.46	82.40	82.39	82.23	82.19	82.11	82.10	82.07	82.07	81.97	81.93	81.48
		DD	DN	IS	IN	UN	MC	RK	FB	DU	kN	ZA	1N
82.46	DD		1.000	.070	**.001**	.066	< **.001**	**.002**	**.007**	.770	< **.001**	.608	< **.001**
82.40	DN			.833	.023	.292	**.003**	**.006**	.027	1.000	< **.001**	.518	< **.001**
82.39	IS				.694	1.000	1.000	1.000	1.000	1.000	.013	1.000	.042
82.23	IN					1.000	1.000	1.000	1.000	1.000	.188	1.000	.387
82.19	UN						1.000	1.000	1.000	1.000	.127	1.000	**.008**
82.11	MC							1.000	1.000	1.000	.190	1.000	1.000
82.10	RK								1.000	1.000	.085	1.000	1.000
82.07	FB									1.000	.135	1.000	**.008**
82.07	DU										.427	.070	< **.001**
81.97	kN											1.000	1.000
81.93	ZA												**.002**
81.48	1N												

Note Bolded values indicate significance at the level of $\alpha = 0.01$

Table 5.18 ERP—statistically significant wins (W) and losses (L) reported by the corrected resampled *t*-test on classification accuracy (across 12,800 splits)

	DD	DN	IS	UN	DU	IN	RK	MC	FB	ZA	kN	1N
W	363	351	217	253	308	177	180	158	186	223	119	130
L	91	97	148	195	258	187	212	190	218	310	278	481
W–L	272	254	69	58	50	−10	−32	−32	−32	−87	−159	−351

first case, it did not even reach 0.011, in the second case it was below 0.014 and in the third case it was slightly lower than 0.029. This can also be seen by the proximity of the points to the point with coordinates (1,1) in Fig. 5.61. Among the other weighting schemes, the most datasets with a difference greater than 0.03 were detected for kN, MC and RK (9), and the least for DN (4). The actual gain differed most from the expected in the *BeetleFly* (8) (MC, DN, DD, IS), *Beef* (7) (kN, RK, UN) and *Rock* (96) (IN) datasets. Except for ZA, DU and FB, *DodgerLoopDay* (26) produced more than 0.03 higher expected than actual gain for all the kNN variants. The average difference in the FP region was, as expected, the smallest for ZA, DU and FB, and the largest for MC, RK and kN.

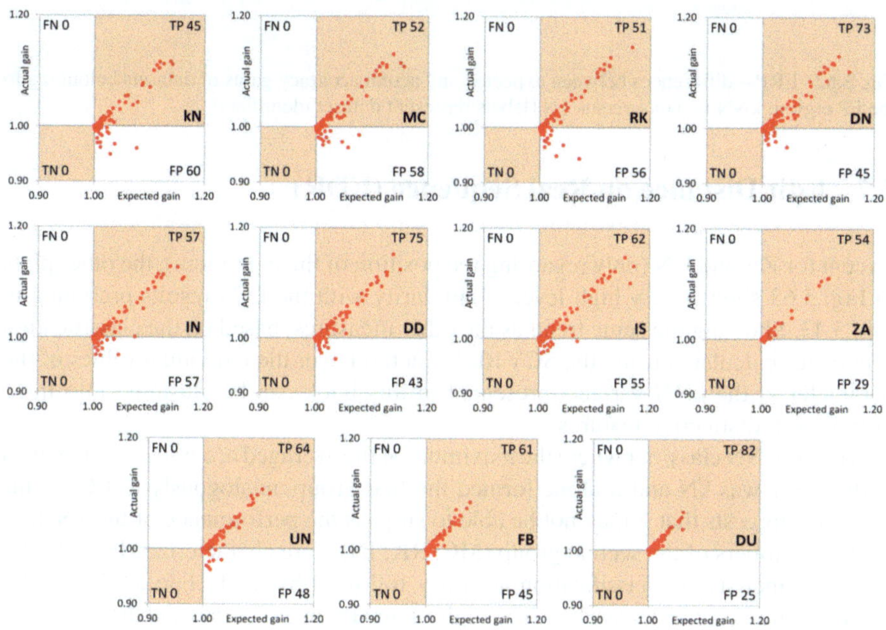

Fig. 5.61 ERP—comparing the actual and expected gain in classification accuracy: kNN variants versus 1N

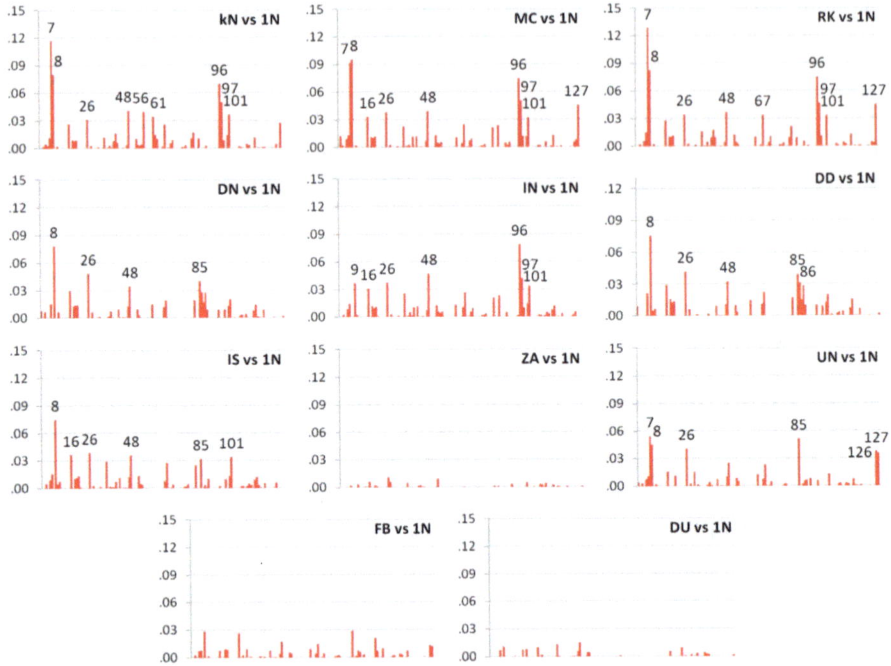

Fig. 5.62 ERP—differences between expected and actual accuracy gains of datasets belonging to the FP region: kNN variants versus 1N (labels represent dataset identifiers)

5.7 Edit Distance on Real Sequence (EDR)

Except for ZA and UN (only regarding the position of the maximum), the other plots in Fig. 5.63 show a very high level of similarity with the LCS results presented in Fig. 5.43. Showing the same trend as the other measures, based on the classification accuracies calculated using the SCV10x10 method over the individual values of the parameter k, the kNN variants presented in this book can be clustered into three groups with distinctive features.

The only NN classifier where the maximum of the averaged accuracy was achieved with $k = 1$ was kN and it alone formed the first group, analogously to LCS. This property suggests that it may not be able to surpass the performance of the 1NN.

The members of the second group (MC, RK, IN) were characterized by a shorter period of increase in classification accuracy followed by its decline until $k = 30$, where the minimum was reached. The highest maximum (77.81%) and the minimum (76.18%) that differed the least from the 1N result (77.06%) was generated by IN, while the performances of MC and RK were very similar. It appears that these weighting schemes may outperform the 1NN classifier.

The third group was the most numerous with EDR, it included as many as seven of the eleven discussed kNN techniques: DN, DD, IS, ZA, UN, FB, DU. In their case,

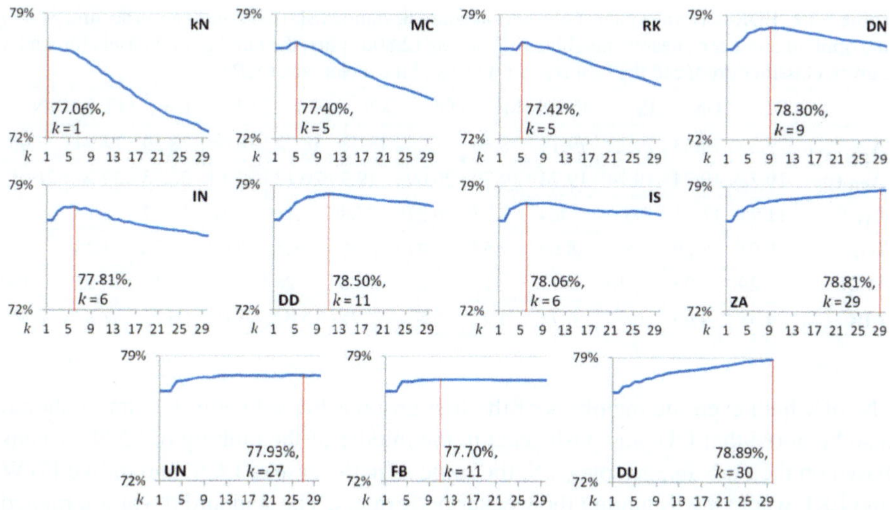

Fig. 5.63 EDR—average classification accuracy across all datasets for different numbers of nearest neighbors

every value of the parameter k (at least up to 30) achieved a higher average accuracy than 1NN, making them good candidates for overriding it. However, it is easy to recognize that they differed by the trends in the change of their performance with the growth of the number of nearest neighbors. In DN, DD and IS, the period of growth was followed by a period of decline, but, unlike the members of the second group, the result obtained with $k = 1$ retained the status of the minimum. After reaching the maximum, UN and FB produced almost constant average accuracies just below the maximum. Unlike them, the results of ZA and DU increased until $k = 29$ and $k = 30$, respectively.

In examining the relationship between the performance of different weighting schemes and the simple 1NN classifier, we relied on the SCV10x10 algorithm, whereby we tuned the number of nearest neighbors of the kNN variants over the training subset with the SCV1x9 method in each of the 100 splits into testing and training subsets. A summary of the obtained results is given in Table 5.19.

The advantage of a larger number of neighbors over favoring the first nearest one, in terms of average classification accuracy calculated over the 128 datasets of the UCR repository, was also confirmed in combination with the EDR distance measure: all kNN variants achieved somewhat better results than 1N. With the exception of ZA (where LCS generated a slightly higher gain over 1N than EDR) and DU (where LCS and EDR were tied), for all other weighting schemes EDR achieved the largest differences over 1N among the elastic measures considered in this the book. As with DTW, LCS and ERP, the best weighting schemes proved to be DD, DN and IS, achieving approximately 2.33, 2.27 and 2.16 percentage points higher average accuracy than 1N, respectively. Contrary to expectations based on the preliminary results, UN (1.23 pp) and FB (0.88 pp) showed the smallest advantage compared to

Table 5.19 EDR—average value (AV) and standard deviation (SD) of classification accuracy (AV) and optimal number of nearest neighbors (k) across 12,800 splits; the number of datasets for which a given classifier produced the best result (WIN) and the mean rank (MR)

	DD	DN	IS	IN	MC	DU	kN	ZA	RK	UN	FB	1N
$\mathbf{A_{AV}}$ (%)	79.39	79.33	79.22	79.14	78.96	78.8	78.78	78.72	78.68	78.29	77.94	77.06
$\mathbf{A_{SD}}$ (pp)	19.72	19.74	19.74	19.74	19.79	20.09	19.9	20.12	20.14	20.75	20.94	21.01
$k_{\mathbf{AV}}$	11.79	11.21	9.05	8.4	7.73	16.27	7.08	7.13	8.58	10.47	5.16	
$k_{\mathbf{SD}}$	9.03	8.89	8.9	8.69	8.53	10.47	8.46	8.5	9.08	9.56	4.25	
WIN	29	23	16	13	13	27	8	28	11	19	13	18
MR	4.35	4.65	6.39	6.71	6.92	5.82	7.75	5.87	7.07	6.77	7.32	8.36

1N, of which even the members of the first group achieved better results. It should also be noted that DU was positioned in the middle of the ranking of kNN variants based on their advantages over 1N, the same as in the case of LCS, and unlike DTW and ERP where it was among the weaker performers. The standard deviation ranged in a 0.65 pp wide band around 20.37 pp. It was the largest for 1N, FB and UN and the smallest for IS, IN and DD.

In accordance to the initial analysis based on individual values of k, the average of the optimal numbers of nearest neighbors was the highest with DU (16.27), followed by DD (11.79) and DN (11.21), in the same order as with LCS and ERP. Although it might not be expected based on the preliminary results, FB and ZA, in addition to kN, relied on the smallest neighborhood, as in the case of the other elastic measures. The standard deviation in the optimal values of k was the smallest with these three weighting schemes and the largest with DU, UN and RK.

EDR visibly stood out among the elastic measures in terms of the number of kNN variants that surpassed 1N with regard to the number of datasets on which they achieved the highest accuracy. As many as 5 out of 11 of them were better than 1N in this context: DD, ZA, DU, DN and DU. In this competition, IN, MC, FB, RK and kN were the least successful. According to the mean rank, DD, DN and DU topped the list, in the same way as with LCS and ERP. The lowest mean ranks were obtained for FB, kN and finally for 1N.

Regarding the proportions of the 12,800 splits where the discussed weighting schemes preferred $k = 1$ (Fig. 5.64), EDR showed very similar results with LCS (Fig. 5.44). Members of the first group (which were outperformed by 1N for $k > 2$ in the initial analysis) were among the schemes that relied the most on $k = 1$. Besides them, in more than 30% of the splits, both ZA and FB opted for only one neighbor, which is quite unexpected considering the initial experiment, especially when it comes to ZA. In accordance with the anticipations, the first nearest neighbor was chosen as the optimal solution the least often by DU, DN and DD. In their case, the proportion of such splits did not even reach 20%.

The number of datasets with the feature that all the kNN variants in each of the 100 iterations within the SCV10x10 evaluation procedure chose $k = 1$ was the smallest in combination with EDR among the elastic measures. There were only two

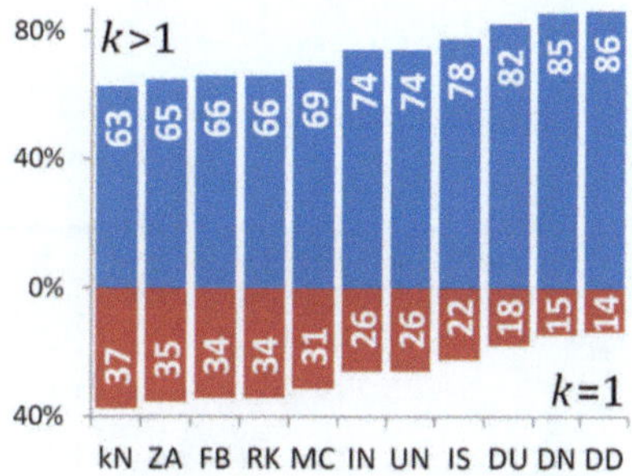

Fig. 5.64 EDR—percentages of splits in which SCV1x9 chose $k = 1$ and $k > 1$ as the optimal number of nearest neighbors on the training set when evaluating classification accuracy using SCV10x10

such datasets here: *Fungi* (46) and *PigAirwayPressure* (86). Looking at individual weighting schemes, the number of datasets in which the optimal number of neighbors for a given classifier was one varied from scheme to scheme. They were present in the largest number when applying ZA (13), FB (11) and UN (8) and in the smallest number with DN, DD and IS (2). At the same time, EDR produced the largest number of datasets for which the average value of the parameter k was greater than 20 for each kNN variant (except for FB). Those marked with numbers 36, 70, 73, 74, 75, 83, 92, 94 and 109 had this attribute. The datasets with the specified feature were the most numerous with DU (53), DD (27), DN (24) and UN (24) and the least numerous with kN (15), MC (16) and IN (16).

As with the other elastic measures, when choosing the optimal number of nearest neighbors, unlike the other weighting techniques, the DU scheme noticeably more often opted for higher values (Fig. 5.65). Among the k-values greater than 1, FB the most preferred 4, in even slightly more than 22% of the splits. In more than 10% of the cases, $k = 4$ was also chosen by MC, RK, DN, IN, IS and UN and $k = 3$ was strongly relied upon by kN, IN, IS and ZA, IS. The frequency of different optimal numbers of neighbors was most unevenly distributed with FB, ZA and kN, where the standard deviation was the highest and amounted to 4.9 pp, 3.3 pp and 3.2 pp, respectively. Deviations in frequency distribution were the most moderate in DU (1.7 pp), DD (1.9 pp) and DN (2.2 pp).

The apparent high degree of similarity between EDR and LCS is easily seen by comparing Figs. 5.66 and 5.46 which show the percentages of those splits where the optimal value of k was less than or equal to the given number on the horizontal axis of the plots. In order to cover more than 90% of the splits, each weighting scheme had to consider the same number of nearest neighbors as in the case of LCS.

As can be seen in Fig. 5.67, from the point of view of the analyzed weighting schemes, the ratio of wins and losses was the most favorable in the case of DU (4.8:1), ZA (4.4:1), DD (3.5:1) and DN (3.5). However, similar to LCS, the win-loss ratio was greater than 1 in the case of the other kNN variants as well. Furthermore, except for kN, MC, RK and IS with which LCS achieved better win-loss ratios, the other schemes achieved the best ratio with EDR. As a result, EDR produced the highest average win-loss ratio (around 2.5), while the deviation was also the highest here (around 1.3).

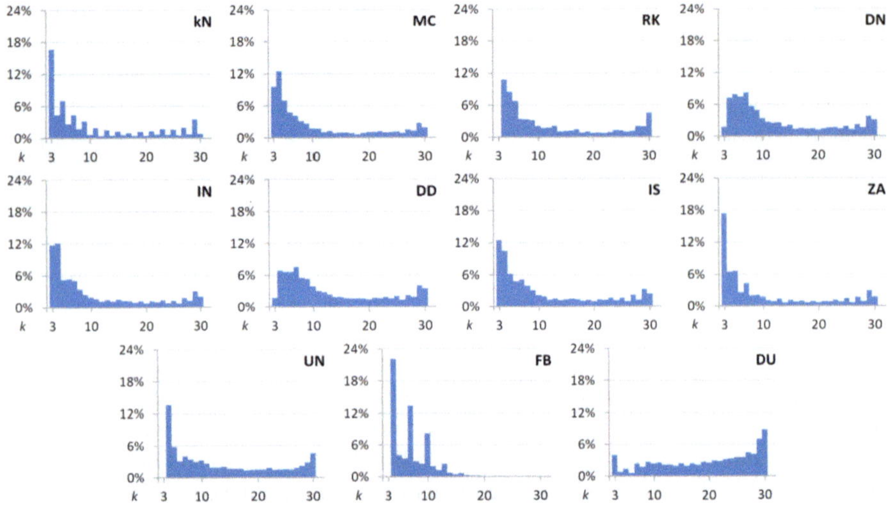

Fig. 5.65 EDR—frequencies (in percentage relative to the 12,800 splits) of different values of the parameter k (≥ 3) chosen by the SCV1x9 algorithm over the training set when evaluating classification accuracy using SCV10x10

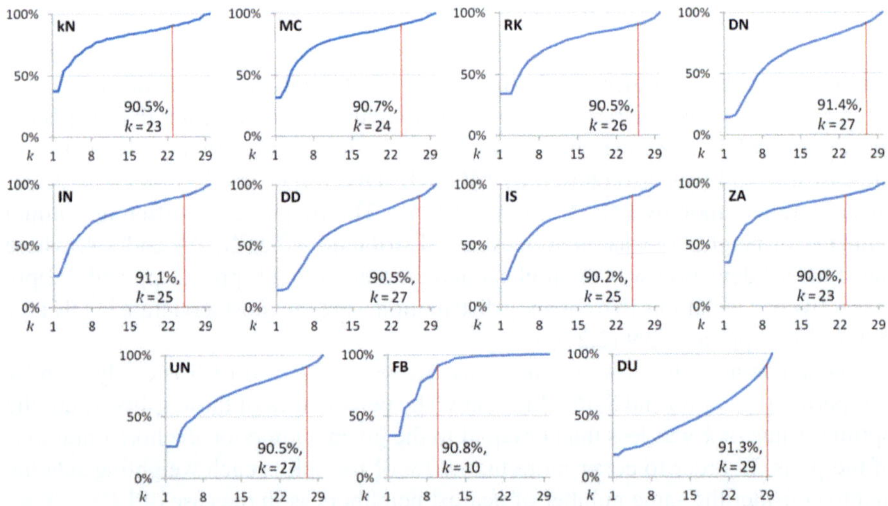

Fig. 5.66 EDR—cumulative display of the frequencies of optimal numbers of nearest neighbors chosen by the SCV1x9 algorithm over the training set when evaluating classification accuracy using SCV10x10

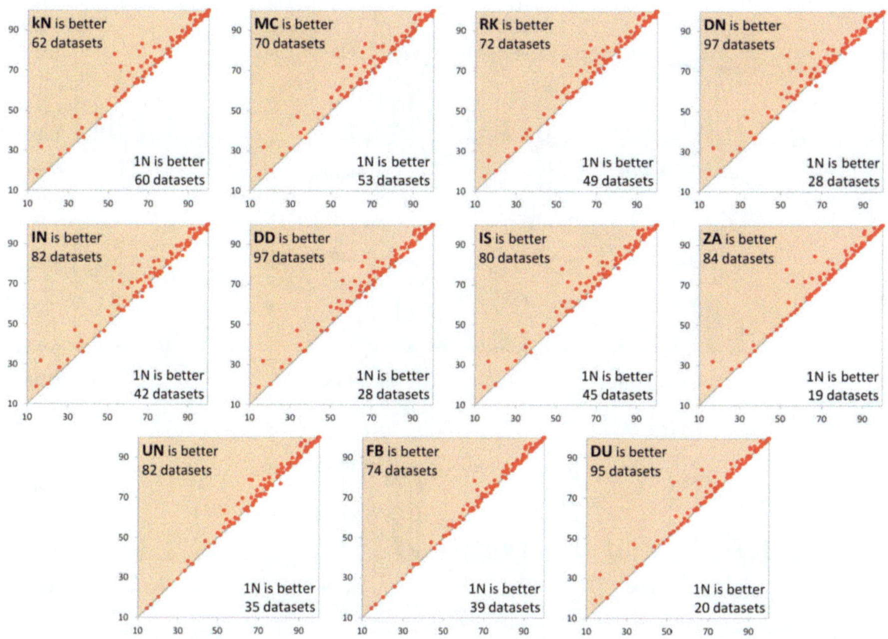

Fig. 5.67 EDR—comparison of the kNN variants with the 1N classifier based on the number of datasets where they achieved higher/lower classification accuracy

It should also be noted that among the elastic measures, DU achieved the highest number of victories and the lowest number of defeats with EDR. In doing so, its positive average difference (Fig. 5.68) was more than twice as large as when using DTW (Fig. 5.38) and ERP (Fig. 5.58), which means that with EDR it achieved a significantly higher gain over 1N on a larger number of datasets. Visually, this can be seen by the greater distance of the points from the diagonal in Fig. 5.67 and by the greater number of higher positive (green) lines in Fig. 5.69. It behaved similarly in combination with LCS as well.

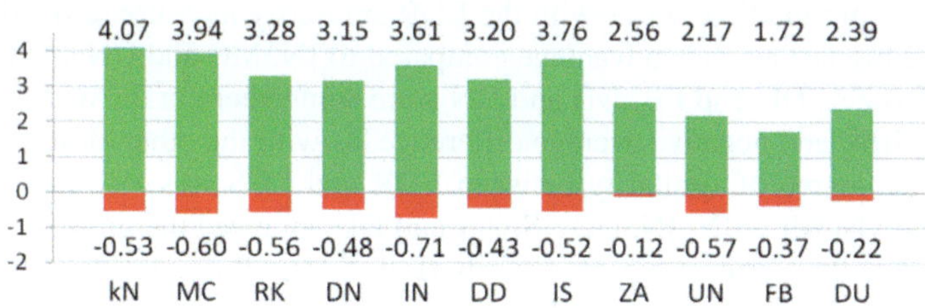

Fig. 5.68 EDR—average difference in classification accuracy of various kNN variants and 1N over the datasets where the performance of a given kNN variant was better (green) and worse (red) than 1N

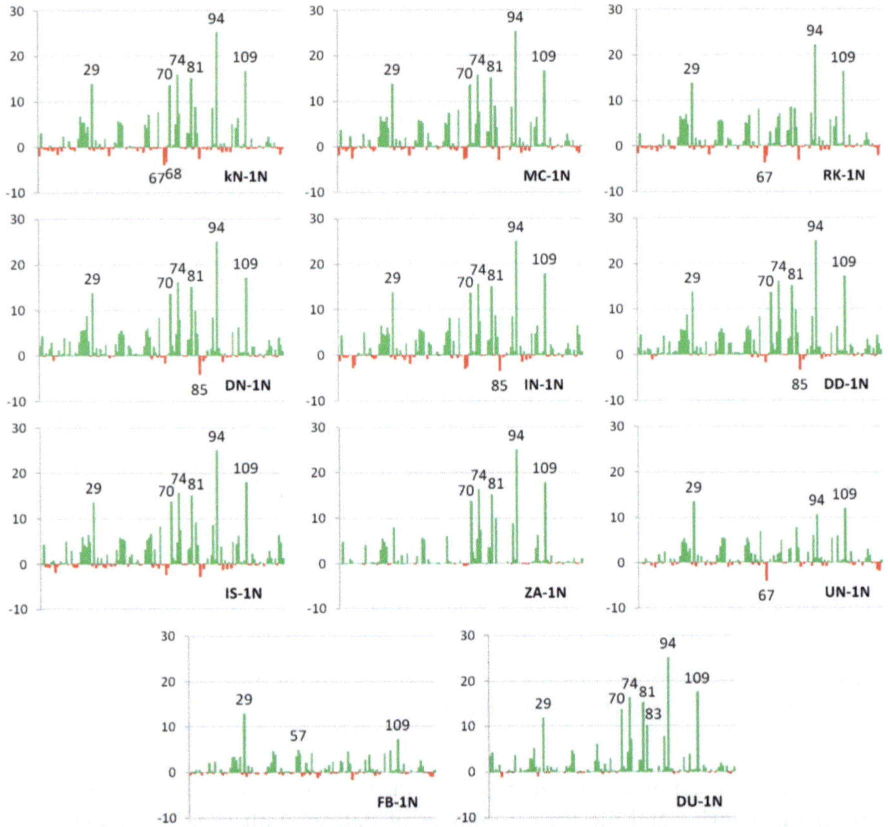

Fig. 5.69 EDR—differences between the classification accuracies of various kNN variants and 1N by datasets (labels represent dataset identifiers)

The largest positive average differences were achieved by kN, MC, IS and IN and the smallest by ZA, DU, UN and FB (Fig. 5.68). Along with that, the average of 11 positive average differences was the highest among all 7 analyzed distance measures when comparing the time series with the EDR similarity measure (around 3.1). 1N showed the highest average advantage compared to IN, MC and UN and the lowest compared to ZA, DU and FB. DD and DN were somewhere in the middle in terms of both positive and negative average difference, as with the other elastic measures.

The high degree of similarity between EDR and LCS was also manifested by the fact that kN, MC, DN, IN, DD, IS, ZA and DU achieved the greatest advantage over 1N with the same group of datasets (Fig. 5.69): *ProximalPhalanxTW* (94), *Meat* (70), *MiddlePhalanxOutlineCorrect* (74), *OliveOil* (81), *Strawberry* (109). Moreover, even the gains over the members of this group were very close: the largest difference between EDR and LCS was 1.22 pp in favor of EDR in the case of the *Strawberry* (109) dataset when using the ZA weighting scheme. The advantage over 1N in the case of *ProximalPhalanxTW* (94) was greater than 25 pp for the listed kNN variants and for the other members of the group it ranged between 13.6 pp and 18 pp, similar to LCS. EDR and LCS differed noticeably in their results on the *Earthquakes* (29) dataset: combined with EDR, almost all weighting schemes (with the exception of ZA and DU) outperformed 1N by more than 13 pp, and when using LCS the difference was less than 5.5 pp.

The performance difference in favor of 1N on no dataset was greater than 4.1 pp over any considered kNN variant. An advantage greater than 3 pp was achieved by 1N only over the following datasets: *Lightning2* (67) (kN, RK, UN), *Lightning7* (68) (kN) and *PickupGestureWiimoteZ* (85) (DN, IN, DD). LCS differed from EDR in this respect as it crossed the 3 pp limit over other datasets (8, 9, 67 and 127). Looking at Fig. 5.69, it is easy to see that the classification accuracy benefited the least from 1N when relying on ZA, DU and FB schemes, where the upper limit of the difference was the smallest (0.44 pp, 1.0 pp and 1.6 pp, respectively).

The statistical significance of the difference in the performance of the analyzed NN classifiers was examined and supported by the Friedman test ($\chi^2(11) = 157.67$, $p < .001$). Nemenyi's post-hoc test (Fig. 5.70) revealed that 1N was significantly different from DD, DN, DU, ZA and IS, as well as kN from DD, DN, DU and ZA and finally DD and DN also from UN, MC, RK and FB.

According to the Wilcoxon signed-rank test (adjusted with the Holm-Bonferroni scheme), all the kNN variants showed a statistically significant difference compared to 1N, except for kN (Table 5.20). In addition, DD was not significantly different from DN alone and vice versa (with the difference between DN and IS, IN and ZA being significant only at the $\alpha = 0.05$ level). Statistical significance was also revealed in the performance of kN in comparison with MC and at the $\alpha = 0.05$ level in comparison with IS and IN. Also, FB differed from IS and UN (at the $\alpha = 0.05$ level) as well.

The results of the corrected resampled *t*-test shown in Table 5.21 confirmed the superiority of DD, DN and IS, especially in relation to RK, kN, FB and 1N.

According to Fig. 5.71, the prediction that 1N would be outperformed on the test data based on the fact that the kNN variants achieved higher classification accuracy over the training data turned out to be wrong to the greatest extent when using the kN, MC and RK weighting schemes: 45%, 40% and 38% of cases, respectively. The number of datasets where the prediction was wrong was the lowest with DN (21%),

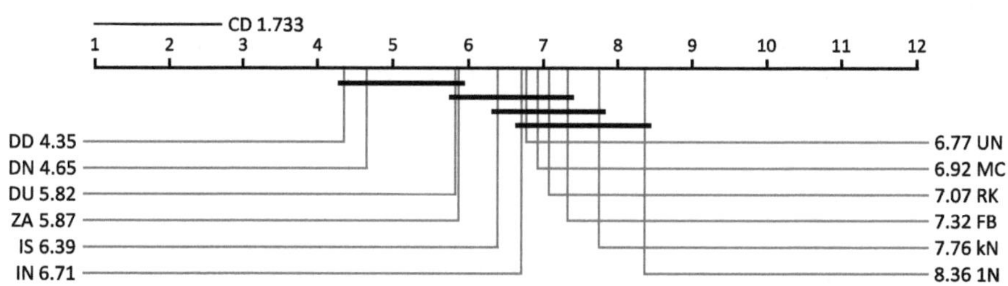

Fig. 5.70 EDR—critical difference diagram comparing average accuracy ranks based on Nemenyi's test: NN variants connected by a solid bar did not differ statistically significantly

Table 5.20 EDR—Holm-Bonferroni-adjusted *p*-values of the pairwise Wilcoxon signed-rank test on classification accuracy

A		79.39	79.33	79.22	79.14	78.96	78.80	78.78	78.72	78.68	78.29	77.94	77.06
		DD	DN	IS	IN	MC	DU	kN	ZA	RK	UN	FB	1N
79.39	DD		.191	**< .001**	**< .001**	**< .001**	**< .001**	**< .001**	**.002**	**< .001**	**< .001**	**< .001**	**< .001**
79.33	DN			.014	.011	**< .001**	**.007**	**< .001**	.012	**< .001**	**< .001**	**< .001**	**< .001**
79.22	IS				1.000	1.000	1.000	.015	1.000	.260	.260	**.006**	**< .001**
79.14	IN					1.000	1.000	.013	1.000	.424	.960	.076	**< .001**
78.96	MC						1.000	**.001**	1.000	1.000	1.000	.990	**.001**
78.80	DU							.443	1.000	.743	1.000	.053	**< .001**
78.78	kN								.076	.433	1.000	1.000	.076
78.72	ZA									.960	1.000	.159	**< .001**
78.68	RK										1.000	.666	**.001**
78.29	UN											.026	**< .001**
77.94	FB												**< .001**
77.06	1N												

Note Bolded values indicate significance at the level of $\alpha = 0.01$

Table 5.21 EDR—statistically significant wins (W) and losses (L) reported by the corrected resampled *t*-test on classification accuracy (across 12,800 splits)

	DD	DN	IS	IN	DU	ZA	MC	UN	RK	kN	FB	1N
W	504	479	342	306	366	337	235	258	204	166	192	98
L	71	100	192	226	293	280	259	345	317	368	410	626
W–L	433	379	150	80	73	57	−24	−87	−113	−202	−218	−528

DD (20%), DU (14%) and ZA (13%). The rate of correctness of the assessment of the actual gain based on the expected was the highest with DD (77%), DN (77%), DU (76%) and the lowest when using FB (59%), RK (57%), MC (56%) and kN (50%).

Figure 5.72 testifies to the fact that, in the FP region, the expected gain on no dataset was greater than the actual one by more than 0.03 when relying on the ZA weighting scheme and only on one dataset each when using the FB and DU algorithms. The difference greater than 0.03 was most frequent with MC and IN (in both cases with 8 datasets). As many as seven kNN variants underestimated the performance of the 1N classifier on the *PickupGestureWiimoteZ* (85) dataset to the greatest extent: kN, MC, RK, DN, IN, DD, IS. In the case of UN, FB and DU, the largest difference in gains was generated for *Lightning2* (67), *Herring* (59) and *Beef* (7), respectively. Additionally, excluding ZA, FB and DU, the expected gain was more than 0.03 higher than the actual gain for all other weighting schemes with the *Lightning2* (67) and

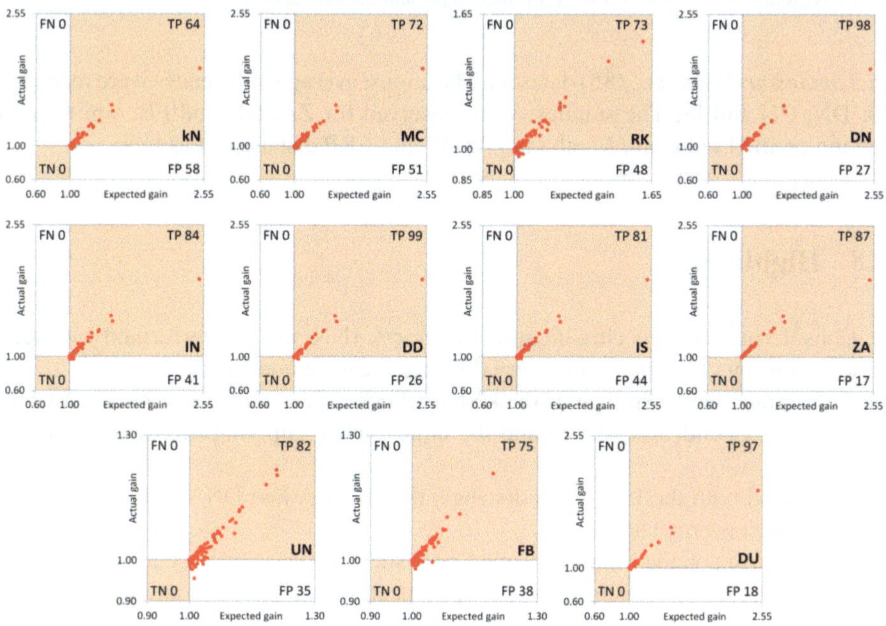

Fig. 5.71 EDR—comparing the actual and expected gain in classification accuracy: kNN variants versus 1N

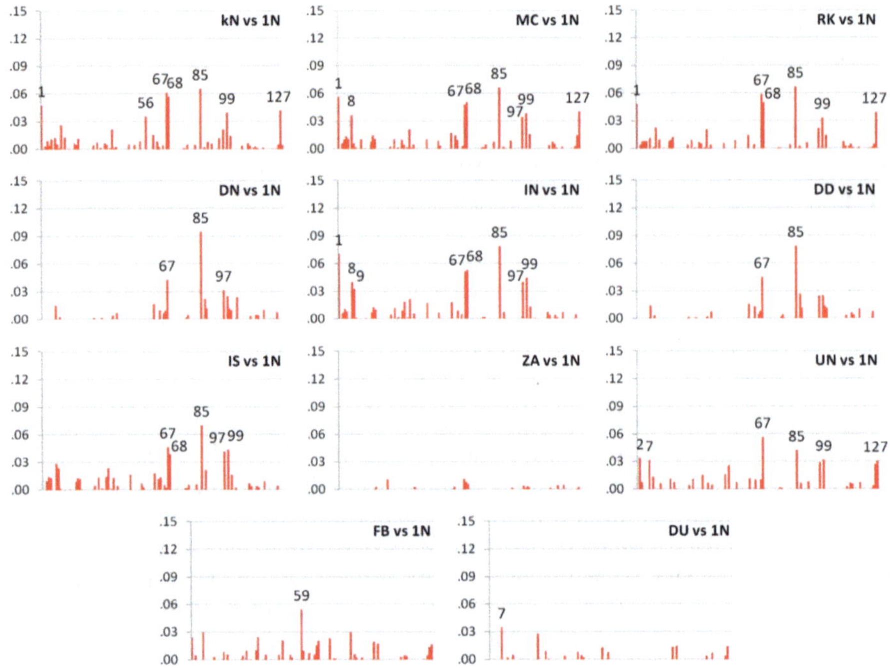

Fig. 5.72 EDR—differences between expected and actual accuracy gains of datasets belonging to the FP region: kNN variants versus 1N (labels represent dataset identifiers)

PickupGestureWiimoteZ (85) datasets. The largest average differences were recorded for DN, UN and IS. The smallest were observed for ZA, DU and FB, similarly to relying on the Euclidean, Manhattan, DTW and ERP distance measures.

5.8 Highlights

- Considering average classification accuracy, 1NN was outperformed by every kNN variant combined with every distance measure considered.
- On average, the advantage of the kNN variants over 1NN was the smallest when using the Euclidean distance and the largest when applying the EDR similarity measure.
- Combined with the Euclidean distance, the DD, ZA and DN schemes produced the highest average accuracies.
- When relying on the Manhattan and Chebyshev distance and the elastic measures, the highest average accuracies were produced by the DD, DN and IS weighting schemes.
- FB generated one of the lowest accuracies with every measure considered. Except when using L_∞ and EDR, kN's average accuracy was also among the lowest.

- On average, the standard deviation of accuracies was highest for the Chebyshev distance and lowest for ERP.
- The average value of the optimal numbers of nearest neighbors was the highest with the DU scheme regardless of the distance measure used—it ranged between 14.35 (L_2) and 16.27 (EDR).
- The average value of the optimal nearest neighbor numbers of DD, DN and UN was higher than all the other schemes, except DU, combined with every measure analyzed in this chapter. It ranged between 7.93 (UN with L_2) and 11.83 (DD with LCS).
- The average value of the optimal numbers of nearest neighbors was the lowest with the FB scheme (except in combination with the L_1 and ERP measure) and ranged between 4.29 (L_∞) and 5.27 (LCS). The second place always belonged to kN whose average k-values ranged between 4.41 (ERP) and 7.08 (EDR).
- Regardless of the distance measure used, the standard deviation of the optimal numbers of nearest neighbors was the highest with DU and UN followed by DD and DN (except with EDR).
- Regardless of the distance measure used, the standard deviation of the optimal numbers of nearest neighbors was the lowest with FB followed by kN (L_2, L_∞, ERP, EDR) or ZA (L_1, DTW, LCS).
- Considering the number of datasets for which it produced the best accuracy, 1NN was hard to beat. It achieved the best result in combination with L_2 and ERP (tied with DU), with L_∞ and DTW it was outmatched only by DD and with LCS by both DD and DU. It was defeated by a larger number of kNN classifiers only when using the L_1 (DD, ZA, DU and UN) and EDR (DD, ZA, DU, DN and UN) measures.
- The lowest mean ranks were produced by the DD and DN schemes independent of the distance measure. The highest mean ranks were generated by kN (L_2, L_1, DTW, ERP), 1N (LCS, EDR) and UN (L_∞).
- The dominance of the first nearest neighbor was, on average, the most pronounced when using the Chebyshev distance and the least present in combination with the LCS measure.
- $k = 1$ was the most relied upon by kN (L_2, DTW, LCS, EDR) and ZA (L_1, L_∞, ERP), in between about 36.6% (kN with LCS) and 63.0% (ZA with L_1) of the 12,800 splits and the least by DD, in between 13.7% (EDR) and 25.3% (L_∞) of splits. DN and DU were also among the schemes least reliant on $k = 1$.
- For several datasets, the optimal value of k was 1 in each of the 100 iterations. On average, they were the most numerous with L_∞ and the least frequent with EDR.
- In several datasets, the average of the optimal values of k was greater than 20. On average, they were the most numerous with LCS and EDR and the least frequent with ERP.
- The win-loss ratio of each kNN variant (except kN and FB) was the lowest with the Chebyshev distance, and if only elastic measures are considered, then with ERP (except for the DU scheme which generated the lowest ratio with DTW).

- The positive average difference in accuracy was the lowest for all kNN variants when applying the Manhattan distance, except for the MC scheme, which produced the lowest value when using the Euclidean distance. If only elastic measures are considered, it was the lowest with ERP, with the exception of IN and IS which produced an even lower average with DTW.
- The negative average difference in accuracy was the highest in combination with DTW, with the exception of ZA (LCS), UN (EDR) and DU (LCS).
- With L_2, L_1, DTW and ERP, the advantage over the accuracy of 1NN was below 9.5 percentage points for each kNN variant.
- With L_∞, except for FB and DU, the other kNN variants outperformed 1NN by more than 10 pp over a small number of datasets (their number was not higher than 3 with any weighting scheme).
- With LCS and EDR, the kNN variants (excluding UN and FB) achieved advantages ranging between 13 and 18 percentage points over the accuracy of the 1NN classifier on several datasets. Moreover, in the case of *ProximalPhalanxTW* (94) the gain was greater than 25 percentage points (except for RK with LCS, for which it was slightly higher than 22 pp).
- Apart from the Chebyshev distance, according to the Friedman test, the difference in the performance of the considered classifiers was statistically significant at the $\alpha = 0.01$ level.
- According to Nemenyi's post-hoc test, when it comes to the Euclidean and Chebyshev distances, no kNN variant was significantly better than 1NN. When it comes to the Manhattan distance and the elastic measures, only the DD and DN (except ERP) scheme outperformed 1NN significantly in combination with each of them.
- According the Holm-Bonferroni-corrected Wilcoxon signed-rank test, when it comes to the Euclidean and Chebyshev distances, only the DD and DU schemes outperformed 1NN significantly at the $\alpha = 0.01$ level (and DN in combination with L_∞). When it comes to the Manhattan distance and the elastic measures, DN, DD, ZA, UN and DU schemes significantly outperformed 1NN combined with all of them.
- On average, datasets were the most numerous in the FP region when using the Euclidean distance. If only elastic measures were considered, then when relying on the ERP measure.
- On average, datasets were the least numerous in the TP region when using the Chebyshev distance. If only elastic measures were considered, then when relying on the ERP measure.
- In the FP region, regardless of the similarity measure, the gain of kN was overestimated the most frequently, and the gain of the DU scheme the least often (except for EDR where ZA, and L_1 where both ZA and FB produced less FP datasets than DU).
- Regardless of the distance measure used, the most datasets were contained in the TP regions of the DD, DN and DU schemes, and, except LCS, the least in the TP regions of the kN, RK and MC schemes (where, combined with L_1, ZA generated even fewer TP datasets than these three schemes).

References

1. Fix, E., Hodges, J.L.: Discriminatory analysis: nonparametric discrimination: consistency properties. Int. Stat. Rev. **57**, 238–247 (1989). https://doi.org/10.2307/1403797
2. Cover, T., Hart, P.: Nearest neighbor pattern classification. IEEE Trans. Inf. Theory **13**, 21–27 (1967). https://doi.org/10.1109/TIT.1967.1053964
3. Dudani, S.A.: The distance-weighted k-nearest-neighbor rule. IEEE Trans. Syst. Man. Cybern. **6**, 325–327 (1976). https://doi.org/10.1109/TSMC.1976.5408784
4. Macleod, J., Luk, A., Titterington, D.: A re-examination of the distance-weighted k-nearest neighbor classification rule. IEEE Trans. Syst. Man Cybern. **17**, 689–696 (1987). https://doi.org/10.1109/TSMC.1987.289362
5. Mitchell, T.M.: Machine learning. McGraw-Hill Inc., New York, NY (1997)
6. Zavrel, J.: An empirical re-examination of weighted voting for k-NN. In: Proceedings on 7th belgian-cutch conference on machine learning, pp. 139–148 (1997)
7. Pao, T.-L., Chen, Y.-T., Yeh, J.-H., Cheng, Y.-M., Lin, Y.-Y.: A comparative study of different weighting schemes on KNN-based emotion recognition in Mandarin speech. In: Huang, D.-S., Heutte, L., Loog, M. (eds.) Advanced intelligent computing theories and applications with aspects of theoretical and methodological issues, pp. 997–1005. Springer, Berlin (2007). https://doi.org/10.1007/978-3-540-74171-8_101
8. Gou, J., Xiong, T., Kuang, Y.: A novel weighted voting for K-nearest neighbor rule. J. Comput. **6**, 833–840 (2011). https://doi.org/10.4304/jcp.6.5.833-840
9. Gou, J., Du, L., Zhang, Y., Xiong, T.: A new distance-weighted k-nearest neighbor classifier. J. Inf. Comput. Sci. **9**, 1429–1436 (2012)
10. agrawal, R., Faloutsos, C., Swami, A.: Efficient similarity search in sequence databases. In: Lomet, D.B. (ed.) Proceedings on 4th international conference foundations of data organization and algorithms (FODO '93), pp. 69–84. Springer, Berlin Heidelberg (1993). https://doi.org/10.1007/3-540-57301-1_5
11. Berndt, D., Clifford, J.: Using dynamic time warping to find patterns in time series. In: Usama, R.U., Fayyad, M. (ed.) Knowledge discovery in databases paper from 1994 AAAI work, pp. 359–370. AAAI Press, Seattle, WC (1994). http://dblp.uni-trier.de/rec/bib/conf/kdd/BerndtC94
12. Das, G., Gunopulos, D.: Time series similarity and indexing. In: Ye, N. (ed.) Handbook of data mining, pp. 279–304. Lawrence Erlbaum Associates, Mahwah, NJ (2003)
13. Chen, L., Ng, R.: On the marriage of Lp-norms and edit distance. In: Nascimento, M.A., Özsu, M.T., Kossmann, D., Miller, R.J., Blakeley, J.A., Schiefer, K.B. (eds.) Proceedings 2004 VLDB Conference, pp. 792–803. Elsevier (2004). https://doi.org/10.1016/B978-012088469-8.50070-X
14. Chen, L., Özsu, M.T., Oria, V.: Robust and fast similarity search for moving object trajectories. In: Proceeding of the 2005 ACM SIGMOD international conference management data—SIGMOD '05, pp. 491–502. ACM Press, New York (2005). https://doi.org/10.1145/1066157.1066213
15. Anh Dau, H., Keogh, E., Kamgar, K., Michael Yeh, C.-C., Zhu, Y., Gharghabi, S., Ann Ratanamahatana, C., Chen, Y., Hu, B., Begum, N., Bagnall, A., Mueen, A., Batista, G., Hexagon, M.L.: The UCR time series classification archive (2019). https://www.cs.ucr.edu/~eamonn/time_series_data_2018/
16. Paparrizos, J.: 2018 UCR time-series archive: backward compatibility, missing values, and varying lengths (2019). https://github.com/johnpaparrizos/UCRArchiveFixes
17. Geler, Z., Kurbalija, V., Radovanović, M., Ivanović, M.: Impact of the Sakoe-Chiba band on the DTW time series distance measure for kNN classification. In: Buchmann, R., Kifor, C.V., Yu, J. (eds.) 7th International conference on knowledge science, engineering and management KSEM 2014, pp. 105–114. Springer international publishing, Cham (2014). https://doi.org/10.1007/978-3-319-12096-6_10
18. Geler, Z.: Role of similarity measures in time series analysis. University of Novi Sad, Serbia (2015). (PhD thesis) https://nardus.mpn.gov.rs/handle/123456789/1703

19. Geler, Z., Kurbalija, V., Radovanović, M., Ivanović, M.: Comparison of different weighting schemes for the kNN classifier on time-series data. Knowl. Inf. Syst. **48**, 331–378 (2016). https://doi.org/10.1007/s10115-015-0881-0
20. Kurbalija, V., Radovanović, M., Geler, Z., Ivanović, M.: A framework for time-series analysis. In: Dicheva, D., Dochev, D. (eds.) Artificial intelligence on methodology system applications SE—5, pp. 42–51. Springer, Berlin Heidelberg (2010). https://doi.org/10.1007/978-3-642-15431-7_5
21. Kurbalija, V., Ivanović, M., Geler, Z., Radovanović, M.: Two faces of the framework for analysis and prediction, part 1—education. Inf. Technol. Control. **47**, 249–261 (2018). https://doi.org/10.5755/j01.itc.47.2.18746
22. Kurbalija, V., Ivanović, M., Geler, Z., Radovanović, M.: Two faces of the framework for analysis and prediction, part 2—research. Inf. Technol. Control. **47**, 489–502 (2018). https://doi.org/10.5755/j01.itc.47.3.18747
23. Witten, I.H., Frank, E., Hall, M.A., Pal, C.J.: Data mining: practical machine Learning tools and techniques, 4th ed. Morgan Kaufmann (2017). https://doi.org/10.1016/C2015-0-02071-8
24. Friedman, M.: The use of ranks to avoid the assumption of normality implicit in the analysis of variance. J. Am. Stat. Assoc. **32**, 675 (1937). https://doi.org/10.2307/2279372
25. Friedman, M.: A comparison of alternative tests of significance for the problem of m rankings. Ann. Math. Stat. **11**, 86–92 (1940). https://doi.org/10.1214/aoms/1177731944
26. Nemenyi, P.B.: Distribution-free multiple comparisons. Princeton University, Princeton (1963) (PhD thesis)
27. Demšar, J.: Statistical comparisons of classifiers over multiple data sets. J. Mach. Learn. Res. **7**, 1–30 (2006)
28. Benavoli, A., Corani, G., Mangili, F.: Should we really use post-hoc tests based on mean-ranks? J. Mach. Learn. Res. **17**, 1–10 (2016)
29. Wilcoxon, F.: Individual comparisons by ranking methods. Biometrics Bull. **1**, 80 (1945). https://doi.org/10.2307/3001968
30. Holm, S.: A simple sequentially rejective multiple test procedure. Scand. J. Stat. **6**, 65–70 (1979)
31. Bouckaert, R.R., Frank, E.: Evaluating the replicability of significance tests for comparing learning algorithm. In: Dai, H., Srikant, R., Zhang, C. (eds.) Advances in knowledge discovery and data mining, pp. 3–12. Springer, Berlin Heidelberg (2004). https://doi.org/10.1007/978-3-540-24775-3_3
32. Batista, G.E.A.P.A., Wang, X., Keogh, E.J.: A complexity-invariant distance measure for time series. In: Proceedings of the 2011 SIAM international conference on data mining, pp. 699–710. Society for Industrial and Applied Mathematics, Philadelphia, PA (2011). https://doi.org/10.1137/1.9781611972818.60

Chapter 6
Combined Influence of Weighting Schemes and Global Constraints on Classification Accuracy

The final chapter examines in detail the combined impact of six selected weighting schemes (see Chap. 3) and global constraints on classification accuracy of the four fundamental elastic similarity measures (DTW [1], LCS [2], ERP [3] and EDR [4]) introduced in Chap. 2 (and summarized in Appendix A) in comparison with the performance of the simple 1NN classifier paired with the corresponding unconstrained measure (1N). Of the 11 kNN variants analyzed in the previous chapter, we will consider the following six here (their brief descriptions are given in Appendix B): majority-voting rule (kN), Dudani's weighting function (DN) [5], inverse distance weight (IN) [5], inverse of the squared distance weight (IS) [6], dual distance-weighted function (DD) [7] and dual-uniform weighting function (DU) [8]. Among them, kN represents a natural extension of the 1NN rule, DN, IN, IS and DD produced the highest average classification accuracies when paired with the unconstrained elastic measures and DU achieved the best win-loss ratio over 1N.

In the rest of this chapter, the combinations of kNN variants and global constraints will be denoted with labels obtained by combining the acronyms of the corresponding classifiers and the applied type of global constraints. For example, kNSC will represent the kN (kNN) classifier in combination with the Sakoe-Chiba band [9] and kNIT with the Itakura parallelogram [10].

Analogously to the previous chapters, the optimal size of the neighborhood (k) was chosen from the interval between 1 and 30 and the optimal warping-window width (r) from the set $R = \{0, 1, 2, \ldots, 48, 49, 50, 55, 60, \ldots, 90, 95, 100\}$. As in previous chapters, we utilized the 128 publicly available datasets from the UCR archive [11] in our experiments and they were preprocessed using Paparizzo's script [12]. Given that the focus of the study was not on comparing the performance of the examined similarity measures, the value of the matching threshold parameter ε of LCS and EDR was fixed at 0.25 based on the recommendations in [4].

The setup of the experiments was based on our earlier papers [13–15] related to the topic of this chapter and they were performed (except for statistical tests) using our *Framework for Analysis and Prediction* [16]. Beyond its research applications (as

Z. Gellér et al., *Recent Advances in Time-Series Classification—Methodology and Applications*, Intelligent Systems Reference Library 264,
https://doi.org/10.1007/978-3-031-77527-7_6

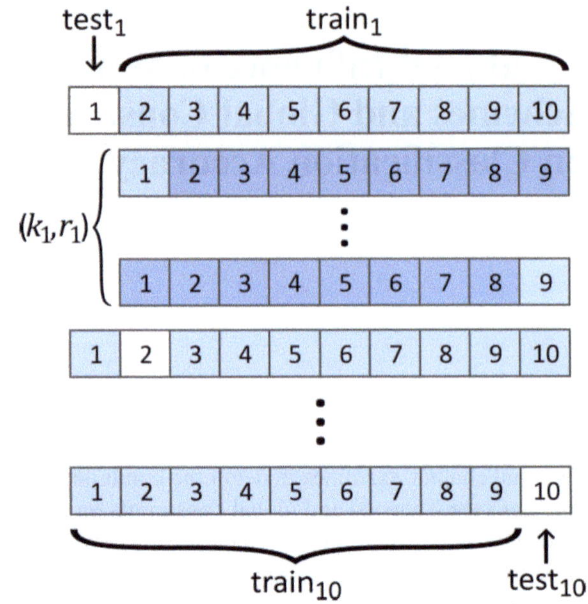

Fig. 6.1 Calculating classification accuracy using tenfold cross-validation where the optimal combination of the parameters k and r is chosen using ninefold cross-validation on the training subset

documented in [17]), this open-source library has also become a valuable auxiliary tool for university teaching [18].

The applied classifier performance evaluation procedure based on nested stratified cross-validation [19] is illustrated in Fig. 6.1. In each iteration of 10 times repeated tenfold stratified cross-validation (SCV10x10), the given dataset was divided into 10 stratified disjoint subsets of approximately the same size. One of them was used for testing and the union of the other 9 for training (i.e. parameter tuning). The optimal combination of parameters k and r was chosen so that the classification accuracy calculated over the training subset using the ninefold stratified cross-validation (SCV1x9) method was maximal. After finding the optimal number of nearest neighbors and the width of the warping window, the accuracy of the model was evaluated by classifying the time series of the testing subset relying on the training subset. After each of the 10 folds was used (exactly once) for testing, the data set was shuffled and the entire procedure was repeated 9 more times. In this way, 100 models of each combination of kNN classifiers and (globally constrained) similarity measures were obtained for each of the 128 UCR dataset and the average of their accuracies represents an estimate of their performance.

The averages of the obtained classification accuracies, optimal numbers of nearest neighbors and optimal warping-window widths of the considered classifiers are presented in Tables 6.1, 6.5, 6.9 and 6.13, along with the corresponding standard deviations, the numbers of datasets on which the highest accuracy was achieved by a given classifier and with the mean ranks.

In order to investigate the effect of global constraints on the preference of one or more nearest neighbors by the analyzed kNN variants, Figs. 6.2, 6.20, 6.38 and 6.56

show the percentages of those splits (out of a total of 12,800) in which the optimal value of the parameter k was equal to 1 or greater than 1. Comparative presentations of the frequencies of different k-values greater than 2 (see the discussion about $k = 2$ and $k = 3$ in Appendix D) obtained by applying the Sakoe-Chiba band and Itakura parallelogram are given in Figs. 6.3, 6.21, 6.39 and 6.58. Tables 6.2, 6.6, 6.10 and 6.14 show the numbers of those datasets where the optimal number of nearest neighbors was 1 in each of the 100 iterations of the SCV10x10 evaluation.

As part of an effort to examine the effect of increasing the considered number of nearest neighbors and applying different weighting schemes to the optimal width of the warping window, Figs. 6.4, 6.22, 6.40 and 6.58 present the frequencies of different values of the parameter r on the example of the DN method.

The numbers of datasets on which a given kNN variant paired with the Sakoe-Chiba band beat and was beaten by the 1NN classifier combined with the corresponding unconstrained measure in terms of classification accuracy are presented in Figs. 6.5, 6.23, 6.41 and 6.59. The results with the Itakura parallelogram are given in Figs. 6.6, 6.24, 6.42 and 6.60. Focusing on the differences between the two types of global constraints, the numbers of datasets where SC outperformed IT and vice versa are shown in Figs. 6.7, 6.25, 6.43 and 6.61. The plots in these figures also depict individual datasets—in the form of 2D points whose x coordinate represents the classification accuracy of one classifier and y coordinate the classification accuracy of the other.

By visualizing the positive and negative average performance differences (see Appendix C for details) between the kNN variants with the constrained and the 1NN classifier with the unconstrained measures, Figs. 6.8, 6.26, 6.44 and 6.62 provide a deeper undestanding of the nature of their (dis)advantage. The positive (negative) average was obtained by summing up the differences in classification accuracy between a given kNN variant and the 1NN method that were greater (less) than zero and dividing by the number of such datasets. A comparison of the SC and IT constraints in this regard is given in Figs. 6.9, 6.27, 6.45 and 6.63.

In order to make it easier to recognize the datasets on which a given kNN variant combined with the Sakoe-Chiba constraint achieved the greatest advantage over 1N, as well as those on which it suffered the greatest defeat from it, Figs. 6.10, 6.28, 6.46 and 6.64 show the differences in their performance at the level of individual datasets. The findings obtained in combination with the Itakura constraint are shown in Figs. 6.11, 6.29, 6.47 and 6.65 and contrasting the two types of global constraints is given in Figs. 6.12, 6.30, 6.48 and 6.66.

We examined the statistical significance of the difference in classifier performance using the Friedman test [20, 21], as recommended in [22], followed by Nemenyi's post-hoc test [23]. The results of Nemenyi's post-hoc test are shown in Figs. 6.13, 6.31, 6.49 and 6.67 in the form of critical difference diagrams. As recommended in [24], we also checked the statistical significance of the difference in the performance of pairs of classifiers using the Wilcoxon signed-rank test [25], whereby we adjusted the obtained p-values in accordance with the Holm-Bonferroni approach [26]. The results of the Wilcoxon test are shown in Tables 6.3, 6.7, 6.11 and 6.15. As the third approach to comparison, we relied on the numbers of statistically significant wins and losses reported by the corrected resampled t-test [27] in Tables 6.4, 6.8, 6.12 and 6.16. As in previous chapters, the results of statistical tests were interpreted at the significance level of $\alpha = 0.01$.

In the final step of the analysis, we examined to what extent the assessment of the advantage of the kNN variants (paired with global constraints) over the 1NN classifier (combined with the corresponding unconstrained measure) based on the training data was misleading. As recommended in [28], we performed this examination using *Texas Sharpshooter* (TS) plots. Accordingly, in Figs. 6.14, 6.32, 6.50 and 6.68, the datasets are shown in the form of two-dimensional points whose x-coordinate represents the expected and y-coordinate the actual gain of the kNN variants paired with the Sakoe-Chiba band over 1N. The gain of a given kNN variant over 1N is defined as the quotient between the accuracy of kNN and 1N. The expected gain was calculated over the training subset and the actual gain over the testing subset of a given iteration of the SCV10x10 procedure. If the gain was greater than 1, the kNN variant performed better than 1N and if it was less than 1, it performed worse. By comparing the x (expected gain) and y (actual gain) coordinates of the points, four cases can be distinguished:

- $x > 1$ and $y > 1$—Expected and actual gains are aligned: we expected the given kNN variant to outperform 1N and it really outperformed it. The given dataset belongs to the TP (*true positive*) region of the TS plot.
- $x < 1$ and $y < 1$—Expected and actual gains are aligned: we expected a given kNN variant to be outperformed by 1N and indeed it was outperformed. The given dataset belongs to the TN (*true negative*) region of the TS plot.
- $x > 1$ and $y < 1$—Expected and actual gains are not aligned: we expected the given kNN variant to outperform 1N but actually 1N performed better. The given dataset belongs to the FP (*false negative*) region of the TS chart. In its case, we chose the worse option based on the training data and worsened the classification accuracy compared to 1N.
- $x < 1$ and $y > 1$—Expected and actual gains are not aligned: we expected the given kNN variant to be outperformed by 1N but it actually performed better. The given dataset belongs to the FN (*false negative*) region of the TS plot. In its case, based on the results on the training data, we missed the opportunity to improve the classification accuracy compared to 1N.

Figures 6.15, 6.33, 6.51 and 6.69 show Texas Sharpshooter plots comparing the expected and actual gains of kNN variants paired with the Itakura parallelogram versus the 1NN classifier combined with the corresponding unconstrained measure, and Figs. 6.16, 6.34, 6.52 and 6.70 contrast the Itakura parallelogram. with the Sakoe-Chiba band when classifying with kNN methods.

To identify those datasets of the FP region where the difference between the expected and the actual gain was the largest, Figs. 6.17, 6.35, 6.53 and 6.71 visualize the difference between the gains when comparing the kNN variants paired with the Sakoe-Chiba band and the 1NN classifier combined with the corresponding unconstrained measure. Figures 6.18, 6.36, 6.54 and 6.72 show that difference for the kNN variants when limiting elasticity with the Itakura parallelogram, also compared to

1N. The differences between the expected and actual gains of IT over SC when using the six selected kNN classifiers are presented in Figs. 6.19, 6.37, 6.55 and 6.73.

The results shown on the TS plots should be interpreted taking into account the following remarks:

- The accuracy gains represent the averages of the gains calculated within 100 iterations of the SCV10x10 evaluation method.
- Since the set of considered warping-window widths contained $r = 100\%$, the SCV1x9 algorithm took it into account when choosing the optimal (k, r) pair, so the expected gain (when comparing kNN variants paired with an constrained measure with the 1NN classifier combined with the corresponding constrained measure) cannot be less than 1. It follows that the TN and FN regions of the respective TS plots are empty.
- Combined with the unconstrained DTW, 1NN misclassified each time series of the testing subset in 3 of the 100 splits of the *Beef* (7) dataset, therefore gains were calculated on the remaining 97.
- Combined with the SC-constrained DTW, the kNN variants misclassified each time series of the testing subset in 1 (IN, DD), 2 (kN, IS) and 3 (DN, DU) out of 100 splits of the *Beef* (7) dataset, therefore gains were calculated based on the remaining 99 (IN, DD), 98 (kN, IS) and 97 (DN, DU).
- Combined with the unconstrained ERP, 1NN misclassified each time series of the testing subset in 5 of the 100 splits of the *Beef* (7) dataset, therefore gains were calculated on the remaining 95.
- Combined with the SC-constrained ERP, the kNN variants misclassified each time series of the testing subset in 2 (kN, IN, IS, DU), 3 (DN) and 4 (DD) out of 100 splits of the *Beef* (7) dataset, therefore gains were calculated based on the remaining 98 (kN, IN, IS, DU), 97 (DN) and 96 (DD).

In the rest of this chapter, we give a detailed overview of the results of the analysis of the combined impact of weighting schemes and global constraints on classification accuracy when relying on the four basic elastic similarity measures.

6.1 Dynamic Time Warping (DTW)

In the competition regarding the average (AV) classification accuracy calculated with the SCV10x10 technique, where the optimal warping-window width and the optimal number of nearest neighbors were selected over the corresponding training subset using the SCV1x9 evaluation technique, the IS, DD, IN and DN classifiers in combination with the SC constraint proved to be the most successful (Table 6.1) and the least successful were kNSC, DUIT, kNIT and finally 1N (the 1NN classifier paired with the unconstrained DTW measure).

The advantage over 1N ranged from 2.55 (kNIT) to 3.42 (ISSC) percentage points (pp), which represents a noticeable increase compared to the results in Chap. 5

Table 6.1 DTW—average value (AV) and standard deviation (SD) of classification accuracy (AV), optimal number of nearest neighbors (*k*) and optimal warping-window width (*r*) across 12,800 splits; the number of datasets for which a given classifier produced the best result (WIN) and the mean rank (MR)

	ISSC	DDSC	INSC	DNSC	ISIT	DDIT	DNIT	DUSC	INIT	kNSC	DUIT	kNIT	1N
A_{AV} (%)	82.66	82.55	82.53	82.48	82.30	82.27	82.23	82.20	82.19	82.10	81.88	81.79	79.23
A_{SD} (pp)	18.51	18.65	18.50	18.68	18.35	18.42	18.44	18.93	18.35	18.87	18.63	18.68	19.45
k_{AV}	7.02	8.27	6.16	7.53	7.01	8.24	7.58	14.70	6.18	4.43	14.68	4.52	
k_{SD}	7.52	7.67	7.12	7.36	7.49	7.60	7.33	10.86	7.08	6.26	10.78	6.19	
r_{AV}(%)	6.84	6.95	6.86	6.94	21.08	21.61	21.68	6.66	21.29	7.11	21.21	21.39	
r_{SD}(pp)	9.68	10.22	9.76	10.19	28.81	29.43	29.50	10.12	28.65	10.56	29.28	28.75	
WIN	23	26	15	14	14	24	14	20	13	8	12	10	20
MR	5.91	5.39	6.41	6.13	7.18	5.98	6.31	6.66	7.18	8.06	7.47	8.84	10.16

(Table 5.10). Specifically, by applying the SC constraint, the advantage of kNN variants over 1N increased by about between 2.19 (DD) and 2.35 (IN) pp and by applying the IT constraint by about between 1.92 (DD) and 2.0 (IN) pp. This improvement was the largest among the elasticity measures discussed in this book.

All six considered kNN classifiers achieved slightly better results when using the Sakoe-Chiba band than when applying the Itakura parallelogram, which reflects the findings of Chap. 4 based on the 1NN classifier.

The standard deviation (SD) ranged between 18.35 (INIT) and 19.45 (1N), and it is noticeable that the kNN variants generated a bit higher SD values in combination with SC than with IT. Furthermore, the application of global constraints resulted in lower SD values compared to the unconstrained measure (Table 5.10 in Chap. 5).

Looking at the average values of the optimal numbers of nearest neighbors in Table 6.1, we can observe the same (decreasing) order of kNN classifiers as when relying on the unconstrained DTW (Chap. 5): DU, DD, DN, IS, IN, kN. The (descending) order in terms of standard deviation is very similar, the only difference is that DN and IS have switched places: DU, DD, IS, DN, IN, kN. It is also noticeable that both the average values and standard deviations obtained applying global constraints were somewhat lower than those observed when using the unconstrained measure (Table 5.10 in Chap. 5). Similar to the case of the unconstrained DTW, the DU scheme produced noticeably higher average k-values than the others.

The higher frequency of smaller k-values (in relation to the unconstrained DTW) was most pronounced for $k = 1$, which can be seen by comparing Figs. 6.2 and 4.34 (the only exception was ISIT which relied on $k = 1$ about 0.27 pp less often than when using the unconstrained DTW). At the same time, kNN variants relied somewhat more strongly on $k = 1$ with SC than with IT, which is most noticeable in the case of the majority-voting approach. As for the frequencies of $k > 2$ (Fig. 6.3), apart from the fact that the AVs were slightly lower (except for ISIT) and the SDs were slightly higher than with the unconstrained DTW, the general traits of the kNN variants did not deviate from those presented in Chap. 5 (Fig. 5.35). No noticeable difference was observed between SC and IT and DU preferred larger numbers of nearest neighbors, the same as with the unconstrained DTW.

Applying global constraints increased the number of datasets for which the optimal number of nearest neighbors was 1 in each of the 100 iterations of the SCV10x10 algorithm (Table 6.2). This feature was not present in the *BME* (10) (except for kN), *CBF* (12), *DodgerLoopWeekend* (28) and *Fungi* (46) datasets when using unconstrained DTW, but did occur with both SC and IT. There were also a few datasets that lost this feature when global constraints were applied, for example, *DiatomSizeReduction* (22) when using SC (except with the DU scheme) and *PigAirwayPressure* (86) when using both types of constraints (the only exceptions were kNSC and DUSC).

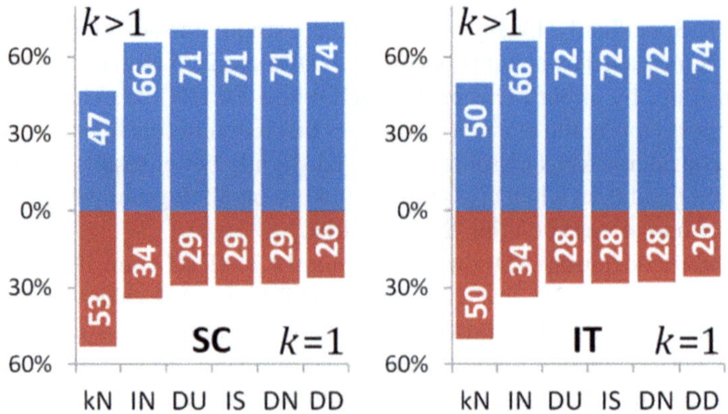

Fig. 6.2 DTW—percentages of splits in which SCV1x9 chose $\boldsymbol{k} = 1$ and $\boldsymbol{k} > 1$ as the optimal number of nearest neighbors on the training set when evaluating classification accuracy using SCV10x10

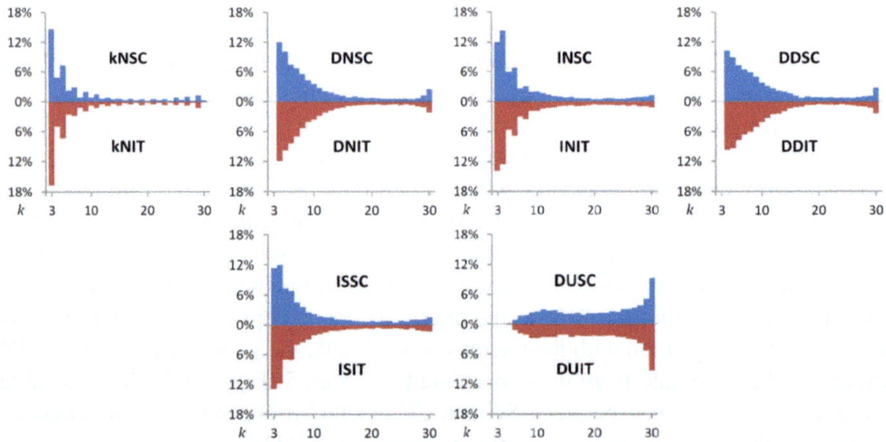

Fig. 6.3 DTW—frequencies (in percentage relative to the 12,800 splits) of different values of the parameter k (≥ 3) chosen by the SCV1x9 algorithm over the training set when evaluating classification accuracy using SCV10x10

Table 6.2 DTW—number of datasets for which the optimal number of nearest neighbors was 1 in each of the 100 iterations of the SCV10x10 algorithm

	kN	DN	IN	DD	IS	DU
DTW	20	6	8	6	8	8
SC	22	9	10	9	9	12
IT	23	10	12	10	11	13

The switch from the 1NN classifier (see Table 4.1 in Chap. 4) to kNN variants was accompanied by a slight increase in the average values of the optimal warping-window widths (by between about 1.29 (IS) and 1.89 (DN) pp for IT and between approximately 0.16 (DU) and 0.62 (kN) pp for SC), but the finding that IT relied on considerable higher r-values than SC (about 3.0 (kN) to 3.2 (DU) times) was still valid (Table 6.1).

In the case of SC, the increase in the number of considered neighbors was also reflected in a minor increase of the warping-window width needed to cover more than 90% of the splits: while using the 1NN rule it was sufficient to consider r-values up to 16% of the length of the time series (see Fig. 4.9b), with the kNN classifiers it was achieved at $r = 17\%$ (DU), $r = 18\%$ (DN, IN, DD, IS) and $r = 19\%$ (kN). In addition, there was also a decrease in the frequency of $r = 0\%$ as the optimal warping-window width for both types of global constraints (from about 19.4% (SC) to between 18.8% (DUSC) and 16.6% (INS) and from about 16.3% (IT) to between 14.4% (DUIT) and 12.8% (kNIT)).

kNN variants affected standard deviation of optimal warping-window widths differently for SC and IT. For IT, considering more than one nearest neighbor led to somewhat larger deviations compared to 1NN (by between about 0.27 (IN) and 1.12 (DN) pp), while SC produced marginally smaller SDs (by between approximately −0.0 (kN) and −0.89 (IS) pp). Still, the observation in Chap. 4 that IT generated significantly higher SD than SC was confirmed by the kNN variants as well. In this case, it was between about 2.7 (kN) and 3.0 (IS) times greater. Besides that, the general features observed when analyzing the impact of global constraints on the 1NN classifier were also present in combination with all the six kNN methods discussed in this chapter, as can be seen in the example of the DN weighting scheme in Fig. 6.4.

When using the kNN variants, similar to the results of the 1NN classifier, the optimal warping-window width was 0% in each of the 100 iterations of splitting *Crop* (21) and *MelbournePedestrian* (72) into testing and training subsets for both SC and IT. When classifying with the kNN variants, *FordA* (42) and *SmoothSubspace* (105) stood out in particular. For the former, the optimal value of the parameter r was 1% in each of the 100 splits with both SC and IT (except for DUIT) and for the latter it was 7% with SC (except for DUSC). This phenomenon did not characterize the 1NN classifier.

1N (20) was surpassed only by DDSC (26), DDIT (24) and ISSC (23) in the number of datasets over which a given classifier achieved the best accuracy (Table 6.1). In this category, DUIT (12), kNIT (10) and kNSC (8) proved to be the least successful. On the other hand, 1N was outperformed by every kNN variant in mean rank. Combined with the unconstrained DTW, only DD surpassed 1N in terms of number of wins and kN had a lower mean rank than 1N (see Table 5.10 in Chap. 5).

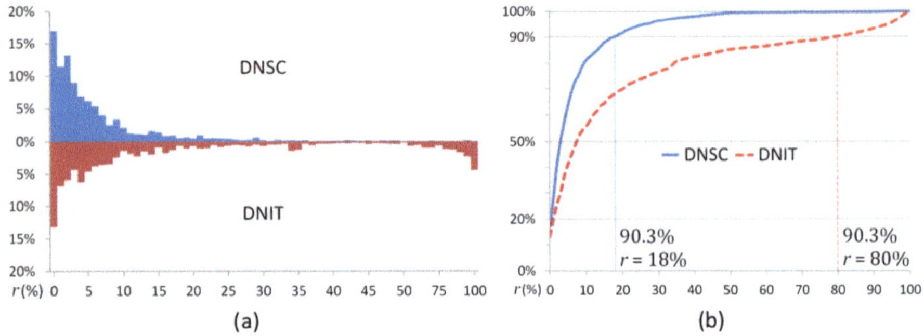

Fig. 6.4 DTW—frequencies (in percentage relative to the 12,800 splits) of different warping-window widths selected by the SCV1x9 algorithm over the training sets when evaluating classification accuracy using SCV10x10 (**a**); cumulative display of frequencies of optimal warping-window widths (**b**)

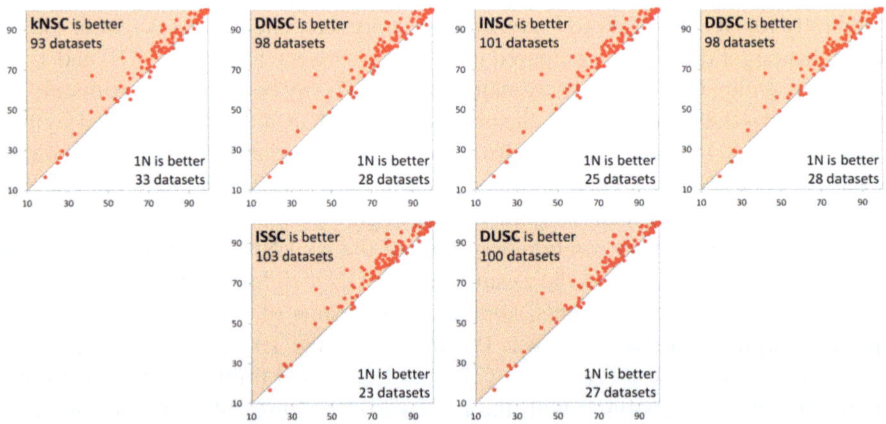

Fig. 6.5 DTW—comparison of the kNN variants (combined with the SC constrained measure) with the 1NN classifier (combined with the unconstrained measure) based on the number of datasets where they achieved higher/lower classification accuracy

Contrasting different kNN variants paired with the SC and IT constraints versus the 1NN classifier with the unconstrained DTW is visualized in Figs. 6.5 and 6.6. The win-loss ratio was the highest for ISSC (4.5:1) and INSC (4.0:1) and the lowest for kNSC (2.8:1) and kNIT (2.2:1). Similar to the 1NN results in Chap. 4 (Fig. 4.11), SC generated slightly better win-loss ratios than IT, but its advantage was much milder when combined with kNN variants. In this respect, SC showed the greatest advantage over IT in the case of the IN and the smallest in the case of the DD weighting scheme.

The noticeably higher win-loss ratios in favor of the kNN classifiers compared to the results in Chap. 5 (Fig. 5.37) reinforced the positive effect of applying global constraints classification performance. The degree of improvement was the highest in the case of kN and the lowest when using the DU classifier. By applying SC, the win-loss ratio became 3.5 and 1.2 times higher for kN and DU, respectively. The degree of improvement when applying the IT constraint was somewhat lower and amounted to 2.8 (kN) and 1.1 (DU). Moreover, in combination with the unconstrained DTW, the number of kN wins over 1N was less than the number of losses (47:58), which was here reversed owing to applying global constraints.

Figure 6.7 shows that the performance of SC and IT was more similar when using the kNN than the 1NN classifier. Namely, the win-loss ratio of SC over IT was approximately 1.6:1 in Chap. 4 (Fig. 4.11c), but with the kNN variants analyzed in this chapter it was between about 1.5:1 (DU) and 1.1:1 (kN).

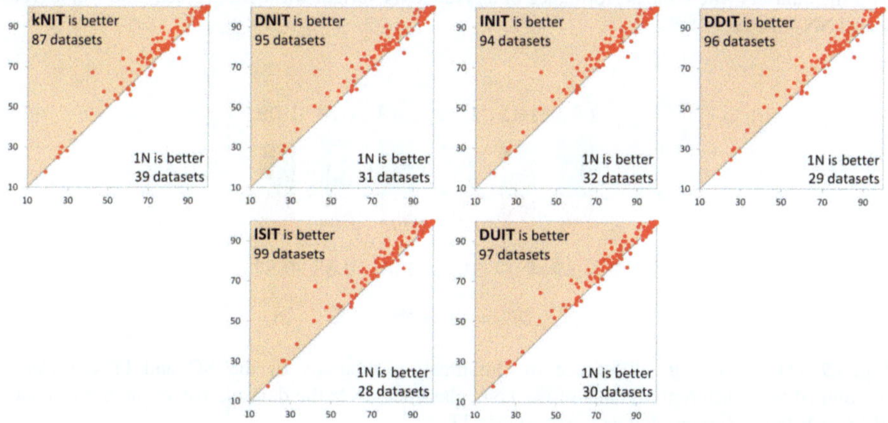

Fig. 6.6 DTW—comparison of the kNN variants (combined with the IT constrained measure) with the 1NN classifier (combined with the unconstrained measure) based on the number of datasets where they achieved higher/lower classification accuracy

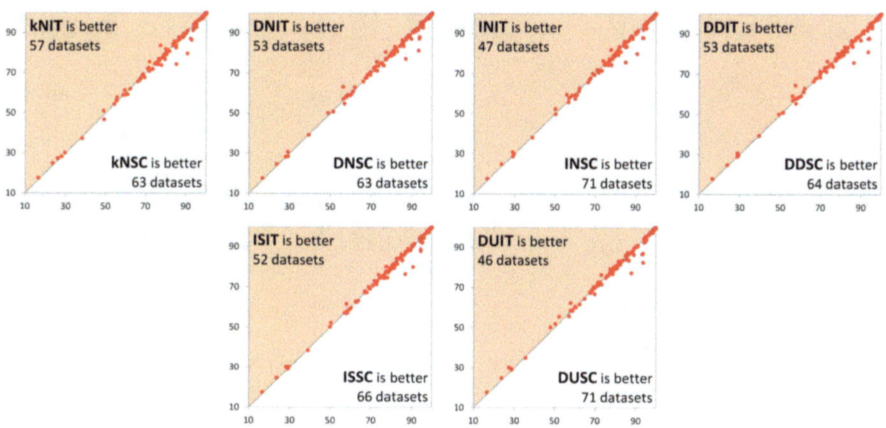

Fig. 6.7 DTW—comparison of the SC and IT constraints combined with the given kNN classifier variant based on the number of datasets where they achieved higher/lower classification accuracy

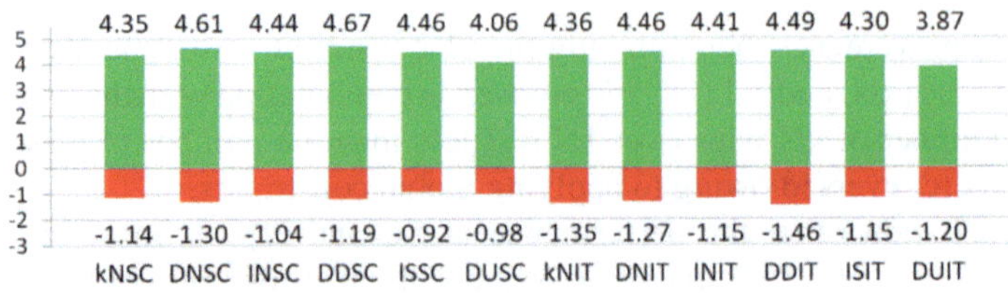

Fig. 6.8 DTW—average difference in classification accuracy of various kNN variants (combined with the SC and IT constrained measure) and 1NN (combined with the unconstrained measure) over the datasets where the performance of a given kNN variant was better (green) and worse (red) than 1NN

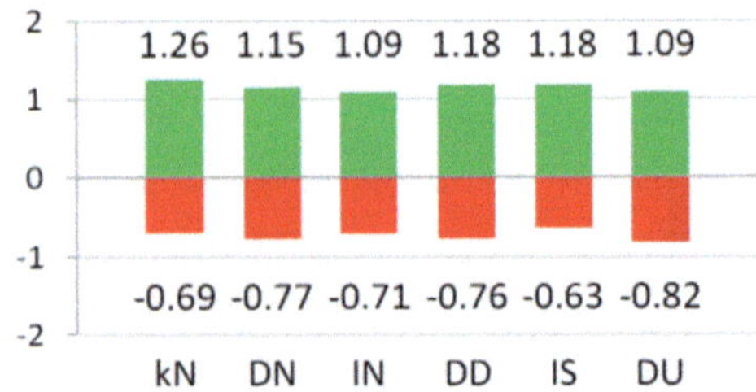

Fig. 6.9 DTW—average difference in classification accuracy of the SC and IT constraints (combined with different variants of the kNN classifier) over the datasets where the performance of SC was better (green) and worse (red) than IT

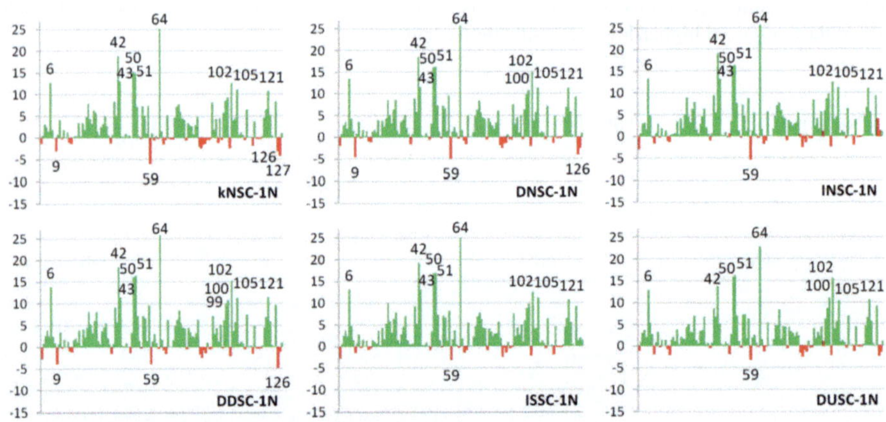

Fig. 6.10 DTW—differences between the classification accuracies of various kNN variants (combined with the SC constrained measure) and 1NN (combined with the unconstrained measure) by datasets (labels represent dataset identifiers)

The positive average differences between the kNN variants and 1N ranged between 4.67 and 3.87, whereby the highest were detected with DDSC and DNSC and the lowest with DUSC and DUIT (Fig. 6.8). These values were higher than those presented in Fig. 5.38 in Chap. 5 for the unconstrained DTW, where without applying global constraints, IS, IN and kN overpowered DD and DN. Except for kNSC and ISIT, the results obtained with DTW were the highest among the elastic measures. Furthermore, when considering the amount of increase relative to the findings with the unconstrained measure, DTW outperformed all other measures in all combinations of schemes and global constraints.

In contrast to the 1NN results in Chap. 4 (Fig. 4.12), except for kN, the other kNN variants generated slightly larger positive average differences with SC than with IT.

The negative average differences were greater than those observed with the unconstrained DTW and ranged between −1.46 and −0.92. 1N achieved the highest average advantage over DDIT and kNIT and the lowest over DUSC and ISSC. With the exception of DN, the other schemes produced slightly higher negative averages with IT than with SC, similar to the 1NN findings. Moreover, except for kNSC, these values were higher than those detected with other elastic measures.

Combined with the kNN classifiers shown in Fig. 6.9, the average advantage of SC (calculated over the datasets where SC outperformed IT) was between 1.3 (DU) and 1.9 (IS) times higher than the average advantage of IT (over datasets where IT outperformed SC). In this regard, the advantage of SC over IT when using the 1NN classifier was close to 1.3 times greater than the other way around (see Fig. 4.12 in Chap. 4). These results further support the dominance of SC over IT. It is also worth noting that the difference in the average advantages of global constraints was, for all six weighting schemes, the largest with DTW among the elastic measures considered, which is also consistent with the findings in Chap. 4.

The most favorable effect of combining kNN classifiers with global constraints occurred on the *InsectWingbeatSound* (64) dataset, where DU achieved an advantage of more than 22 percentage points over 1NN paired with the unconstrained DTW (Figs. 6.10 and 6.11). The average accuracy of the other kNN variants was even more than 25 pp higher than that achieved with 1N. All kNN variants achieved an advantage greater than 10 pp also with *ArrowHead* (6), *FordA* (42) and *UWaveGestureLibraryY* (121). *ArrowHead* (6) and *FordA* (42) generated an average accuracy more than 10 pp higher than the average accuracy of the unconstrained DTW when using the 1NN classifier (see Fig. 4.13 in Chap. 4) as well. However, none of these three datasets were among those over which the kNN variants achieved the greatest advantages over 1NN when using the unconstrained DTW in Chap. 5 (Fig. 5.39). When constraining the width of the warping window employing the Sakoe-Chiba band, the kNN variants outperformed 1N by more than 10 pp also in the case of the datasets labeled 50, 51, 43 (except DU), 102 and 105. Given that when considering the unconstrained DTW, the advantage of the kNN classifier over 1NN did not reach 9 pp, these results strongly confirm the beneficial impact of applying global constraints.

The lag of each discussed SC-constrained kNN variant was greater than 3 pp only with the *Herring* (59) dataset. When it comes to IT, the advantage of 1N was not less than 8 pp for any kNN variant on the *BeetleFly* (8) dataset.

In accordance with the results of the 1NN classifier in Chap. 4 (Fig. 4.13c), SC outperformed IT to the greatest extent on the *BeetleFly* (8), *GesturePebbleZ1* (50), *GesturePebbleZ2* (51) and *ShapeletSim* (102) datasets (Fig. 6.12). However, in

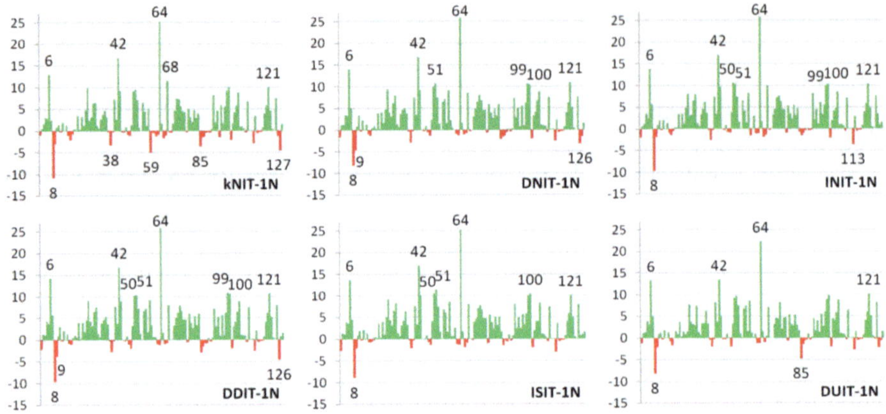

Fig. 6.11 DTW—differences between the classification accuracies of various kNN variants (combined with the IT constrained measure) and 1NN (combined with the unconstrained measure) by datasets (labels represent dataset identifiers)

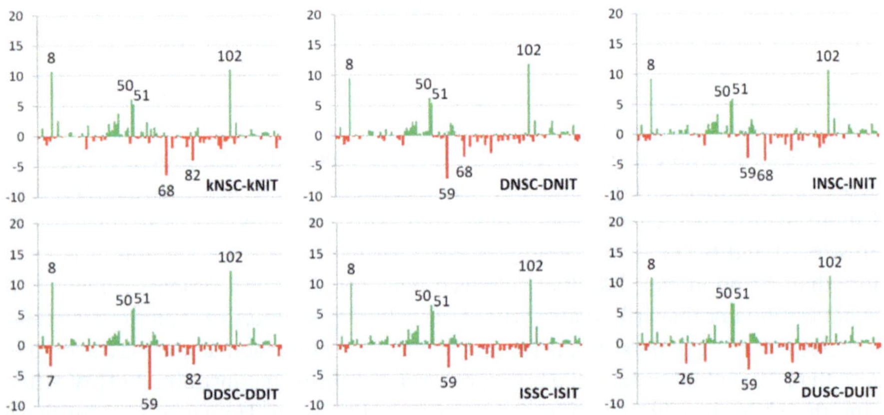

Fig. 6.12 DTW—differences between the classification accuracies of the SC and IT constrained measure (combined with different variants of the kNN classifier) by datasets (labels represent dataset identifiers)

contrast to the 1NN classifier, SC suffered the biggest defeat from IT most often on the *Herring* (59) dataset.

The statistical significance of the difference in the performance of the considered classifiers was confirmed by the result of the Friedman test ($\chi^2(12) = 189.87$, $p < .001$). From Fig. 6.13, it can be seen that according to Nemenyi's post-hoc test, 1N was not significantly different from kNIT alone. Given that when relying on the unconstrained DTW, only DD and DN differed significantly from 1N (see Fig. 5.40 in Chap. 5), this result can be primarily attributed to the application of global constraints. As reported by the less conservative Wilcoxon signed-rank test (with the Holm-Bonferroni correction), the difference between 1N and kNIT was also significant (Table 6.3). The inferiority of 1NN combined with the unconstrained DTW compared to kNN variants paired with global constraints (and especially in relation to DD, DN, IS and IN with SC) was also confirmed with the numbers of statistically significant wins and losses shown in Table 6.4.

The estimate of the advantage of kNN variants over 1N based on the expected accuracy gain was wrong in between about 15% (IS) and 22% (kN) of the datasets when using the SC constraint and between about 18% (IS) and 28% (kN) with the application of the IT constraint (Figs. 6.14 and 6.15). It can also be observed that in combination with IT, each kNN variant had a higher number of FP and a lower number of TP datasets than in combination with SC.

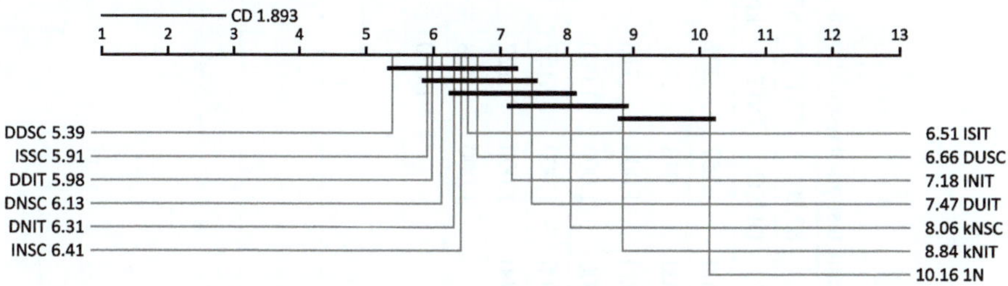

Fig. 6.13 DTW—critical difference diagram comparing average accuracy ranks based on Nemenyi's test: classifiers connected by a solid bar did not differ statistically significantly

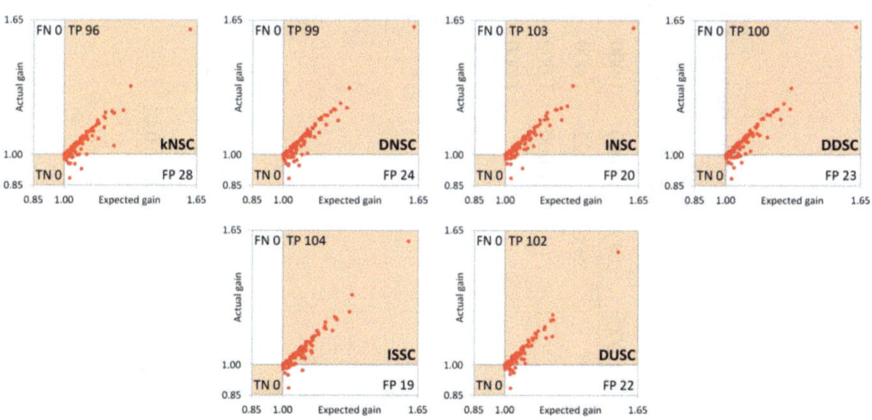

Fig. 6.14 DTW—comparing the actual and expected gain in classification accuracy: kNN variants (combined with the SC constrained measure) versus 1NN (combined with the unconstrained measure)

Table 6.3 DTW—Holm-Bonferroni-adjusted p-values of the pairwise Wilcoxon signed-rank test on classification accuracy

A		82.66	82.55	82.53	82.48	82.30	82.27	82.23	82.20	82.19	82.10	81.88	81.79	79.23
		ISSC	DDSC	INSC	DNSC	ISIT	DDIT	DNIT	DUSC	INIT	kNSC	DUIT	kNIT	1N
82.66	ISSC		1.000	.043	1.000	1.000	1.000	1.000	.157	.222	**< .001**	.025	**< .001**	**< .001**
82.55	DDSC			.506	.022	1.000	1.000	1.000	.018	.277	**< .001**	**.007**	**< .001**	**< .001**
82.53	INSC				1.000	1.000	1.000	1.000	1.000	.716	**< .001**	.240	**< .001**	**< .001**
82.48	DNSC					1.000	1.000	1.000	1.000	1.000	**< .001**	.066	**< .001**	**< .001**
82.30	ISIT						1.000	1.000	1.000	.157	.812	.191	**< .001**	**< .001**
82.27	DDIT							1.000	1.000	.122	.667	**.005**	**< .001**	**< .001**
82.23	DNIT								1.000	.428	.735	.206	**< .001**	**< .001**
82.20	DUSC									1.000	.067	1.000	.342	**< .001**
82.19	INIT										1.000	1.000	**< .001**	**< .001**
82.10	kNSC											1.000	1.000	**< .001**
81.88	DUIT												.417	**< .001**
81.79	kNIT													**< .001**
79.23	1N													

Note Bolded values indicate significance at the level of $\alpha = 0.01$

Table 6.4 DTW—statistically significant wins (W) and losses (L) reported by the corrected resampled *t*-test on classification accuracy (across 12,800 splits)

	DDSC	DNSC	ISSC	INSC	DDIT	ISIT	DNIT	DUSC	INIT	kNSC	DUIT	kNIT	1N
W	449	406	407	352	340	324	303	351	261	260	240	174	163
L	132	158	160	197	213	233	234	289	285	342	355	427	1005
W-L	317	248	247	155	127	91	69	62	−24	−82	−115	−253	−842

Applying the Sakoe-Chiba band reduced the number of datasets in the FP region by approximately between 1.1 (from 25 to 22 for DU) and 2.3 (from 46 to 20 for IN) times compared to the results in Chap. 5 (Fig. 5.41). At the same time, there was an increase in the number of datasets in the TP region by approximately between 1.3 (from 80 to 100 for DD) and 2.0 (from 49 to 96 for kN) times. The impact of constraining elasticity with the Itakura parallelogram was more moderate, the biggest drop in the number of FP datasets was also observed with the IN scheme (by about 1.8 times), but in combination with the DU scheme it even led to an increase (from 25 to 27). The largest increase in the number of TP datasets was detected, the same as with SC, for the kN (from 49 to 88) and the smallest for the DD and DU (from 80 to 97) schemes.

Both SC and IT made the biggest gains by switching from 1NN to IS: the number of datasets in the FP region decreased from 23 and 37 to 19 and 23 and their number in the TP region increased from 99 and 85 to 104 and 101 (see Fig. 4.15 in Chap. 4). It should also be mentioned that for SC, the transition from 1NN to majority-voting kNN resulted in an increase in the number of FP datasets and a decrease in those in the TP region. Furthermore, looking at the average of all six kNN variants, according to the degree of benefit from taking into account more than one neighbor, only ERP achieved a slightly weaker result than DTW, but only for IT.

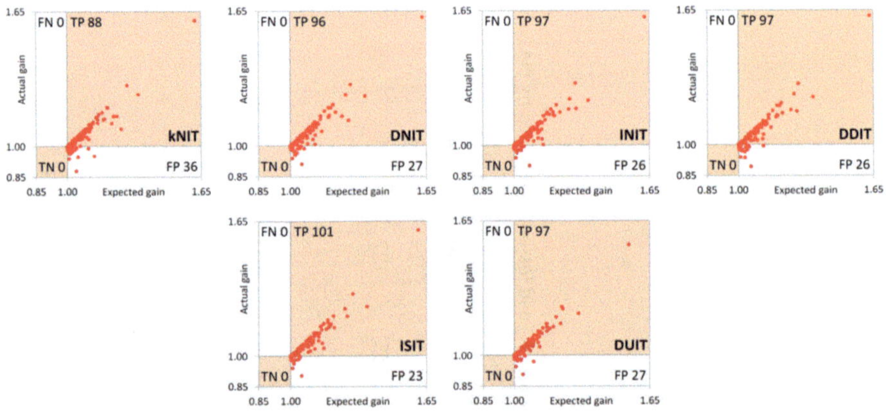

Fig. 6.15 DTW—comparing the actual and expected gain in classification accuracy: kNN variants (combined with the IT constrained measure) versus 1NN (combined with the unconstrained measure)

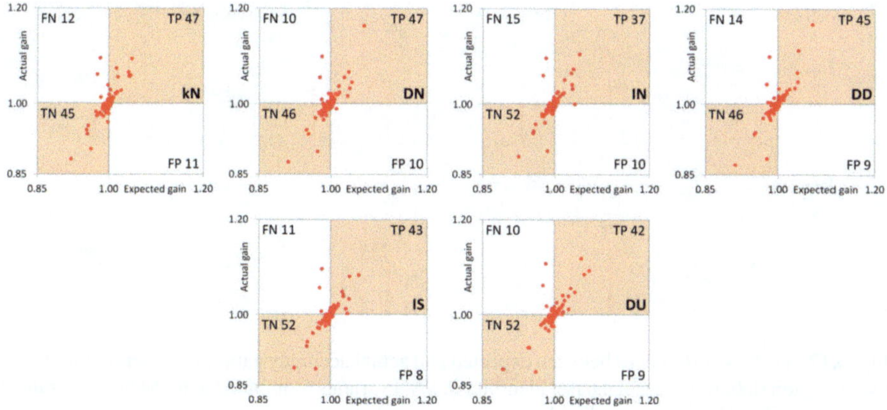

Fig. 6.16 DTW—comparing the actual and expected gain in classification accuracy: IT versus SC (combined with different variants of the kNN classifier)

As can be seen from Fig. 6.16, the number of datasets where, based on the expected gain, the better type of global constraints was successfully identified was the highest with IS (95) and the lowest with IN (89), which represents an improvement compared to the 1NN classifier (84), the results of which are presented in Chap. 4 (Fig. 4.15c). In addition, the total number of datasets in the FN and FP regions (for which, based on the training set, we failed to correctly identify which of the two types of global constraints would be more successful on the test set) was reduced from 30 when using the 1NN classifier to between 19 (IS and DU) and 25 (IN). These results suggest that, when it comes to DTW, the kNN variants make less mistakes in choosing the better type of global constraints based on the training data. Averaging the six kNN methods, among the elastic measures considered, DTW produced the most significant improvement over the results with 1NN in this category.

When comparing the SC constrained kNN variants with the 1NN classifier paired with the unconstrained DTW the expected gain was greater than the actual gain in the FP region by more than 0.05 for the *Computers* (17), *Herring* (59), *PigAirway-Pressure* (86) and *PigCVP* (88) datasets (Fig. 6.17). Relying on the IT constraint, in the overestimation of the kNN variants over 1N that threshold was exceeded with *BeetleFly* (8), *GestureMidAirD2* (48) and *PigAirwayPressure* (86), as shown in Fig. 6.18. These datasets were not the only ones where the difference between expected and actual gain was greater than 0.05, but they were common to all kNN classifiers analyzed in this chapter. In addition, they were all among the datasets for which the deviation between expected and actual gain was the largest when evaluating the performance of constrained DTW variants against unconstrained DTW using the 1NN classifier in Chap. 4 (see Fig. 4.16).

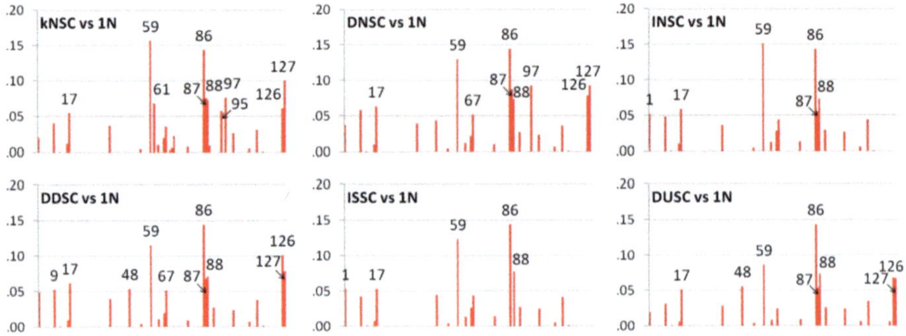

Fig. 6.17 DTW—differences between expected and actual accuracy gains of datasets belonging to the FP region (labels represent dataset identifiers): kNN variants (combined with the SC constrained measure) versus 1NN (combined with the unconstrained measure)

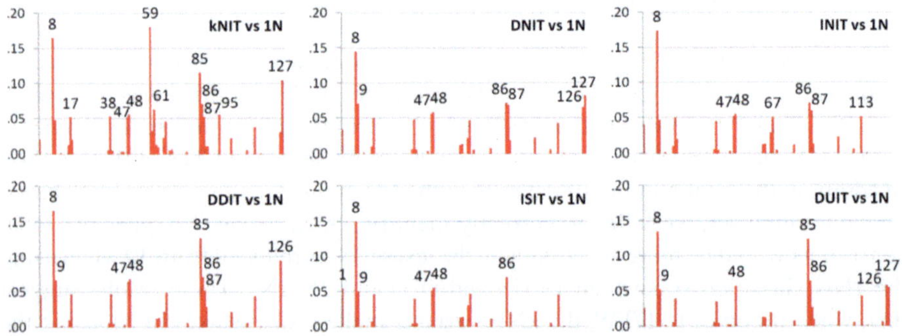

Fig. 6.18 DTW—differences between expected and actual accuracy gains of datasets belonging to the FP region (labels represent dataset identifiers): kNN variants (combined with the IT constrained measure) versus 1NN (combined with the unconstrained measure)

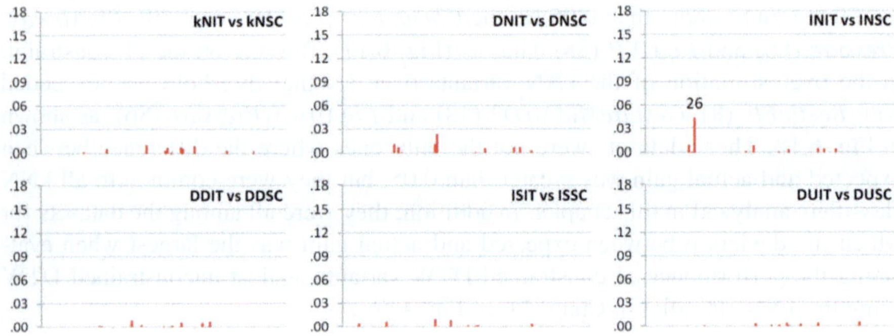

Fig. 6.19 DTW—differences between expected and actual accuracy gains of datasets belonging to the FP region (labels represent dataset identifiers): IT versus SC (combined with different variants of the kNN classifier)

Analogous to the results with the 1NN classifier in Chap. 4, the difference between the expected and actual gain was less than 0.03 when comparing IT and SC in the case of kNN variants as well (Fig. 6.19). The only exception was *DodgerLoopDay* (26) with the IN scheme.

6.2 Longest Common Subsequence (LCS)

The best classification accuracy obtained utilizing the SCV10x10 evaluation algorithm (where in each of the 100 splits into testing and training subsets, the optimal value of the parameters k and r was chosen over the training subset using the SCV1x9 algorithm) was achieved by the DD weighting scheme with the application of the SC-constrained LCS (Table 6.5), while the combination of 1NN and the unconstrained LCS (denoted as 1N) proved to be the least successful in this regard. The advantage of the considered kNN variants over 1N ranged between 2.57 and 1.83 percentage points (pp). Applying global constraints, these performance differences were by approximately 0.09 (DUIT) to 0.5 (ISIT) pp greater than the results obtained with the unconstrained LCS in Chap. 5 (see Table 5.13). Based on these findings, two conclusions can be drawn: SC and IT showed very similar performances and constraining the elasticity of LCS only very moderately improved the average classification accuracy of the considered weighting schemes.

With the exception of IS, the other schemes achieved consistent results with the 1NN classifier (Table 4.4 in Chap. 4): SC generated somewhat higher accuracy than IT.

The largest standard deviation (SD) was observed for 1N (20.76 pp) and the smallest for ISIT (19.16 pp). Similar to DTW, SC produced a slightly larger deviation than that observed with IT (with 1N, in Chap. 4, it was the other way around). Moreover, the application of global constraints marginally reduced the SDs compared to the values obtained with the unconstrained LCS in Chap. 5.

The employment of global constraints did not change the (descending) order of the weighting schemes with regard to the average value of the optimal numbers of nearest neighbors (Table 6.5): DU, DD, DN, IS, IN, kN (see Chap. 5). However, it did increase both the average values and the standard deviations to a small extent. Furthermore, it can be observed that both AV and SD were slightly higher with IT than with SC for each analyzed kNN variant (only the IN scheme was an exception in terms of AV). Similar to when combined with the unconstrained LCS, DU stood out by generating significantly higher average optimal k-values than the other schemes.

Comparing Figs. 6.20 and 5.44 (Chap. 5), it is clear that constraining the warping window did not noticeably change the proportion of splits in which various kNN variants chose $k = 1$ as the best option. According to the degree of reliance on the first nearest neighbor, kN stood out the most, followed by IN, IS, DU, DN and DD, in the same order as when using the unconstrained LCS. At the same time, the preference for $k = 1$ was marginally higher with IT than with SC (the difference ranged between 0.28 pp (kN) and 1.19 pp (DN)). Regarding the frequencies of higher

Table 6.5 LCS—average value (AV) and standard deviation (SD) of classification accuracy (AV), optimal number of nearest neighbors (k) and optimal warping-window width (r) across 12,800 splits; the number of datasets for which a given classifier produced the best result (WIN) and the mean rank (MR)

	DDSC	DDIT	DNSC	ISIT	ISSC	DNIT	INSC	INIT	kNSC	kNIT	DUSC	DUIT	1N
$\mathbf{A_{AV}}$ (%)	80.15	80.15	80.07	80.05	80.04	80.03	80.01	79.95	79.58	79.56	79.52	79.42	77.59
$\mathbf{A_{SD}}$ (pp)	19.29	19.18	19.31	19.16	19.30	19.20	19.28	19.18	19.49	19.32	19.56	19.49	20.76
$\boldsymbol{k_{AV}}$	11.99	12.03	11.40	8.90	8.89	11.43	8.67	8.65	7.11	7.19	16.28	16.31	
$\boldsymbol{k_{SD}}$	9.21	9.31	9.12	8.98	8.89	9.21	8.91	8.94	8.59	8.71	10.59	10.62	
$\boldsymbol{r_{AV}}$(%)	6.92	27.95	6.95	26.30	7.05	27.77	7.18	26.45	6.92	26.66	6.36	26.76	
$\boldsymbol{r_{SD}}$(pp)	7.10	31.40	7.15	29.80	7.22	31.05	7.31	29.81	6.99	29.87	6.53	31.09	
WIN	23	19	14	18	16	9	10	9	9	8	20	7	22
MR	4.95	5.56	5.42	7.10	5.98	6.37	6.70	7.50	8.38	8.93	6.91	7.93	9.28

k-values (> 2) shown in Fig. 6.21, apart from a small increase in standard deviations (for both SC and IT), the kNN variants maintained the general features they had when comparing the time series with the unconstrained LCS (Fig. 5.45). It is also evident that, in this matter, there were no visible differences between the SC and IT constraints.

As can be seen from Table 6.6, constraining the warping window had no significant effect on the number of datasets that exhibited the property that the optimal value of the parameter k was 1 in each of the 100 iterations of the SCV10x10 algorithm. In general, the same datasets were involved. Among them, for *BME* (10), *Fungi* (46), *PigAirwayPressure* (86) and *TwoPatterns* (117), this feature was present for all weighting schemes, analogous to the unconstrained LCS.

The result of Chap. 4 that, on average, IT relied on about 4 times higher values of the parameter r than SC was also confirmed in combination with the kNN variants

Fig. 6.20 LCS—percentages of splits in which SCV1x9 chose $k = 1$ and $k > 1$ as the optimal number of nearest neighbors on the training set when evaluating classification accuracy using SCV10x10

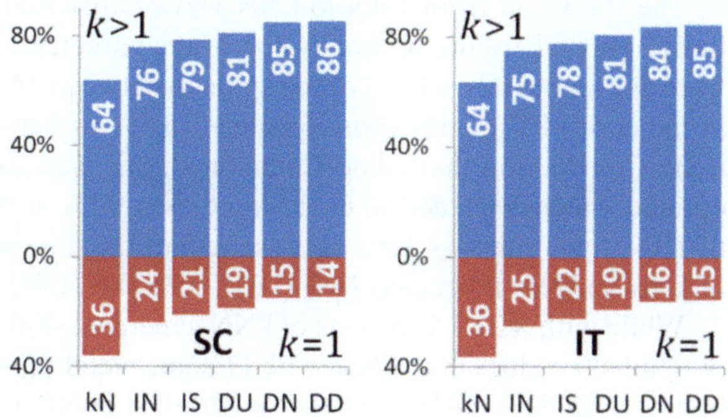

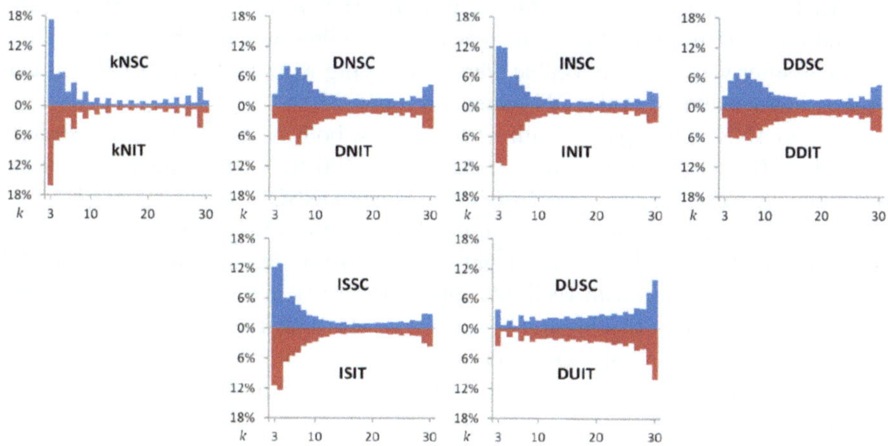

Fig. 6.21 LCS—frequencies (in percentage relative to the 12,800 splits) of different values of the parameter k (≥ 3) chosen by the SCV1x9 algorithm over the training set when evaluating classification accuracy using SCV10x10

Table 6.6 LCS—number of datasets for which the optimal number of nearest neighbors was 1 in each of the 100 iterations of the SCV10x10 algorithm

	kN	DN	IN	DD	IS	DU
LCS	12	4	6	4	6	6
SC	12	5	5	5	5	6
IT	11	4	5	4	5	7

discussed in this chapter (Table 6.5). Namely, the quotient between the average *r*-values was the lowest for IN (almost 3.7) and the highest for DU (a little more than 4.2). However, considering more than one neighbor led to an increase in AV for both types of global constraints (when it comes to IT, the difference ranged between approximately 1.50 (IS) and 3.16 (DD) pp and in the case of SC between about 0.35 (DU) and 1.16 (IN) pp).

The transition from 1NN to kNN variants resulted in a decreased proportion of those splits where the optimal value of the parameter *r* was 0% (in the case of SC, from approximately 17.0% of the splits to between 16.1% (DU) and 12.6% (IS) and in the case of IT, from approximately 15.1% to between 13.6% (DU) and 10.7% (IS)). Considering the Sakoe-Chiba band, this was also manifested through higher optimal *r*-values needed to cover more than 90% of the 12,800 splits: with the use of 1NN, this limit was 15% of the length of the time series and with the application of kNN variants it became 16% (DU), 18% (kN, DN, DD) and 19% (IN, IS).

While with SC all 5 variants of kNN generated slightly higher standard deviations of optimal *r*-values than 1NN, with IT some weighting schemes produced somewhat higher (DD, DU, DN) and some a little lower SD values (kN, IN, IS). However, in no case did the difference reach even 1.5 pp. In addition, similar to Chap. 4, IT produced about 4.1 (IN) to 4.8 (DU) times greater SD than SC. Nevertheless, the general characteristics of the optimal warping-window widths observed with the 1NN classifier were also noticeable when classifying with the kNN variants, as can be seen in the example of the DN scheme in Fig. 6.22.

As with 1NN, when paired with kNN variants, both types of global constraints chose $r = 0\%$ as the best option in each of the 100 SCV10x10 iterations on the following datasets: *FreezerSmallTrain* (45) (except DNSC), *Meat* (70), *Melbourne-Pedestrian* (72) and *OliveOil* (81). While relying on the 1NN classifier this list also included the *Wine* (124) dataset, it only exhibited this feature here with the DU scheme. In combination with 1NN, SC opted for $r = 0\%$ in each iteration also in the case of the *Chinatown* (13), *FreezerRegularTrain* (44) and *Strawberry* (109) datasets, but with kNN classifiers this was observed only for *FreezerRegularTrain* (44). Additionally, when utilizing the Sakoe-Chiba constraint, the value of the parameter *r* was 1% in each split of *FordA* (47) and *FordB* (48) and 3% in each SCV10x10 iteration of the *Crop* (21) dataset, both with 1NN and kNN variants.

Except for DDSC, all other classifiers generated fewer wins than 1N (Table 6.5), which is (except for DU) consistent with the results in Chap. 5 devoted to the analysis

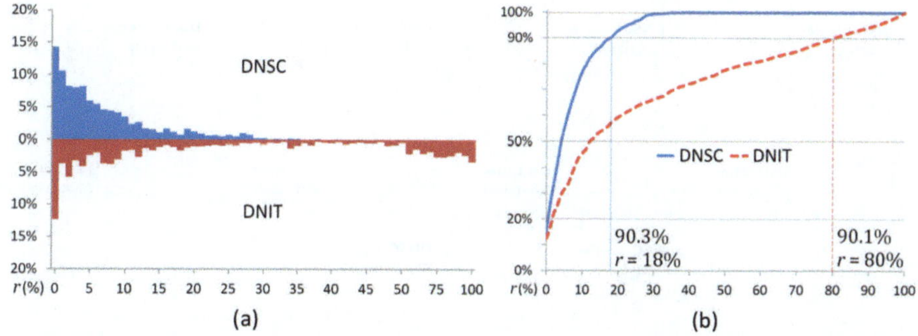

Fig. 6.22 LCS—frequencies (in percentage relative to the 12,800 splits) of different warping-window widths selected by the SCV1x9 algorithm over the training sets when evaluating classification accuracy using SCV10x10 (**a**); cumulative display of frequencies of optimal warping-window widths (**b**)

of the impact of different weighting schemes on classification relying on the unconstrained LCS (Table 5.13). This means that the employment of global constraints did not yield any performance gains in this respect. In terms of mean rank, just as when using unconstrained LCS, 1N was outperformed by all classifiers considered.

The numbers of wins of the kNN variants paired with the SC and IT constrained LCS over the 1NN classifier paired with the unconstrained LCS and vice versa are shown in Figs. 6.23 and 6.24 along with the classification accuracies achieved on individual UCR datasets (represented by dots). The best win-loss ratio was achieved by DDSC (3.2:1) and DDIT (3.1:1) and the worst by kNIT (1.7:1) and kNSC (1.6:1). Apart from kN, the other weighting schemes achieved a slightly better win-loss ratio in combination with the Sakoe-Chiba band than with the Itakura parallelogram, which was also the case in Chap. 4 with the 1NN classifier. However, it should be emphasized that with 1NN, the number of wins of the unconstrained LCS was higher than the number of wins of the constrained variants, which highlights the favorable impact of switching to kNN classifiers.

From limiting the elasticity of the LCS measure, only the kN and IS weighting schemes benefited both when using the SC and when applying the IT constraint. The biggest drawback was observed in the case of the DU method whose win-to-loss ratio decreased from 4.3:1 (see Fig. 5.47 in Chap. 5) to 2.5:1 with SC and even to 1.9:1 with IT.

Comparing the performance of SC and IT through their numbers of wins and losses, it can be concluded that considering more than one neighbor, they seem more similar to each other (see Figs. 6.25 and 4.23c). The DU scheme is an exception here as in its case the win-loss ratio of SC over IT was 1.8:1, therefore the difference between SC and IT was greater than with 1NN (where the ratio was 1.7:1).

Regarding the positive average difference (Fig. 6.26), the greatest advantage in relation to 1N was achieved by INIT (4.38) and ISIT (4.17) and the smallest by DUIT (3.32) and DUSC (3.14). The results obtained with SC and IT were higher than those

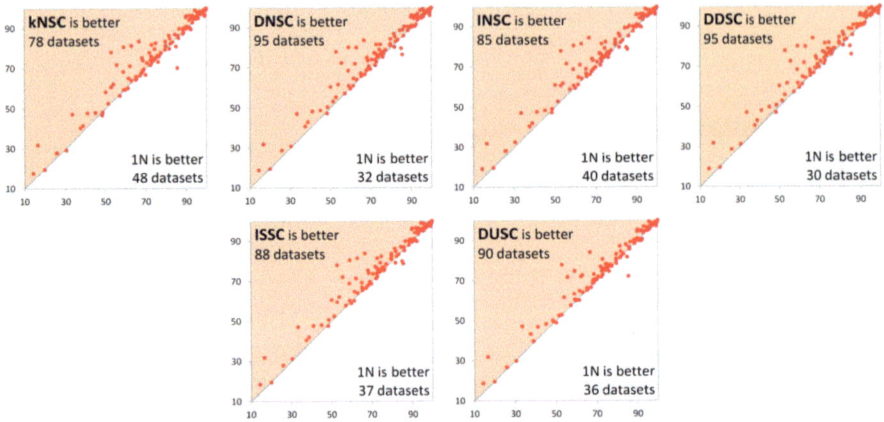

Fig. 6.23 LCS—comparison of the kNN variants (combined with the SC constrained measure) with the 1NN classifier (combined with the unconstrained measure) based on the number of datasets where they achieved higher/lower classification accuracy

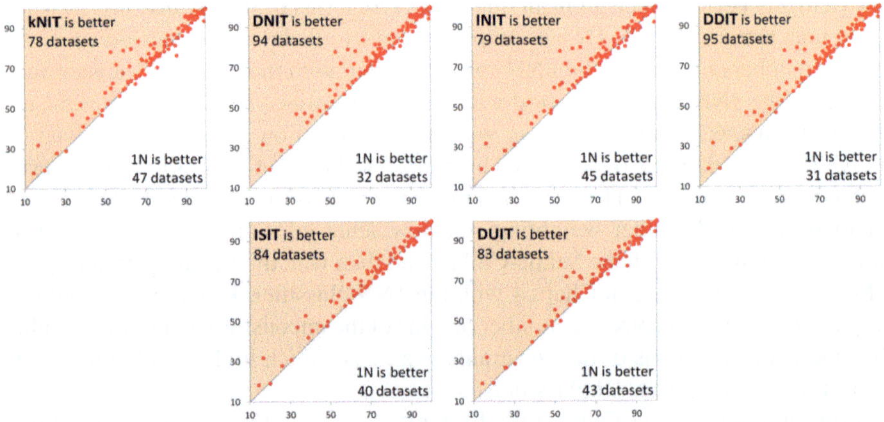

Fig. 6.24 LCS—comparison of the kNN variants (combined with the IT constrained measure) with the 1NN classifier (combined with the unconstrained measure) based on the number of datasets where they achieved higher/lower classification accuracy

with the unconstrained LCS in Chap. 5 (see Fig. 5.48), which favors the application of global constraints.

Contrary to the results in Chap. 4 (Fig. 4.24), kN, IN, IS and DU generated larger average differences with IT than with SC, but the reverse was true for DN and DD.

The average advantage of 1N over the variants of kNN was somewhat higher than in combination with the unconstrained LCS: it was the largest with kNIT (−1.34) and kNSC (−1.18) and the smallest with ISIT (−0.88) and INSC (−0.88). When analyzing the performance of the 1NN classifier in Chap. 4, we saw that IT generated

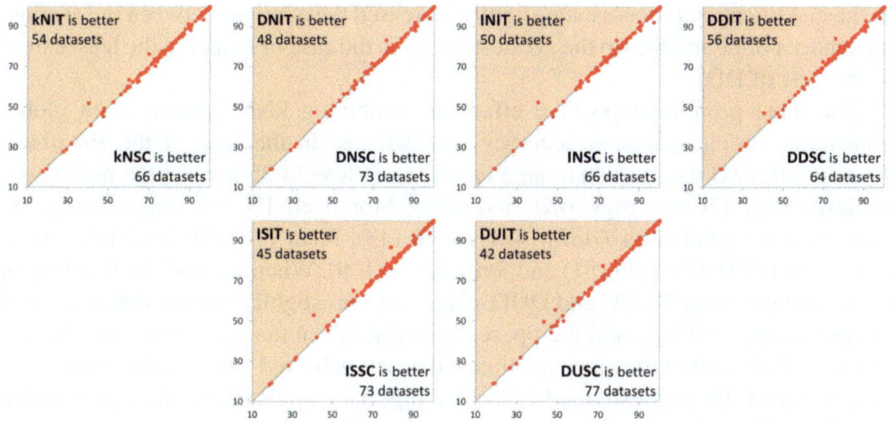

Fig. 6.25 LCS—comparison of the SC and IT constraints combined with the given kNN classifier variant based on the number of datasets where they achieved higher/lower classification accuracy

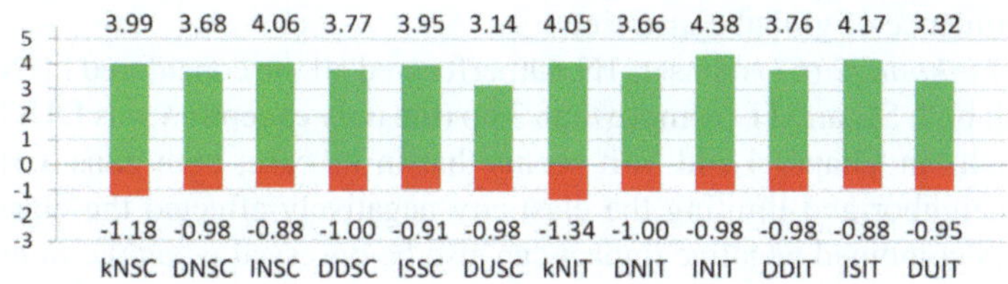

Fig. 6.26 LCS—average difference in classification accuracy of various kNN variants (combined with the SC and IT constrained measure) and 1NN (combined with the unconstrained measure) over the datasets where the performance of a given kNN variant was better (green) and worse (red) than 1NN

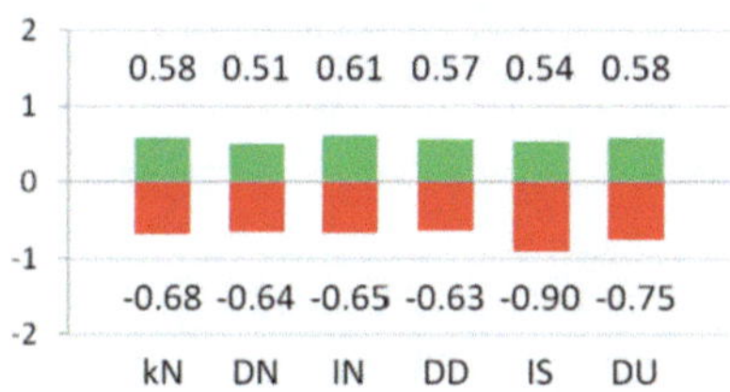

Fig. 6.27 LCS—Average difference in classification accuracy of the SC and IT constraints (combined with different variants of the kNN classifier) over the datasets where the performance of SC was better (green) and worse (red) than IT

a little higher negative average difference than SC, with which only kN, DN and IN matched here.

Figure 6.27 shows that the average advantage of IT was slightly higher than the average advantage of SC, similar to the case of the 1NN classifier in Chap. 4 (Fig. 4.24). Moreover, except for the IN scheme, including more than one neighbor in the classification process even slightly increased the quotient between the average advantages of IT and SC (to the greatest extent in the case of IS and to the least extent in the case of DD).

The most prominent positive effect of combining kNN variants with global constraints on classification accuracy was detected in the case of the *ProximalPhalanxTW* (94) dataset where an advantage of close to 25 percentage points was achieved over 1N (see Figs. 6.28 and 6.29). Moreover, 1N was outperformed by more than 13 pp also on *FordA* (42), *FordB* (43), *Meat* (70), *MiddlePhalanxOutlineCorrect* (74), *OliveOil* (81) and *Strawberry* (109). When it comes to *FordB* (43), the advantage of the DUSC and DUIT classifiers was slightly smaller and amounted to approximately 9.8 pp and 8.2 pp, respectively. Given that, except for *FordA* (42) and *FordB* (43), all of the aforementioned datasets achieved very similar advantages over 1N with the unconstrained LCS (see Fig. 5.49 in Chap. 5), the performance improvement in their case can be attributed solely to the application of different weighting schemes. On the other hand, when relying on the unconstrained LCS, the difference between the average accuracy of 1N and kNN variants on *FordA* (42) and *FordB* (43) was not greater than 5 pp, which, in their case, clearly emphasizes the positive influence of global constraints.

On the *Lightning2* (67) dataset, 1N outperformed all the considered kNN variants paired with both SC and IT by more than 3 pp (the only exception was DUIT). Based on the results in Chaps. 4 and 5, it seems that in its case, both considering more than one neighbor and limiting the elasticity negatively affected the classification accuracy. A combined negative impact can also be observed when the *BeetleFly* (8) and *BirdChicken* (9) datasets are concerned.

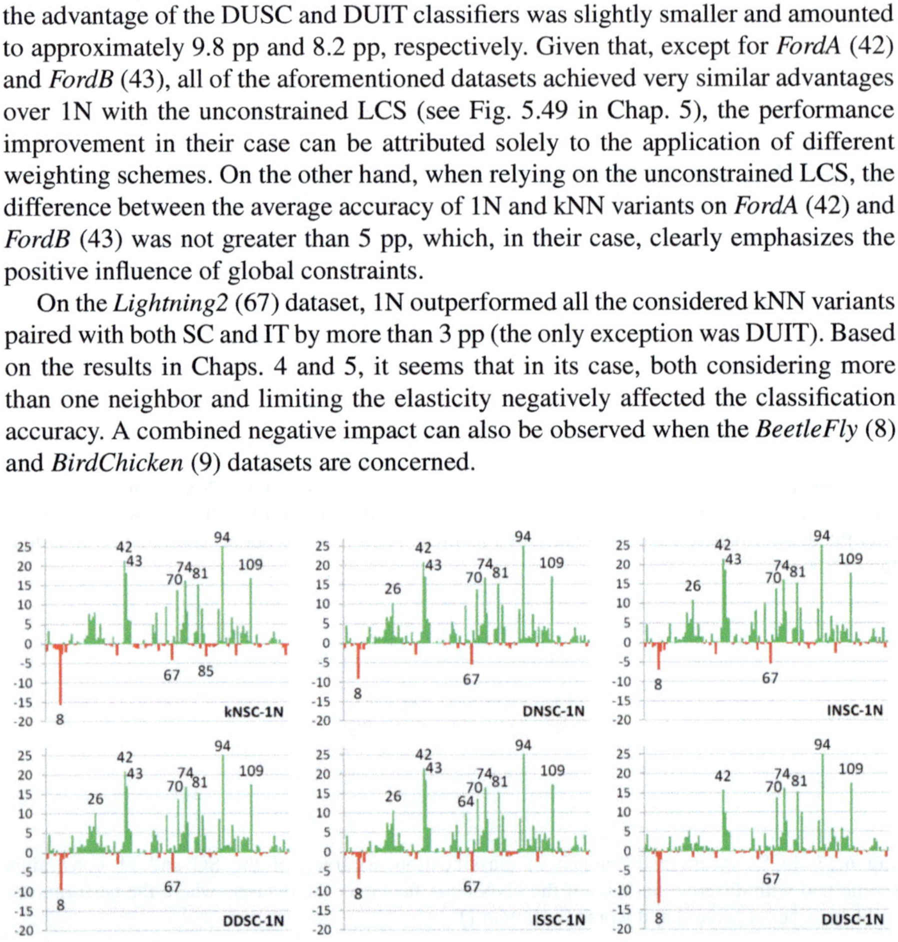

Fig. 6.28 LCS—differences between the classification accuracies of various kNN variants (combined with the SC constrained measure) and 1NN (combined with the unconstrained measure) by datasets (labels represent dataset identifiers)

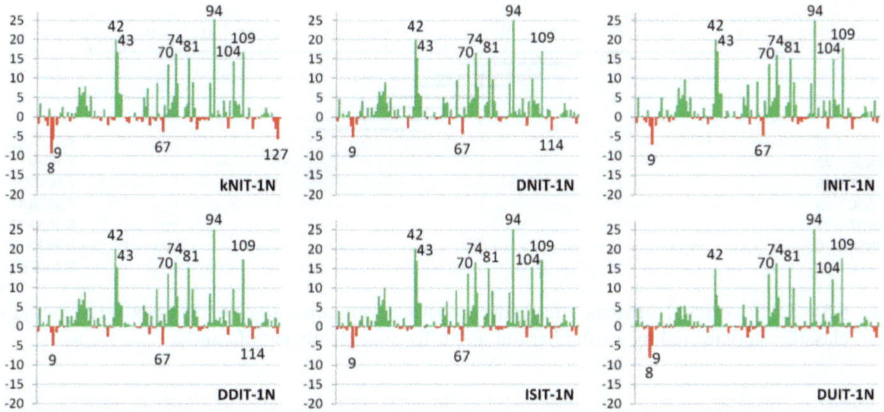

Fig. 6.29 LCS—differences between the classification accuracies of various kNN variants (combined with the IT constrained measure) and 1NN (combined with the unconstrained measure) by datasets (labels represent dataset identifiers)

SC outperformed IT by more than 3 pp only on the *BirdChicken* (9) dataset (Fig. 6.30) and IT's advantage was greater than 4 pp on *BeetleFly* (8) and *SmallKitchenAppliances* (104). When considering the 1NN classifier in Chap. 4, of these three datasets only *BeetleFly* (8) exhibited a similar result (see Fig. 4.25c).

The performance of the discussed classifiers differed significantly, which was verified by applying the Friedman test ($\chi^2(12) = 194.33, p < .001$). According to Nemenyi's post-hoc test (Fig. 6.31), the effectiveness of 1N was significantly lower than the effectiveness of most kNN variants, the only exceptions being kNIT, kNSC, DUIT, INIT. This finding confirms the result with the unconstrained LCS in Chap. 5

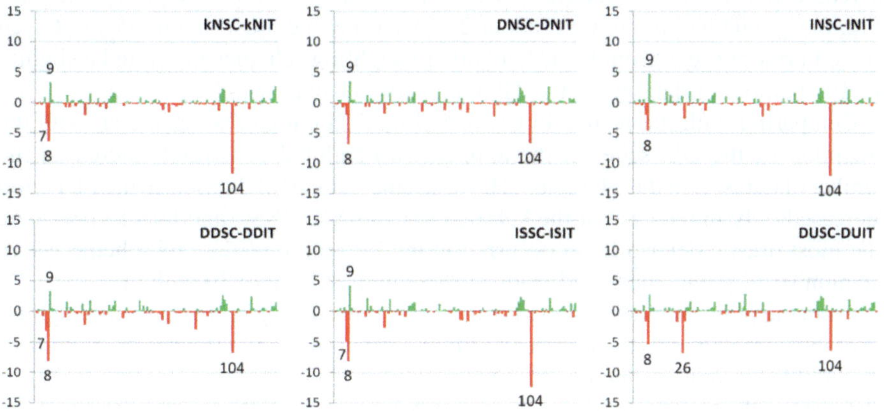

Fig. 6.30 LCS—differences between the classification accuracies of the SC and IT constrained measure (combined with different variants of the kNN classifier) by datasets (labels represent dataset identifiers)

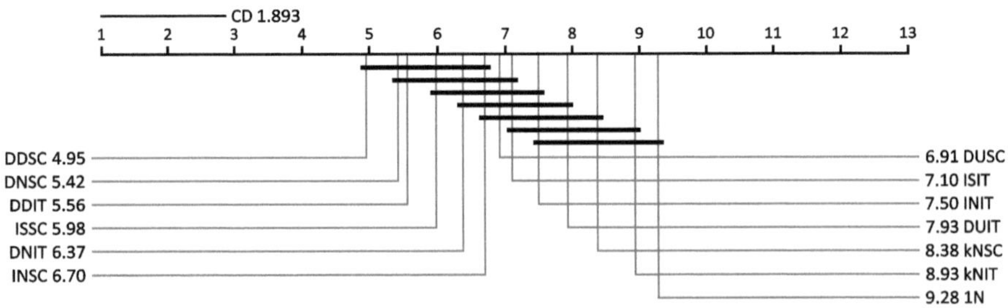

Fig. 6.31 LCS—critical difference diagram comparing average accuracy ranks based on Nemenyi's test: classifiers connected by a solid bar did not differ statistically significantly

(Fig. 5.50) according to which 1N and kN were not significantly different. In addition, it suggests that the kNN variants were slightly more successful in combination with the Sakoe-Chiba band than with the Itakura parallelogram, which deviates from the result with the 1NN classifier in Chap. 4 (Fig. 4.26) where the difference between the SC-constrained and the unconstrained LCS was not statistically significant. The Holm-Bonferroni-adjusted p-values of the pairwise Wilcoxon signed-rank test in Table 6.7 indicate that all kNN variants significantly outperformed the accuracy of 1N, the same as in the case of the unconstrained LCS in Chap. 5 (Table 5.14). Additionally, kN was also significantly outperformed by most of the other kNN classifiers. Based on the numbers of statistically significant wins and losses shown in Table 6.8, the biggest differences were present between DDSC, DNSC and DDIT on the one hand and kNSC, kNIT and 1N on the other.

When constraining the elasticity of the LCS measure using the Sakoe-Chiba band, the expected gain of kNN variants over 1N was misleading in between approximately 20% (DD) and 34% (kN) of cases (Fig. 6.32). Applying the Itakura parallelogram increased the number of datasets in the FP region for all six weighting schemes (Fig. 6.33): DD had the fewest (about 22%) and kN the most (about 35%) of such cases. Furthermore, except for kN, all other weighting schemes generated a slightly smaller number of datasets in the TP region with IT than with SC.

Comparing with the results in Chap. 5 (Fig. 5.51), it can be seen that, with the exception of the DU scheme, the application of global constraints increased the number of datasets in the TP region. However, the number of datasets in the FP region decreased only in the case of the kNSC, INSC, DDSC, ISSC and DDIT classifiers. The most drastic deterioration occurred in the FP region of the DU scheme where the number of points increased from 21 to 33 with SC and to 39 with IT.

From the point of view of global constraints, considering more than one neighbor favorably affected both the number of FP and the number of TP datasets. With both SC and IT, the improvement over 1NN results in Chap. 4 (see 4.27) was the most moderate with the kN and the most pronounced with the DD scheme. The number of FP cases was reduced by between 1.5 and 2.6 times for SC and between 1.6 and 2.6 times for IT. The number of TP cases increased between 1.6 and 2.0 times for SC and between 1.7 and 2.1 times for IT.

Table 6.7 LCS—Holm-Bonferroni-adjusted p-values of the pairwise Wilcoxon signed-rank test on classification accuracy

A		80.15	80.15	80.07	80.05	80.04	80.03	80.01	79.95	79.58	79.56	79.52	79.42	77.59
		DDSC	DDIT	DNSC	ISIT	ISSC	DNIT	INSC	INIT	kNSC	kNIT	DUSC	DUIT	1N
80.15	DDSC		1.000	.021	.025	.177	.037	**.009**	**.004**	**< .001**	**< .001**	**< .001**	**< .001**	**< .001**
80.15	DDIT			1.000	.155	1.000	**< .001**	1.000	**.009**	**.001**	**< .001**	.025	**< .001**	**< .001**
80.07	DNSC				.080	.951	.265	.113	.027	**< .001**	**< .001**	**.002**	**< .001**	**< .001**
80.05	ISIT					.270	1.000	1.000	.215	.074	**< .001**	1.000	.721	**< .001**
80.04	ISSC						1.000	.721	.132	**< .001**	**< .001**	.721	.045	**< .001**
80.03	DNIT							1.000	.356	**.005**	**< .001**	.295	**< .001**	**< .001**
80.01	INSC								.417	**< .001**	**< .001**	1.000	.102	**< .001**
79.95	INIT									.721	**.005**	1.000	.920	**< .001**
79.58	kNSC										1.000	.262	1.000	**.001**
79.56	kNIT											.492	1.000	**.005**
79.52	DUSC												.018	**< .001**
79.42	DUIT													**< .001**
77.59	1N													

Note Bolded values indicate significance at the level of $\alpha = 0.01$

Table 6.8 LCS—statistically significant wins (W) and losses (L) reported by the corrected resampled *t*-test on classification accuracy (across 12,800 splits)

	DDSC	DNSC	DDIT	ISSC	DNIT	INSC	DUSC	ISIT	INIT	DUIT	kNSC	kNIT	1N
W	497	425	399	340	329	313	354	275	254	248	194	168	199
L	103	127	148	192	189	200	316	298	325	427	399	484	787
W-L	394	298	251	148	140	113	38	−23	−71	−179	−205	−316	−588

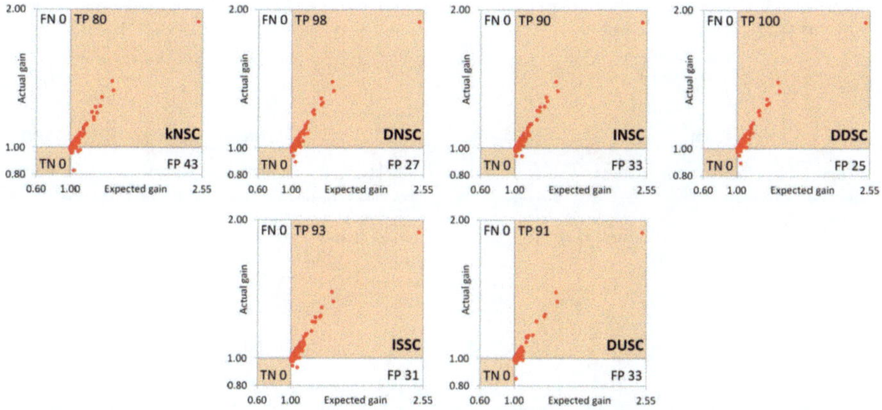

Fig. 6.32 LCS—comparing the actual and expected gain in classification accuracy: kNN variants (combined with the SC constrained measure) versus 1NN (combined with the unconstrained measure)

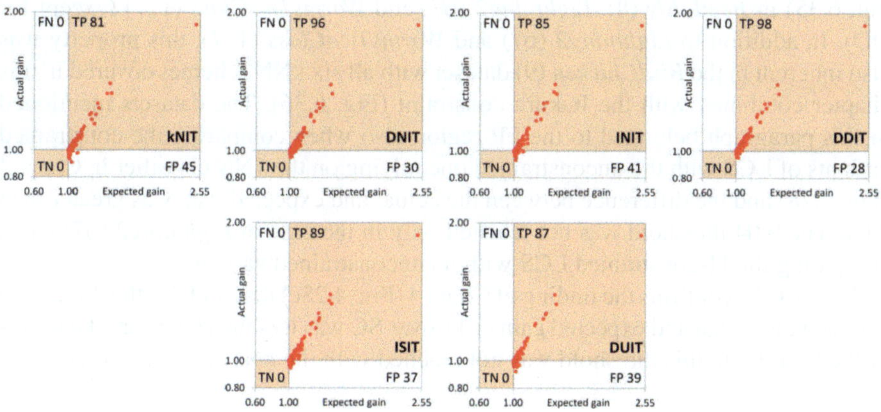

Fig. 6.33 LCS—comparing the actual and expected gain in classification accuracy: kNN variants (combined with the IT constrained measure) versus 1NN (combined with the unconstrained measure)

The number of datasets for which, based on the expected gain, we could not successfully determine which type of global constraints would actually be better (the union of FN and FP regions in Fig. 6.34) was the smallest for kN and DN (24) and the largest for IN and IS (34). The same weighting schemes were the ones where the number of datasets for which the expected and actual gains were matched (the union of TN and TP regions in Fig. 6.34) was the highest (92 for kN and DD) and the lowest (81 for IN and IS). Compared to the results of 1NN in Chap. 4 (Fig. 4.27c), improvement was achieved only with kN and DD.

Among the datasets in the FP region, the actual gain was more than 0.05 less than the expected for all kNN classifiers paired with the Sakoe-Chiba constraint (see

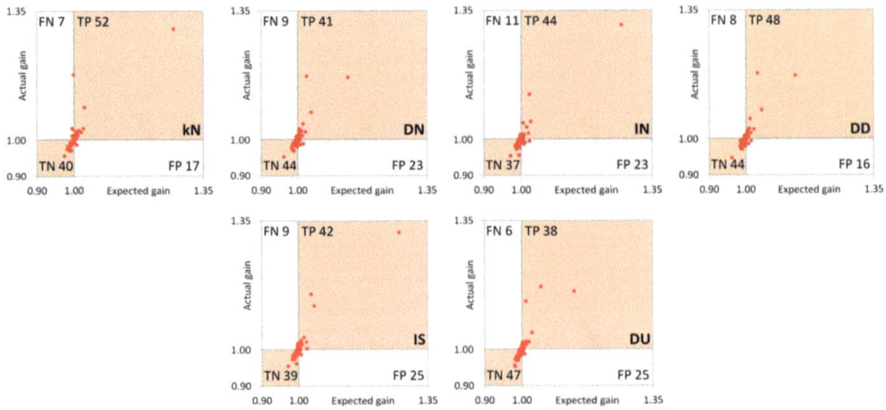

Fig. 6.34 LCS—comparing the actual and expected gain in classification accuracy: IT versus SC (combined with different variants of the kNN classifier)

Fig. 6.35) in *BeetleFly* (8), *Lightning2* (67) and *WormsTwoClass* (127) (except for DD). In addition to *Lightning2* (67) and *WormsTwoClass* (127), this property was also inherent in the *BirdChicken* (9) dataset with all six kNN schemes covered in this chapter combined with the Itakura constraint (Fig. 6.36). The datasets mentioned in this paragraph belonged to the FP region also when comparing the constrained variants of LCS with the unconstrained one relying on the 1NN classifier in Chap. 4 (Fig. 4.28) and the difference between the actual and expected gain was greater than 0.04. The 0.04 threshold was not reached only in the case of *Lightning2* (67) when comparing the IT-constrained LCS with its unconstrained variant.

Figure 6.37 confirms the finding of Chap. 4 (Fig. 4.28c) that for LCS the difference between the actual and expected gain of IT over SC was less than 0.03 for all datasets of the FP region: this threshold was not reached with the kNN variants either.

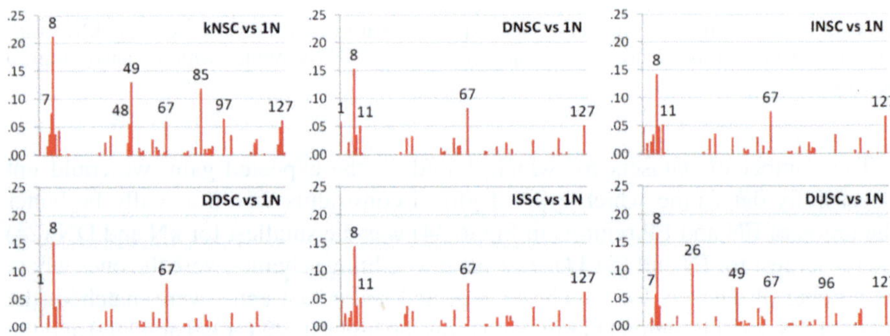

Fig. 6.35 LCS—differences between expected and actual accuracy gains of datasets belonging to the FP region (labels represent dataset identifiers): kNN variants (combined with the SC constrained measure) versus 1NN (combined with the unconstrained measure)

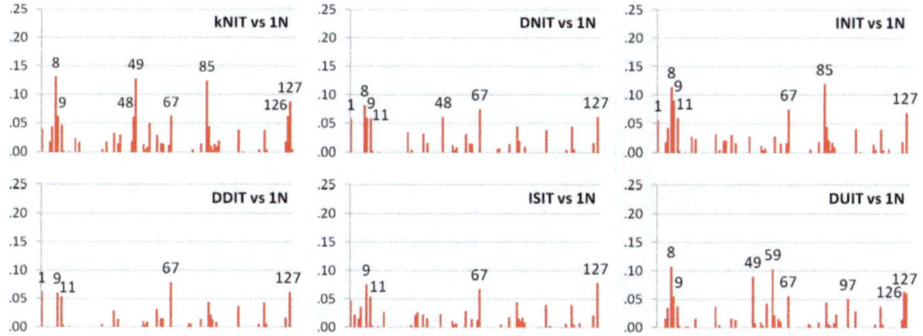

Fig. 6.36 LCS—differences between expected and actual accuracy gains of datasets belonging to the FP region (labels represent dataset identifiers): kNN variants (combined with the IT constrained measure) versus 1NN (combined with the unconstrained measure)

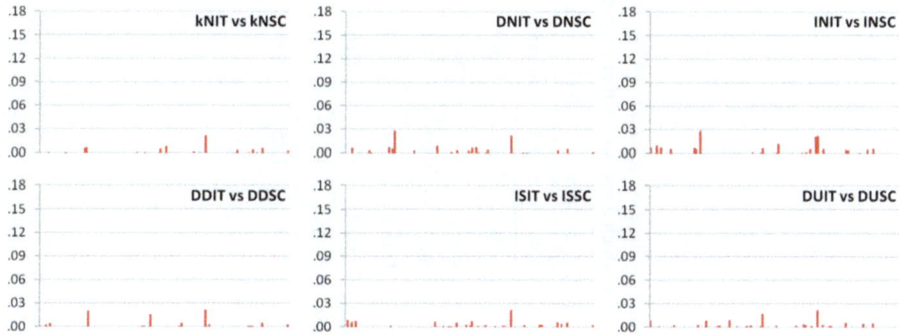

Fig. 6.37 LCS—differences between expected and actual accuracy gains of datasets belonging to the FP region (labels represent dataset identifiers): IT versus SC (combined with different variants of the kNN classifier)

6.3 Edit Distance with Real Penalty (ERP)

According to Table 6.9, the advantage of the discussed kNN variants over 1N (1NN classifier combined with the unconstrained ERP) in terms of average (AV) classification accuracy (calculate with the SCV10x10 evaluation method in which the optimal values of the parameters *k* and *r* were chosen over the appropriate training subsets relying on the SCV1x9 algorithm) ranged between 1.08 (DUIT) and 1.69 (DDSC) percentage points (pp), which was the smallest improvement among the elastic measures analyzed. By comparing with the results of Chap. 5 (Table 5.16), a weak favorable influence of global constraints can be observed. When paired with the Sakoe-Chiba band, the kNN variants improved their advantage over 1N by approximately between 0.57 (DU) and 0.73 (kN) pp and when combined with the Itakura parallelogram by approximately between 0.49 (DU) and 0.65 (kN) pp.

Table 6.9 ERP—average value (AV) and standard deviation (SD) of classification accuracy (AV), optimal number of nearest neighbors (k) and optimal warping-window width (r) across 12,800 splits; the number of datasets for which a given classifier produced the best result (WIN) and the mean rank (MR)

	DDSC	DNSC	DDIT	DNIT	ISSC	ISIT	INSC	INIT	kNSC	DUSC	kNIT	DUIT	1N
A_{AV} (%)	83.17	83.09	83.07	83.04	83.01	82.97	82.93	82.85	82.7	82.64	82.62	82.56	81.48
A_{SD} (pp)	17.01	17.06	16.97	17.02	17.04	16.98	17.06	17.09	17.36	17.41	17.38	17.45	17.97
k_{AV}	9.11	8.70	9.00	8.54	6.19	5.97	5.77	5.67	4.65	15.39	4.52	15.50	
k_{SD}	8.06	7.90	7.99	7.70	7.11	6.82	6.92	6.78	6.27	11.08	6.02	11.09	
r_{AV} (%)	5.92	5.94	20.26	20.24	5.94	20.18	6.01	20.49	5.88	5.63	19.56	19.90	
r_{SD}(pp)	7.72	7.67	27.56	27.55	7.72	27.00	7.79	27.16	7.57	7.07	26.41	27.30	
WIN	34	23	17	15	13	10	7	6	12	20	14	17	26
MR	5.16	5.52	6.00	5.97	6.90	7.29	7.25	7.86	7.82	6.61	8.46	7.22	8.94

Analogous to the 1NN classifier in Chap. 4 (see Table 4.7), the kNN variants also generated slightly better average accuracy when applying the SC than when using the IT constraint.

DDIT generated the smallest (16.97) and 1N the largest (17.97) standard deviation (SD). kN, IN and DU generated slightly higher SD values with IT than with SC and the reverse was true for DN, DD and IS. Applying global constraints led to a modest reduction of SDs compared to the findings in Chap. 5.

The averages of the optimal numbers of nearest neighbors (Table 6.9) were not noticeably different from those reported in Chap. 5, i.e. the application of global restrictions had no considerable effect on this indicator. Consistent with the results of the unconstrained ERP, the DU scheme relied, on average, on approximately 1.7 (DD) to 3.3 (kNSC) and 3.4 (kNIT) times the number of nearest neighbors than the others. As for the standard deviations, similar to LCS and EDR, limiting the width of the warping window led to their slight increase (the only exception was DNIT). Except for DU, all other kNN variants produced slightly higher AV and SD values in combination with the SC than with the IT constraint.

The difference in the degree of dependence of the discussed kNN variants on the first neighbor in terms of whether a global constraint was applied or not, ranged between -2.29 (kNIT) and 0.98 (DUSC) percentage points (compare Fig. 6.38 with Fig. 5.54 in Chap. 5) and the (descending) order remained unchanged: kN, IN, IS, DU, DN, DD. Similar to DTW, kNN variants opted for $k = 1$ in combination with SC a little more often than with IT. As can be seen from Fig. 6.39, there were no visible differences between SC and IT in frequencies of k-values greater than 3. Besides, very similar general characteristics were also observed when relying on the unconstrained ERP in Chap. 5 (see Fig. 5.55).

The application of global constraints did not significantly change the number of datasets with the characteristic that in each of the 100 iterations of performance evaluation of a given kNN classifier, the optimal number of nearest neighbors was 1 (Table 6.10). In most cases, the same datasets were involved both with and without global constraints, but there were some discrepancies as well. *CBF* (12), *ChlorineConcentration* (14) (except with DU), *Fungi* (46), *TwoPatterns* (117) exhibited the above property in combination with every kNN classifier analyzed in this chapter,

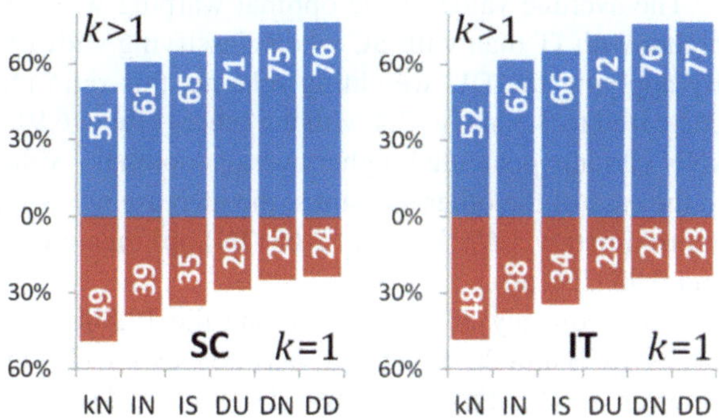

Fig. 6.38 ERP—percentages of splits in which SCV1x9 chose $k = 1$ and $k > 1$ as the optimal number of nearest neighbors on the training set when evaluating classification accuracy using SCV10x10

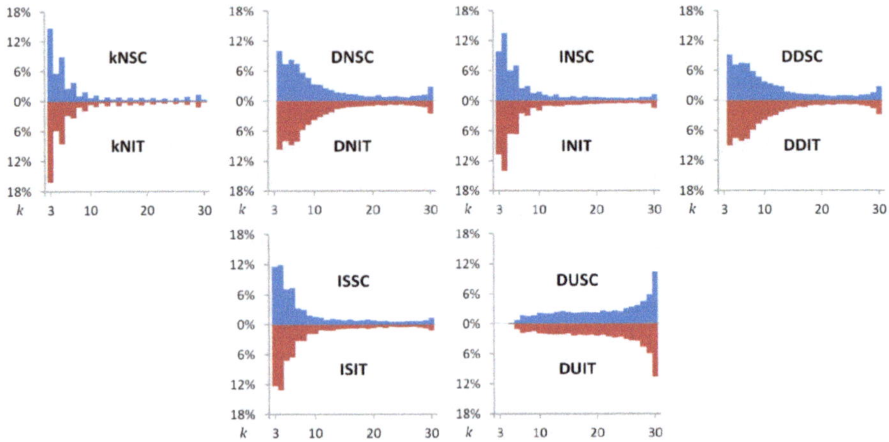

Fig. 6.39 ERP—frequencies (in percentage relative to the 12,800 splits) of different values of the parameter k (≥ 3) chosen by the SCV1x9 algorithm over the training set when evaluating classification accuracy using SCV10x10

Table 6.10 ERP—number of datasets for which the optimal number of nearest neighbors was 1 in each of the 100 iterations of the SCV10x10 algorithm

	kN	DN	IN	DD	IS	DU
ERP	21	5	12	5	10	8
SC	18	5	13	5	9	8
IT	19	5	14	5	10	9

both with the unconstrained ERP and its constrained variants. While *DiatomSizeReduction* (22) possessed the aforementioned property with all kNN variants paired with the unconstrained ERP, in combination with global constraints it retained it only with the DU scheme. *DodgerLoopWeekend* (28), on the other hand, obtained the described feature by constraining the warping window with either SC or IT, with all six weighting schemes.

The average value of the optimal warping-window widths was about 3.5 times higher with IT than with SC when classifying with 1NN (see Table 4.7 in Chap. 4). By applying the DU weighting scheme, this ratio remained the same, with kN it decreased to 3.3 and to 3.4 with the others (Table 6.9). It can also be noticed that the kNN variants generated higher average optimal r-values than the 1NN classifier: as in the case of the other measures, the increase was slightly greater with IT (between approximately 1.69 (kN) and 2.63 (IN) pp) than with SC (between about 0.56 (kN) and 0.93 (IN) pp).

The frequency of $r = 0\%$ among the 12,800 optimal r-values decreased from 23.3% (SC) and 20.4% (IT) to between 18.2% (IS) and 21.7% (DU) for SC and to between 16.4% (IN) and 18.8% (DU) for IT. Furthermore, for both types of global constraints, the threshold to cover more than 90% of the total number of splits into

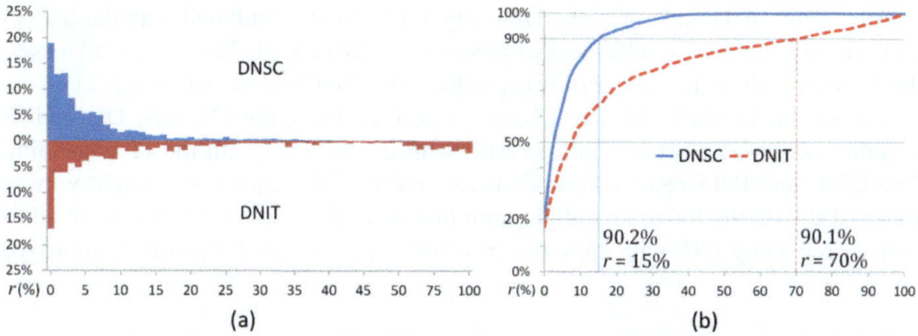

Fig. 6.40 ERP—frequencies (in percentage relative to the 12,800 splits) of different warping-window widths selected by the SCV1x9 algorithm over the training sets when evaluating classification accuracy using SCV10x10 (**a**); cumulative display of frequencies of optimal warping-window widths (**b**)

training and test subsets was increased. In the case of SC from $r = 13\%$ to $r = 15\%$ and when it comes to IT, from $r = 60\%$ to $r = 65\%$ (kN) and $r = 70\%$ (all other weighting schemes).

It can also be concluded that, compared to the findings in Table 4.7, the kNN variants generated somewhat higher standard deviations too. Again, the application of the Itakura parallelogram led to a slightly greater increase than the use of the Sakoe-Chiba band. In the first case, the increase ranged between about 1.10 (kN) and 2.25 (DD) and in the second between approximately 0.53 (DU) and 1.26 (IN) percentage points. Furthermore, similar to the results with the 1NN classifier, IT produced between roughly 3.5 (IN) and 3.9 (DU) times higher SD values than SC. However, as can be seen on the example of the DN scheme in Fig. 6.40, the general characteristics of SC and IT did not differ much from those observed with 1NN.

Crop (21), *DodgerLoopWeekend* (28), *Fungi* (46), *Meat* (70) and *Wine* (124) retained the property (observed with 1NN) that the highest accuracy on each of their 100 training subsets was achieved with $r = 0\%$ with both types of global constraints. *MelbournePedestrian* (72) lost the mentioned feature with kNIT, INIT and ISIT, and *OliveOil* (81) when classifying with kN and IN classifiers (both with SC and IT). When limiting the elasticity with the Sakoe-Chiba band, analogous to 1NN, *CBF* (12) and *FordB* (43) chose $r = 1\%$ as the best option in each of 100 iterations also with the kNN variants analyzed in this chapter. While increasing the number of nearest neighbors *Adiac* (2) lost this feature, it appeared in the case of the *FordA* (42) dataset (except with DUSC).

In terms of the number of datasets on which the given classifier achieved the highest accuracy among those analyzed ones (Table 6.9), only DDSC surpassed 1N (in Chap. 5, only the DU scheme was ahead of 1N in this regard). Nevertheless, 1N had the highest mean rank (unlike in Chap. 5 where the mean rank of IN and kN was higher than 1N's).

By achieving higher classification accuracy than the 1NN classifier paired with the unconstrained ERP similarity measure on 89, and lower accuracy on 35 UCR datasets (Figs. 6.41 and 6.42), the DD weighting scheme combined with the Sakoe-Chiba constraint achieved the best win-loss ratio (2.5:1). With 74 wins and 50 losses, the lowest result in this category was produced by the kNIT classifier (1.5:1).

Unlike the 1NN classifier in Chap. 4 (see Fig. 4.35), the kN, DN, DD and IS schemes achieved a better win/loss ratio with the SC than with the IT constraint. The INSC and INIT results were identical, and the DU scheme was slightly more successful with the Itakura parallelogram than with the Sakoe-Chiba band. In addition, by replacing 1NN with kNN variants, both types of global constraints improved

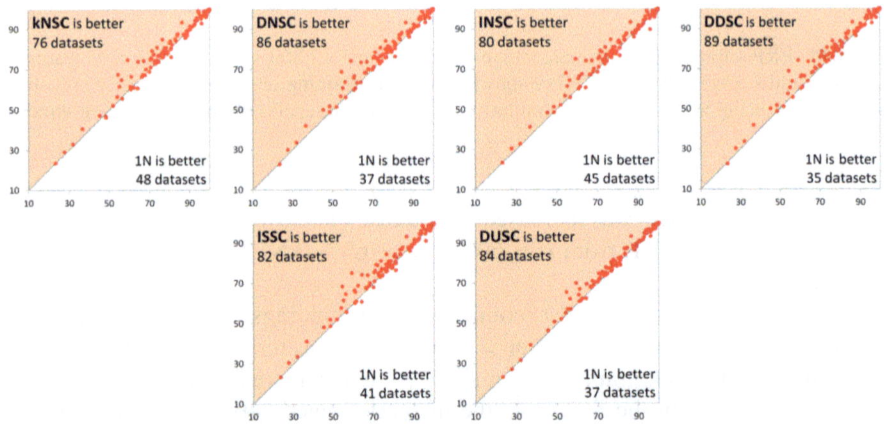

Fig. 6.41 ERP—comparison of the kNN variants (combined with the SC constrained measure) with the 1NN classifier (combined with the unconstrained measure) based on the number of datasets where they achieved higher/lower classification accuracy

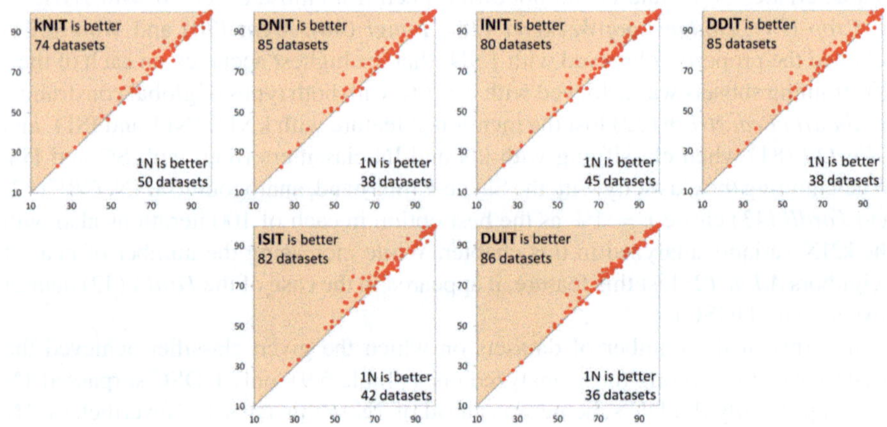

Fig. 6.42 ERP—comparison of the kNN variants (combined with the IT constrained measure) with the 1NN classifier (combined with the unconstrained measure) based on the number of datasets where they achieved higher/lower classification accuracy

their win-loss ratios relative to the unconstrained ERP—only the majority-voting method represented an exception for IT.

The utilization of global constraints positively affected the win-loss ratio of all the kNN variants discussed in this chapter, except for the DU scheme, in which case there was a deterioration from 3.3:1 (Fig. 5.57 in Chap. 5) to 2.3:1 (SC) and 2.4:1 (IT).

The win-loss ratio of SC over IT was approximately 1.5:1 when classifying with 1NN in Chap. 4 (see Fig. 4.35c), but when relying on kNN variants (Fig. 6.43) it ranged between 1:1 (IS) and 1.5:1 (DD). Thus, it can be concluded that their performance has converged to a slight extent by considering more than one nearest neighbor. The closeness of the points to the diagonal in the plots of Fig. 6.43 also supports the resemblance of the performance of SC and IT.

On average (Fig. 6.44), the positive difference between kNN variants combined with the constrained ERP versus 1NN with the unconstrained ERP was highest for ISSC (2.82) and INSC (2.76) and lowest for DUSC (1.99) and DUIT (1.90). The numbers shown here are higher than those presented in Chap. 5 (Fig. 5.58), which favors the application of global constraints. However, comparing with the results of the other analyzed elastic measures, it can also be noted that ERP produced the lowest positive average differences for all combinations of schemes and global constraints.

Except for DD, all other weighting schemes achieved a slightly higher average advantage over 1N with the Sakoe-Chiba band than with the Itakura parallelogram, which is consistent with the findings based on the 1NN classifier in Chap. 4 (Fig. 4.36).

It should also be taken into account that by limiting the elasticity, the average advantage of 1NN was also increased compared to its advantage over kNN variants paired with the unconstrained ERP. The largest negative average difference was detected for ISSC (−0.86) and the smallest for DUSC (−0.49). Paired with the 1NN

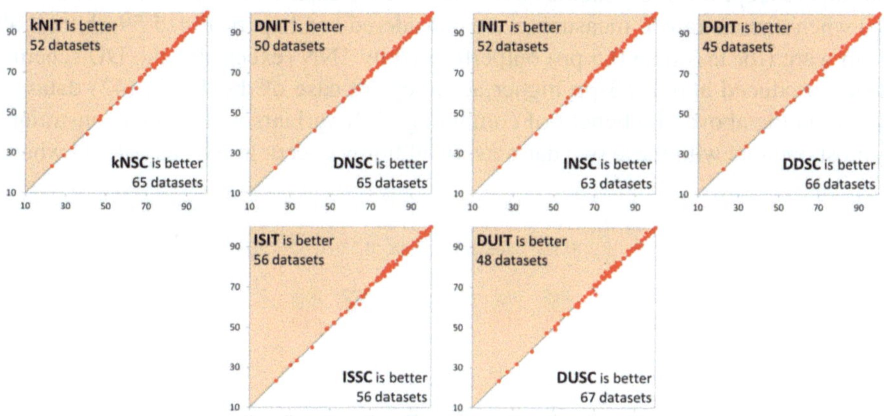

Fig. 6.43 ERP—comparison of the SC and IT constraints combined with the given kNN classifier variant based on the number of datasets where they achieved higher/lower classification accuracy

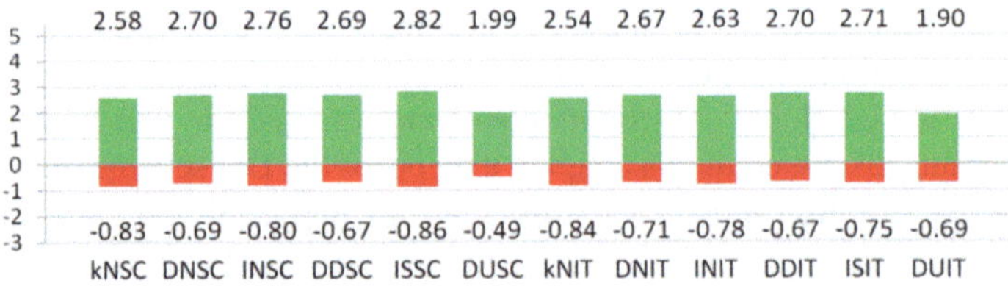

Fig. 6.44 ERP—average difference in classification accuracy of various kNN variants (combined with the SC and IT constrained measure) and 1NN (combined with the unconstrained measure) over the datasets where the performance of a given kNN variant was better (green) and worse (red) than 1NN

classifier, IT was somewhat inferior to SC in this regard in Chap. 4, which was confirmed here with most kNN variants (with the exception of IN and IS). At the same time, except for ISSC, ERP generated the lowest negative average differences among the elastic measures discussed.

As can be seen from Fig. 6.45, the average advantage of SC over IT ranged between 0.53 (IS) and 0.45 (DN) and the average advantage of IT over SC between 0.50 (DU) and 0.43 (kN). Since the positive average difference was slightly higher than the negative one for each kNN variant, in this matter, similarly to the 1NN classifier in Chap. 4 (Fig. 4.36), a slight dominance of SC over IT can be noted.

As evidenced by Figs. 6.46 and 6.47, *FordA* (42) and *InsectWingbeatSound* (64) stood out in particular, as all considered kNN variants with both SC and IT achieved more than 10 percentage points higher classification accuracy than 1N (1NN with the unconstrained ERP). Paired with the 1NN classifier, the constrained variants of ERP achieved the greatest advantage over the unconstrained one also with these two datasets (see Fig. 4.37 in Chap. 4), but the difference in their performance did not reach 9 pp. When it comes to *InsectWingbeatSound* (64), the advantage of kNN variants (except the DU scheme) over 1NN was greater than 5 pp (but less than 7 pp) when unconstrained measures were considered in Chap. 5 (Fig. 5.59), but all of them were (for less than 0.5 pp) outperformed by 1NN (except for the DU scheme which produced about 0.3 pp higher accuracy) in case of the *FordA* (42) dataset. Based on the above, the benefit of combining kNN variants with global constraints is most obvious with these two datasets. In addition, except in the case of DU where

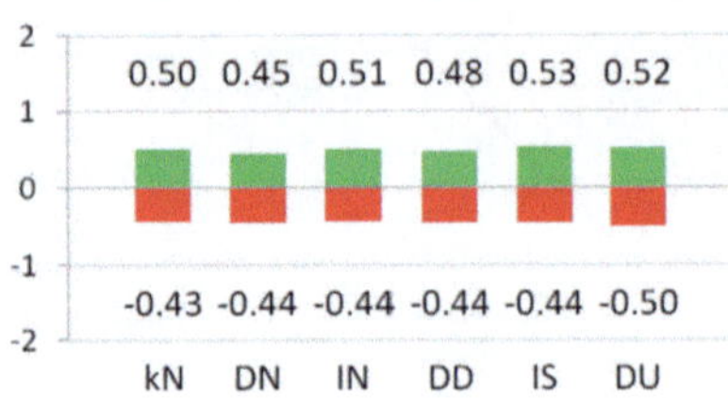

Fig. 6.45 ERP—average difference in classification accuracy of the SC and IT constraints (combined with different variants of the kNN classifier) over the datasets where the performance of SC was better (green) and worse (red) than IT

it remained unchanged, an increase in the number of datasets where the advantage of kNN variants over 1N was greater than 5 pp can be observed, which is also in favor of such combining.

An advantage of more than 3 percentage points was produced by 1N most often only on two datasets: *BeetleFly* (8) and *BirdChicken* (9). In their case, an increased negative effect of pairing kNN with global constraints can be recognized.

When comparing the performance of SC and IT based on the accuracy of the 1NN classifier in Chap. 4 (Fig. 4.37c), SC outperformed IT by more than 3 pp only on *PickupGestureWiimoteZ* (85) and the advantage of IT over SC did not reach that threshold on any dataset. This setup was replicated only with the DU scheme (Fig. 6.48). When it comes to the other analyzed schemes, SC outperformed IT by more than 3 pp only on *SemgHandMovementCh2* (99) (IN and IS) and IT outperformed SC by more than 3 pp only on the *Lightning7* (68) (DN and DD) dataset. The difference in classification accuracy between SC and IT paired with the majority-voting rule did not exceed this threshold on any UCR dataset.

According to the Friedman test ($\chi^2(12) = 134.39, p < .001$), the performance of the tested classifiers was significantly different. Nemenyi's post-hoc test revealed a significant difference between 1N and all classifiers listed on the left side of Fig. 6.49 (DDSC, DNSC, DNIT, DDIT, DUSC and ISSC). Since only DD differed significantly from 1N in the analysis of the unconstrained ERP classification results (see Fig. 5.60 in Chap. 5), the significant difference of other combinations of weighting schemes and constraints can only be attributed to the application of the constraints. It is also easy to see that DDSC differed significantly from all the classifiers listed on the right side of Fig. 6.49. As claimed by the slightly more lenient Wilcoxon signed-rank test (adjusted with the Holm-Bonferroni rule), the average accuracy of 1N was significantly worse than all combinations of schemes and global constraints discussed in this chapter (Table 6.11). When relying on unconstrained ERP (see Table 5.17 in

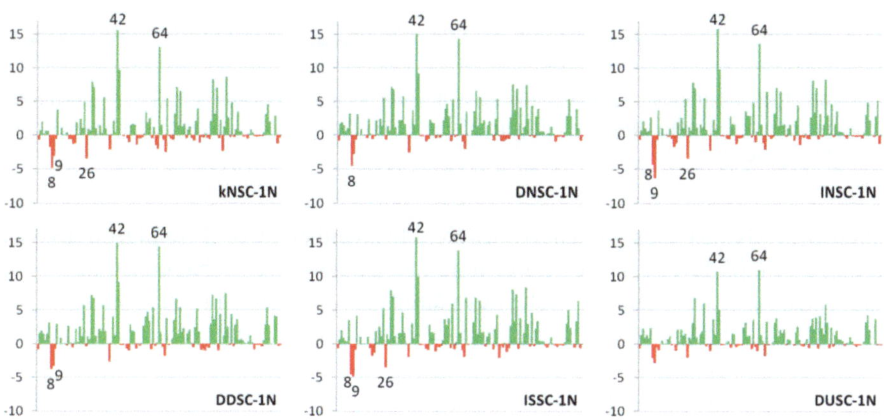

Fig. 6.46 ERP—differences between the classification accuracies of various kNN variants (combined with the SC constrained measure) and 1NN (combined with the unconstrained measure) by datasets (labels represent dataset identifiers)

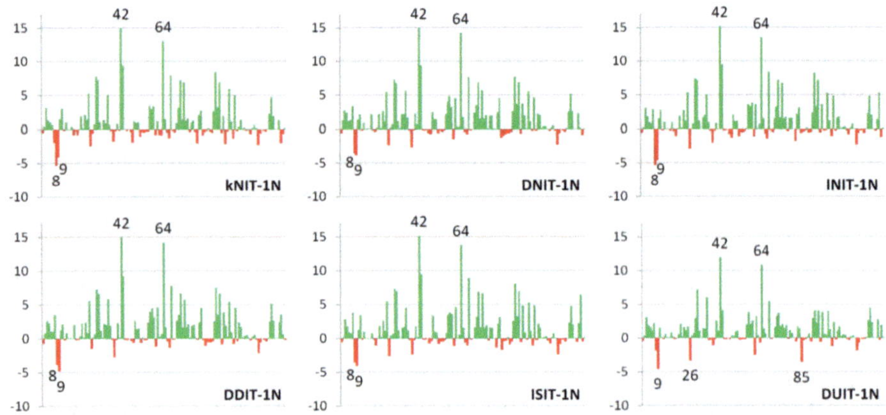

Fig. 6.47 ERP—differences between the classification accuracies of various kNN variants (combined with the IT constrained measure) and 1NN (combined with the unconstrained measure) by datasets (labels represent dataset identifiers)

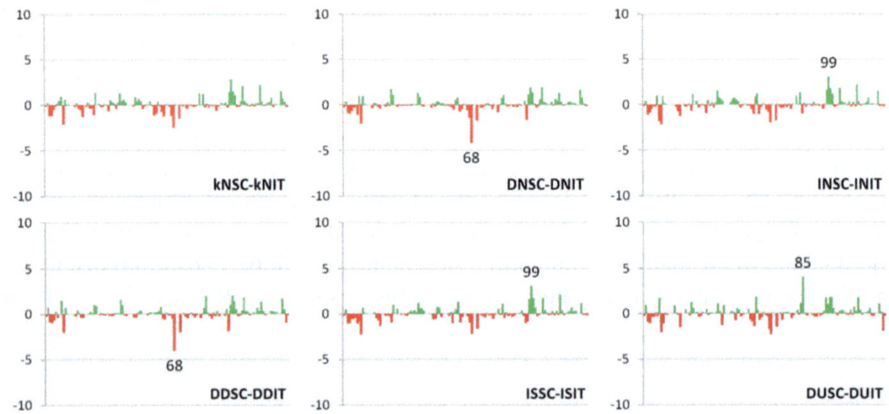

Fig. 6.48 ERP—differences between the classification accuracies of the SC and IT constrained measure (combined with different variants of the kNN classifier) by datasets (labels represent dataset identifiers)

Chap. 5), 1NN was not significantly different from either kN or IN (and from IS only at the $\alpha = 0.05$ level), further confirming the suitability of global constraints. The advantage of the DD and DN schemes paired with both SC and IT over the 1NN classifier with the unconstrained ERP was also confirmed by the corrected resampled *t*-test (Table 6.12).

Relying on the SC constraint, the expected advantage over 1N was most often overestimated by the kN, IN and IS (in about 30% of cases) and least often by the DD (in about 23% of cases) weighting scheme (Fig. 6.50). Applying the Itakura parallelogram, the frequency of false positive estimates was the highest with the

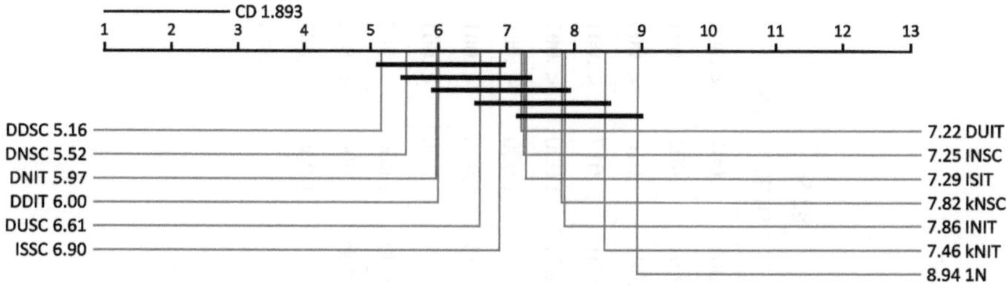

Fig. 6.49 ERP—critical difference diagram comparing average accuracy ranks based on Nemenyi's test: classifiers connected by a solid bar did not differ statistically significantly

kN (approximately 34%) and the lowest with the DU (approximately 26%) method (Fig. 6.51). Moreover, kN, DN and DD produced slightly fewer FP and slightly more TP datasets with SC than with IT, and the reverse was true for IN and IS (i.e., they achieved better results with IT than with SC). DU generated identical FP and TP numbers with both types of constraints.

With the exception of DU, the application of global constraints had a favorable effect on both the number of FP and the number of TP cases compared to the results with the unconstrained ERP in Chap. 5 (Fig. 5.61). SC generated the greatest improvement in both the FP and TP regions for kN, whose number of incorrect estimates was reduced from 60 to 45 and the number of correct ones increased from 45 to 82, thus achieving approximately 1.5 times better results in the first case and 1.8 times in the second case. IT provided the greatest progress in the number of FP datasets to the IN scheme (from 57 to 38, i.e. approximately 1.5 times), and in the number of TP datasets to the kN method (from 45 to 78, i.e. about 1.7 times).

By comparing the plots of Figs. 6.50 and 6.51 with Fig. 4.39a, b in Chap. 4, it can be seen that increasing the number of nearest neighbors did not lead to improvement only in the case of the Itakura parallelogram combined with the majority-voting rule and only in the FP region. When it comes to SC, the biggest improvement came from switching from 1NN to DD: the number of FP datasets was reduced by 1.5 times (from 45 to 30) and the number of TP datasets was increased by about 1.3 times (from 71 to 93). IT benefited the most from switching from the 1NN classifier to the DU scheme: both the number of incorrect and the number of correct predictions improved by nearly 1.2 times (in the first case it was reduced from 40 to 33, and in the second case it was increased from 76 to 88).

Based on Fig. 6.52, the choice between SC and IT was the most successful when classifying time series using DN and the least successful when they were combined with the IN weighting scheme. Compared to the results of the 1NN classifier in Chap. 4 (Fig. 4.39c), all six kNN variants slightly increased the number of correct predictions (total number of datasets in the TP and TN regions), but only DN, IS and DU decreased the number of incorrect predictions (total number of datasets in the FP and FN regions).

Except for the DU scheme, all other kNN variants overestimated their advantage over the 1NN classifier in the FP region the most on the *BeetleFly* (8) dataset both

Table 6.11 ERP—Holm-Bonferroni-adjusted *p*-values of the pairwise Wilcoxon signed-rank test on classification accuracy

A		83.17	83.09	83.07	83.04	83.01	82.97	82.93	82.85	82.70	82.64	82.62	82.56	81.48
		DDSC	DNSC	DDIT	DNIT	ISSC	ISIT	INSC	INIT	kNSC	DUSC	kNIT	DUIT	1N
83.17	DDSC		.514	.618	.578	.065	.091	*.006*	**< .001**	**< .001**	**.003**	**< .001**	**.001**	**< .001**
83.09	DNSC			1.000	1.000	.618	.410	.039	**.002**	**.001**	.018	**< .001**	**.002**	**< .001**
83.07	DDIT				1.000	1.000	.158	.554	**.001**	.047	.869	**< .001**	.016	**< .001**
83.04	DNIT					1.000	.061	.554	**< .001**	.052	1.000	**< .001**	.017	**< .001**
83.01	ISSC						1.000	.368	.580	.018	1.000	.031	1.000	**< .001**
82.97	ISIT							1.000	.258	1.000	1.000	.017	1.000	**< .001**
82.93	INSC								1.000	.055	1.000	.133	1.000	**< .001**
82.85	INIT									1.000	1.000	.078	1.000	**< .001**
82.70	kNSC										1.000	1.000	1.000	**.002**
82.64	DUSC											1.000	1.000	**< .001**
82.62	kNIT												1.000	**.007**
82.56	DUIT													**< .001**
81.48	1N													

Note Bolded values indicate significance at the level of $\alpha = 0.01$

Table 6.12 ERP—statistically significant wins (W) and losses (L) reported by the corrected resampled *t*-test on classification accuracy (across 12,800 splits)

	DDSC	DNSC	DDIT	DNIT	ISSC	INSC	ISIT	DUSC	INIT	kNSC	DUIT	kNIT	1N
W	440	417	333	311	300	255	224	282	197	204	234	147	167
L	102	107	134	134	187	195	237	317	273	320	366	400	739
W-L	338	310	199	177	113	60	−13	−35	−76	−116	−132	−253	−572

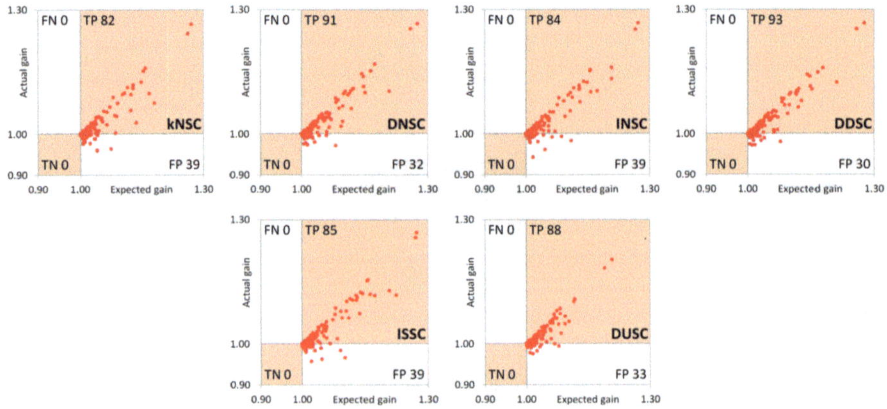

Fig. 6.50 ERP—comparing the actual and expected gain in classification accuracy: kNN variants (combined with the SC constrained measure) versus 1NN (combined with the unconstrained measure)

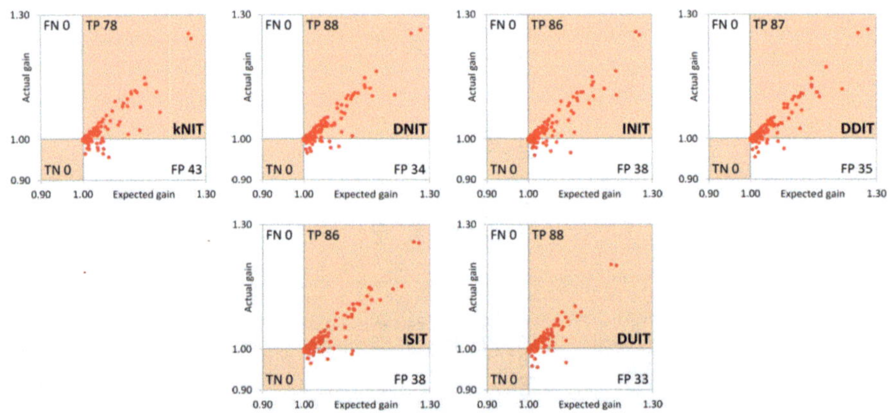

Fig. 6.51 ERP—comparing the actual and expected gain in classification accuracy: kNN variants (combined with the IT constrained measure) versus 1NN (combined with the unconstrained measure)

in combination with SC and paired with IT (see Figs. 6.53 and 6.54). With IT, the expected gain was greater than the actual gain by more than 0.05 on the *DodgerLoopDay* (26) dataset as well. These two datasets were located in the FP region of the kNN classifiers discussed in this chapter (except DU in both cases and IN in the case of *BeetleFly* (8)) also when comparing time series with the unconstrained ERP in Chap. 5 (see Fig. 5.62), with a difference between the expected and actual gain greater than 0.03.

When contrasting the Itakura parallelogram with the Sakoe-Chiba band (Fig. 6.55), the difference between the expected and the actual gain in the FP

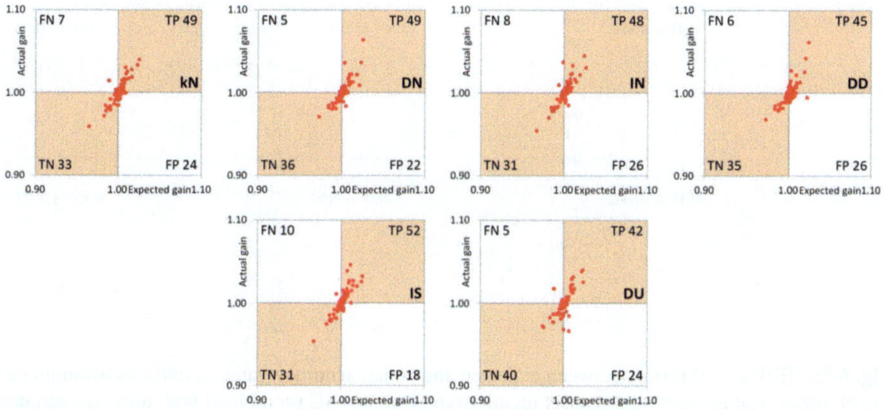

Fig. 6.52 ERP—comparing the actual and expected gain in classification accuracy: IT versus SC (combined with different variants of the kNN classifier)

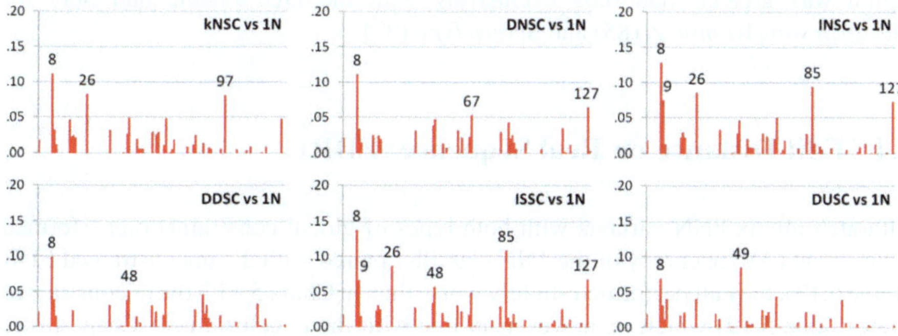

Fig. 6.53 ERP—differences between expected and actual accuracy gains of datasets belonging to the FP region (labels represent dataset identifiers): kNN variants (combined with the SC constrained measure) versus 1NN (combined with the unconstrained measure)

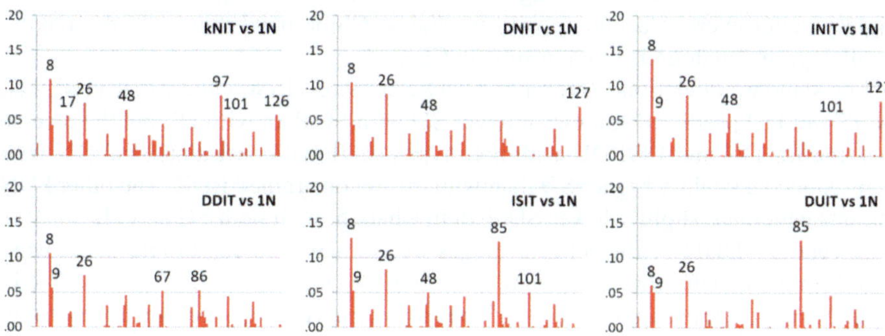

Fig. 6.54 ERP—differences between expected and actual accuracy gains of datasets belonging to the FP region (labels represent dataset identifiers): kNN variants (combined with the IT constrained measure) versus 1NN (combined with the unconstrained measure)

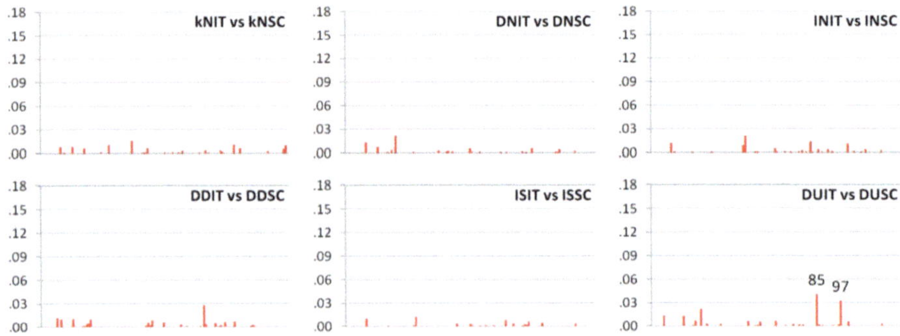

Fig. 6.55 ERP—differences between expected and actual accuracy gains of datasets belonging to the FP region (labels represent dataset identifiers): IT versus SC (combined with different variants of the kNN classifier)

region was greater than 0.03 exclusively with the DU scheme and only for *PickupGestureWiimoteZ* (85) and *ScreenType* (97).

6.4 Edit Distance on Real Sequence (EDR)

Although all six kNN variants with both types of global constraints outperformed the average (AV) accuracy of the 1NN classifier paired with the unconstrained EDR (Table 6.13), they all performed slightly worse than in Chap. 5 with the unconstrained EDR (Table 5.19) which is, however, in line with the conclusions in Chap. 4 (see Table 4.10). The greatest deterioration occurred in the case of the DU scheme using the IT constraint (about 0.24 percentage points (pp)) and the smallest in the case of the IN scheme using the SC constraint (approximately 0.02 pp).

Except for the majority-voting rule, the other weighting schemes achieved slightly better average accuracy with the Sakoe-Chiba band than with the Itakura parallelogram, analogous to the results in Chap. 4.

The standard deviation (SD) ranged between 19.65 (ISSC) and 21.01 (1N). kN, IN, IS and DU produced somewhat lower SD with SC than with IT, and the reverse was true for DN and DD. DU with both types of global constraints, and DN with SC produced a slightly higher SD than with the unconstrained EDR. The other kNN variants generated slightly lower SD than in Chap. 5, with both SC and IT.

Except for DDIT's, the other averages of optimal *k*-values (Table 6.13) were slightly higher than those generated when considering the unconstrained EDR in Chap. 5. However, the application of global constraints did not affect the relationship between DU and the other schemes in this regard: DU's average was noticeably higher here as well. It can also be noted that all six kNN variants produced a little higher AV when limiting the elasticity with the Sakoe-Chiba band than with the Itakura parallelogram. The standard deviations of the optimal numbers of nearest neighbors

Table 6.13 EDR—average value (AV) and standard deviation (SD) of classification accuracy (AV), optimal number of nearest neighbors (k) and optimal warping-window width (r) across 12,800 splits; the number of datasets for which a given classifier produced the best result (WIN) and the mean rank (MR)

	DDSC	DDIT	DNSC	ISSC	DNIT	ISIT	INSC	INIT	kNIT	kNSC	DUSC	DUIT	1N
A_{AV} (%)	79.27	79.24	79.22	79.18	79.17	79.17	79.12	79.08	78.74	78.73	78.67	78.56	77.06
A_{SD} (pp)	19.72	19.70	19.75	19.65	19.72	19.68	19.66	19.69	19.85	19.84	20.10	20.15	21.01
k_{AV}	11.90	11.76	11.35	9.33	11.29	9.14	8.73	8.58	7.24	7.38	16.52	16.35	
k_{SD}	9.16	9.19	9.05	9.10	9.09	9.01	8.91	8.82	8.64	8.74	10.42	10.49	
r_{AV}(%)	5.86	27.41	5.82	6.03	27.29	27.90	5.97	27.23	26.62	5.72	5.60	26.77	
r_{SD}(pp)	5.54	30.55	5.47	5.6	30.03	30.82	5.60	30.56	30.07	5.37	5.38	30.77	
WIN	28	18	12	14	13	9	12	12	10	9	17	12	26
MR	5.11	5.06	5.93	6.89	5.86	7.40	7.18	7.61	8.51	8.60	7.08	7.39	8.39

were, except for DUSC, also slightly larger than those obtained in combination with the unconstrained EDR.

Similar to LCS, when choosing the optimal number of nearest neighbors, the weighting schemes discussed in this chapter slightly more often opted for $k = 1$ in combination with the IT than with the SC constraint (Fig. 6.56). Compared to the results with the unconstrained EDR in Chap. 5 (Fig. 5.64), the application of the Itakura parallelogram marginally increased and the utilization of the Sakoe-Chiba band lightly decreased the degree of first-neighbor dependence. However, the difference compared to the unconstrained EDR did not reach 1 pp. The employment of global constraints did not noticeably affect the frequencies of higher k-values (see Fig. 5.65 in Chap. 5), and as Fig. 6.57 demonstrates, there were no major deviations between SC and IT either.

The application of global constraints had no noticeable effect on the number of datasets with the feature that the optimal number of neighbors was 1 in each of

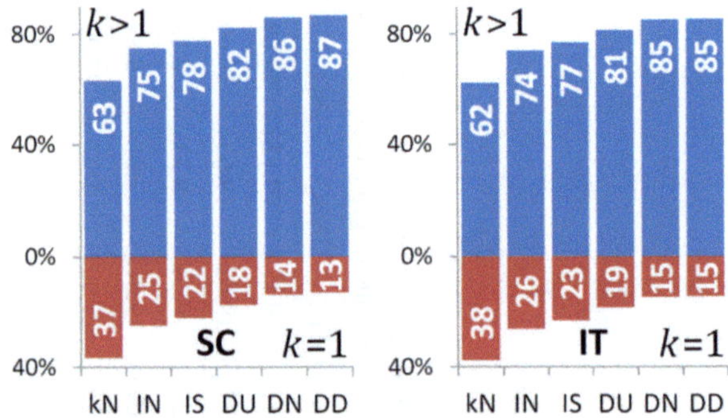

Fig. 6.56 EDR—percentages of splits in which SCV1x9 chose $k = 1$ and $k > 1$ as the optimal number of nearest neighbors on the training set when evaluating classification accuracy using SCV10x10

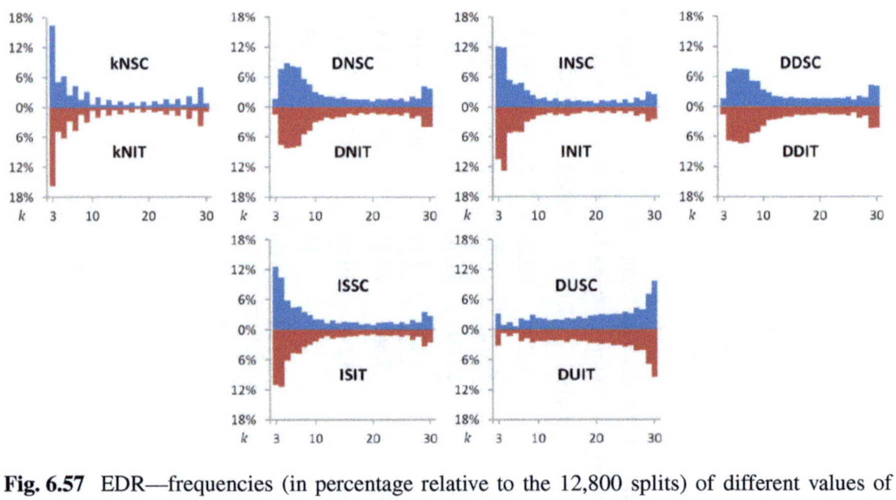

Fig. 6.57 EDR—frequencies (in percentage relative to the 12,800 splits) of different values of the parameter k (≥ 3) chosen by the SCV1x9 algorithm over the training set when evaluating classification accuracy using SCV10x10

the 100 iterations of the SCV10x10 evaluation process (Table 6.14). By comparing Tables 6.2, 6.6, 6.10 and 6.14, it can be seen that EDR has retained the characteristic that among the considered elastic measures it produced the smallest number of datasets with the specified characteristic. As in the Chap. 5, only *Fungi* (46) and *PigAirwayPressure* (86) exhibited this property with all six weighting schemes analyzed in this chapter.

Similar to 1NN in Chap. 4 (Table 4.10), IT relied on significantly larger warping-window widths than SC also in combination with the kNN variants (Table 6.13). The average of the optimal r-values was between about 4.6 (IN) to 4.8 (DU) times higher with the Itakura than with the Sakoe-Chiba constraint. In addition, considering more than one neighbor increased the AV values (compared to findings with 1NN). Analogous to the other analyzed measures, the increase was greater with IT (it ranged between approximately 2.63 (kN) and 3.91 (IS) pp) than with SC (between about 0.38 (DU) and 0.80 (IS) pp).

The impact of switching to the kNN variants, as far as SD is concerned, also manifested itself through an increase compared to 1NN results (by between about 1.12 (DN) and 1.91 (IS) pp for IT, and by between approximately 0.25 (kN) and 0.46 (IN) pp for SC). As with 1NN, among the four elastic similarity measures, the difference in SD values obtained with IT and with SC was the largest precisely with EDR: IT exceeded SC by between about 5.5 (IN) and 5.7 (DU) times.

Similar to LCS, the reliance on $r = 0\%$ has weakend (for SC from approximately 20.2% to between approximately 18.4% (DU) and 14.8% (IS), and for IT from roughly 17.0% to between approximately 15.3% (DU) and 13.0% (IS) of the 12,800 splits into test and training subsets). The accumulation of these changes is also reflected in slightly higher r-value thresholds needed to cover more than 90% of the total number of splits: for the Sakoe-Chiba band from 13% to 14% (kN, DN, DD and DU) and 15% (IN and IS) and for the Itakura parallelogram from 75% to 80%). However, as can be seen from the example of the DN scheme shown in Fig. 6.58, the general traits have not changed significantly compared to 1NN.

The lower degree of reliance on $r = 0\%$ was also reflected in the smaller number of datasets where it represented the best option in each of the 100 iterations of the SCV10x10 algorithm. Several datasets lost this property by increasing the number of nearest neighbors: *ACSF1* (1), *FreezerRegularTrain* (44) (except wtih DNSC), *FreezerSmallTrain* (45), *ProximalPhalanxOutlineAgeGroup* (92), i *Strawberry* (109). *Meat* (70), *MelbournePedestrian* (72), *OliveOil* (81) were invariant in this respect, and *Wine* (124) retained this feature selectively (only with IN, IS and DU

Table 6.14 EDR—number of datasets for which the optimal number of nearest neighbors was 1 in each of the 100 iterations of the SCV10x10 algorithm

	kN	DN	IN	DD	IS	DU
EDR	5	2	3	2	2	6
SC	7	2	3	2	2	6
IT	9	2	4	2	3	5

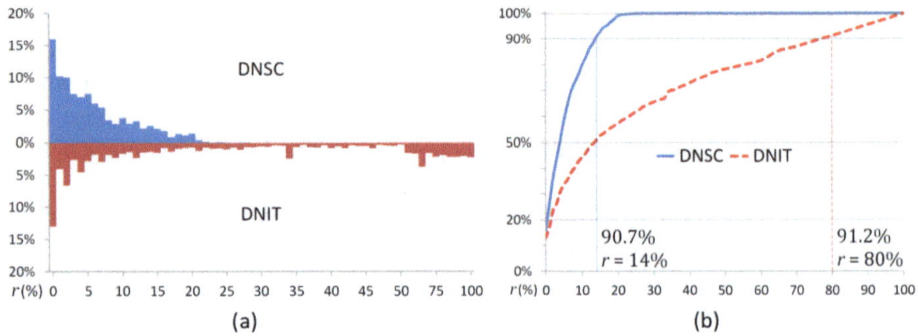

Fig. 6.58 EDR—frequencies (in percentage relative to the 12,800 splits) of different warping-window widths selected by the SCV1x9 algorithm over the training sets when evaluating classification accuracy using SCV10x10 (**a**); cumulative display of frequencies of optimal warping-window widths (**b**)

schemes). In addition, when limiting the elasticity using the SC constraint *FordA* (42) opted for $r = 1\%$ in each of the 100 iterations also with the kNN variants discussed in this chapter (and unlike 1NN, when using the IT constraint as well, where only DU represented an exception). With the exception of DU, in the combination of the Sakoe-Chiba band and the other four weighting schemes, this feature was also exhibited by *FordB* (43).

In terms of the number of datasets on which a given classifier achieved the best accuracy, EDR did not differ from LCS and ERP: only DDSC was better than 1N (Table 6.13). This, however, represents a step backwards for DN and DU as both of them outperformed 1NN when classifying with the unconstrained EDR in Chap. 5 (see Table 5.19). EDR was the only elastic measure where the mean rank of 1N was not the highest—however, only the kN scheme was inferior to it (with both types of global constraints).

Among the four basic elastic similarity measures, the win-loss ratio of the kNN variants paired with the SC and IT constraints compared to the 1NN classifier combined with the unconstrained measure was, except for the DD scheme, the lowest with EDR. As can be seen from Figs. 6.59 and 6.60, it was the highest for DDSC (2.6:1) and DDIT (2.3:1) and the lowest for kNIT (1.0:1) and kNSC (0.9:1). Nevertheless, given that when relying on the 1NN classifier in Chap. 4 (see Fig. 4.47), the unconstrained EDR considerably outperformed both SC and IT, these results represent major improvements (even in the case of kNSC), highlighting the positive impact of different weighting schemes. From the point of view of kNN classifiers, their advantage over 1NN was reduced by limiting the width of the warping window (see Fig. 5.67 in Chap. 5). As with LCS, the biggest decline was registered for the DU scheme, whose win-loss ratio dropped from 4.8:1 to 2.0:1 with SC and to 1.8:1 with IT. Additionally, it is worth mentioning that in the case of kN, the application of the Sakoe-Chiba constraint turned a (very) narrow victory into a narrow defeat.

EDR stood out also in terms of the difference between the Sakoe-Chiba band and the Itakura parallelogram (regarding the number of their mutual wins and losses, see

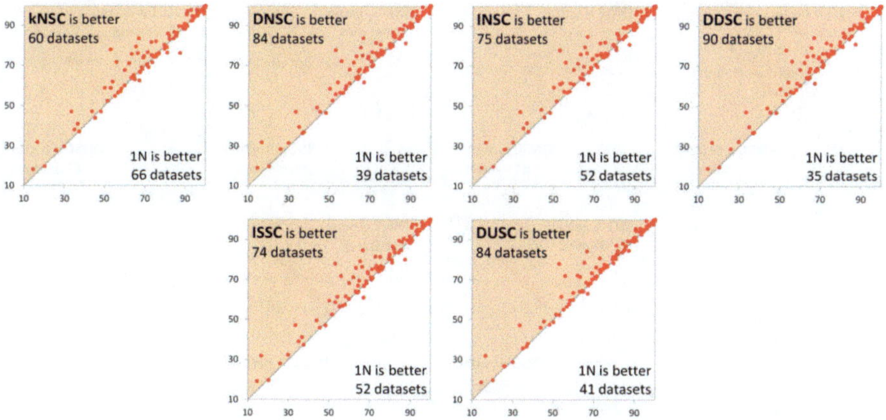

Fig. 6.59 EDR–comparison of the kNN variants (combined with the SC constrained measure) with the 1NN classifier (combined with the unconstrained measure) based on the number of datasets where they achieved higher/lower classification accuracy

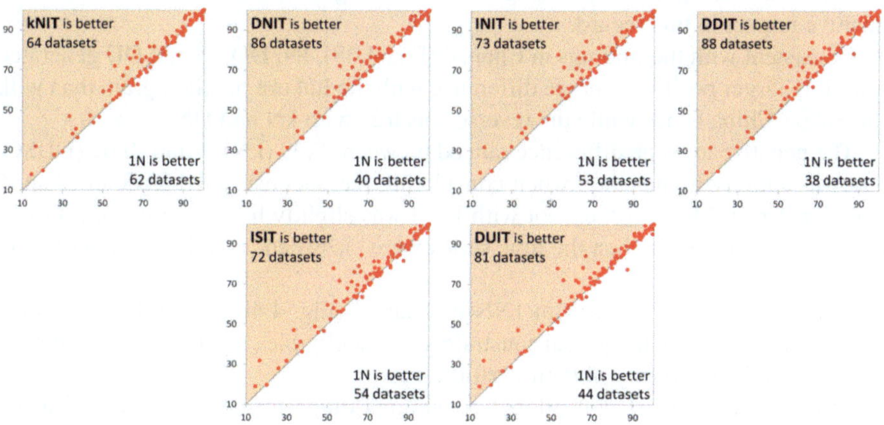

Fig. 6.60 EDR—comparison of the kNN variants (combined with the IT constrained measure) with the 1NN classifier (combined with the unconstrained measure) based on the number of datasets where they achieved higher/lower classification accuracy

Fig. 6.61), which, viewed through the prism of most weighting schemes (DU, IN, DD, IS), was greater than that reported by the 1NN classifier in Chap. 4 (Fig. 4.47c). However, the points representing the datasets in the plots of Fig. 6.61 are still very close to the diagonal, confirming that SC and IT were very similar in terms of classification accuracy when it comes to kNN variants as well.

According to Fig. 6.62, the largest positive average differences were achieved by kNSC (4.55) and ISIT (4.35) and the smallest by DUIT (2.88) and DUSC (2.76). As all the values recorded here were somewhat higher than those achieved with the

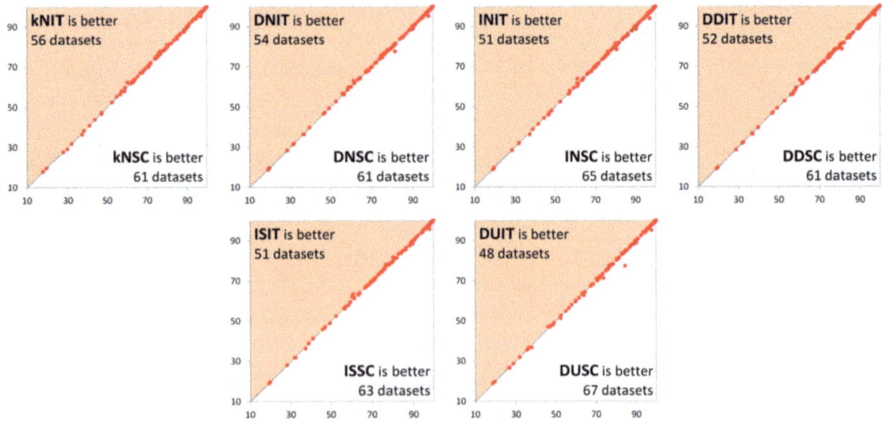

Fig. 6.61 EDR—comparison of the SC and IT constraints combined with the given kNN classifier variant based on the number of datasets where they achieved higher/lower classification accuracy

unconstrained EDR in Chap. 5 (Fig. 5.68), limiting the elasticity had a moderately positive impact in this regard.

Consistent with the findings in Chap. 4 (Fig. 4.48), IN, DD, IS and DU generated a slightly larger positive average difference with the Itakura parallelogram than with the Sakoe-Chiba band, while the reverse was true with kN and DN.

The negative average difference ranged between -0.99 (kNIT) and -0.62 (DUSC) and was slightly higher than when considering the unconstrained EDR in Chap. 5 (see Fig. 5.68). Moreover, except with IS, it was slightly higher with respect to IT than to SC, similar to when the time series were classified with the 1NN method in Chap. 4.

Analogous to the results with 1NN in Chap. 4 (Fig. 4.48), the difference in the average advantages of the global constraints was negligible (Fig. 6.63), which further emphasizes the similarity of their performance.

kNN variants paired with global constraints outperformed the 1NN classifier combined with the unconstrained EDR to the greatest extent on the following

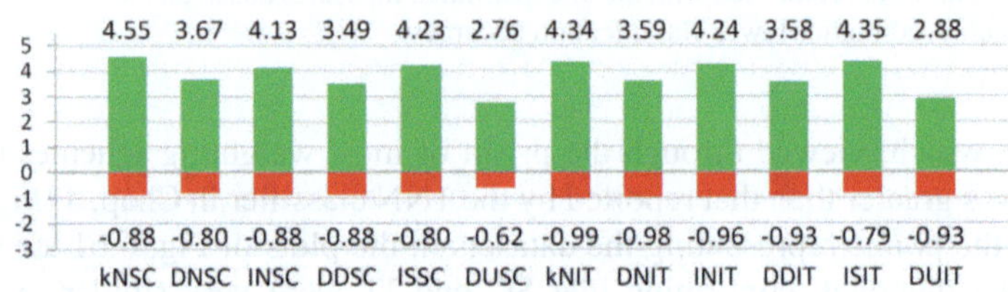

Fig. 6.62 EDR—average difference in classification accuracy of various kNN variants (combined with the SC and IT constrained measure) and 1NN (combined with the unconstrained measure) over the datasets where the performance of a given kNN variant was better (green) and worse (red) than 1NN

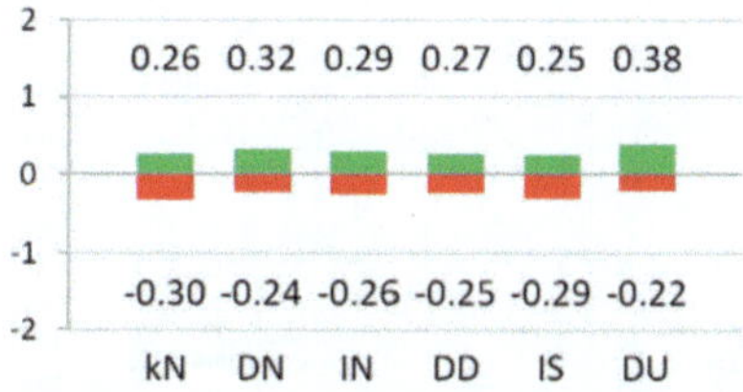

Fig. 6.63 EDR—average difference in classification accuracy of the SC and IT constraints (combined with different variants of the kNN classifier) over the datasets where the performance of SC was better (green) and worse (red) than IT

datasets (see Figs. 6.64 and 6.65): *Earthquakes* (29), *FordA* (42), *Meat* (70), *MiddlePhalanxOutlineCorrect* (74), *OliveOil* (81), *ProximalPhalanxTW* (94) and *Strawberry* (109). Achieving about 25 percentage points higher classification accuracy than with 1N, *ProximalPhalanxTW* (94) had the greatest benefit from combining different weighting schemes with constraining the elasticity of EDR. The difference in performance in case of the other mentioned datasets was greater than 11 pp—with the exception of *FordA* (42) when relying on the DU scheme, with which it amounted to about 8.2 pp with SC and 7.7 pp with IT. It should also be emphasized here that:

- In a descending ranking based on the amount of advantage achieved over 1N, the first four places always belonged to the same datasets in the same order regardless of the weighting scheme used and the type of global constraints: 94, 109, 74, 81.
- With the exception of *Earthquakes* (29), all other datasets mentioned above generated very similar differences in classification accuracy with respect to 1N also when comparing the time series with the LCS similarity measure, which reinforces the previous chapter's findings of a high degree of similarity between LCS and EDR.
- Since, except for *FordA* (42), the advantage of the listed datasets over 1N in Chap. 5 (Fig. 5.69) differed only slightly from the results in this chapter, in their case the application of global constraints did not lead to a noticeable improvement. On the other hand, when it comes to *FordA* (42), the accuracy of the kNN variants in combination with unconstrained EDR was only less than 1 pp higher than the accuracy of the 1NN classifier, which in its case clearly points to a very favorable impact of limiting the elasticity.

BeetleFly (8) and *PickupGestureWiimoteZ* (85) suffered the greatest damage from the transition from 1NN to kNN and from the unconstrained to the constrained EDR (Figs. 6.64 and 6.65). In addition to them, when limiting the width of the warping window using the Itakura parallelogram, 1N also outperformed the classification accuracy of kNN variants by more than 3 percentage points on *BirdChicken* (9). As a very similar situation was noted when examining the influence of global constraints on the accuracy of the 1NN classifier in Chap. 4 (see Fig. 4.49) and within Chap. 5 (Fig. 5.69), out of the mentioned datasets, 1N achieved an advantage of more than 3 pp only with *PickupGestureWiimoteZ* (85) and only over the DN, IN and DD

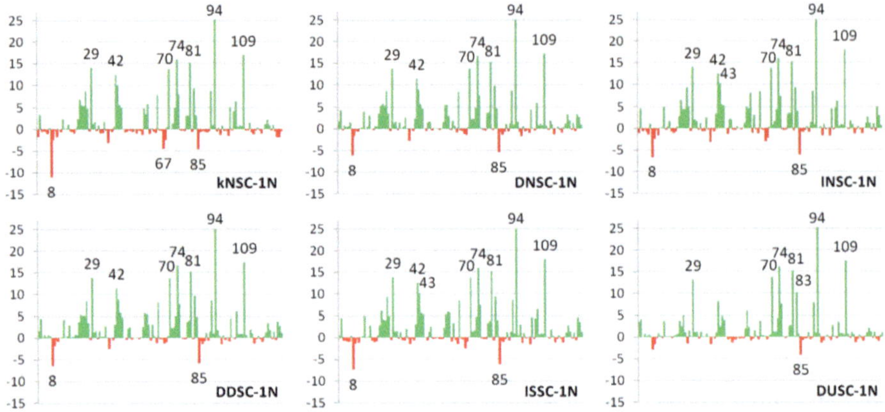

Fig. 6.64 EDR—differences between the classification accuracies of various kNN variants (combined with the SC constrained measure) and 1NN (combined with the unconstrained measure) by datasets (labels represent dataset identifiers)

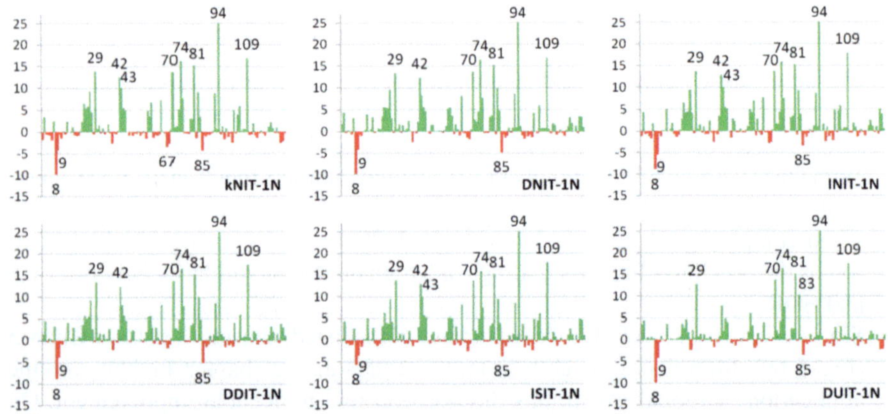

Fig. 6.65 EDR—differences between the classification accuracies of various kNN variants (combined with the IT constrained measure) and 1NN (combined with the unconstrained measure) by datasets (labels represent dataset identifiers)

schemes, these results, in the case of the mentioned datasets, clearly confirm the detrimental influence of SC and IT.

Figure 6.66 clearly testifies that the difference between the Sakoe-Chiba band and the Itakura parallelogram in terms of classification accuracy using the EDR measure to evaluate the similarity between time series is negligible. The difference between their performance in favor of SC was greater than or equal to 3 pp only with *BeetleFly* (8) (DN abd DU) and *BirdChicken* (9) (IN).

A statistically significant difference in the performance of the analyzed classifiers was confirmed using the Friedman test ($\chi^2(12) = 151.35$, $p < .001$). For

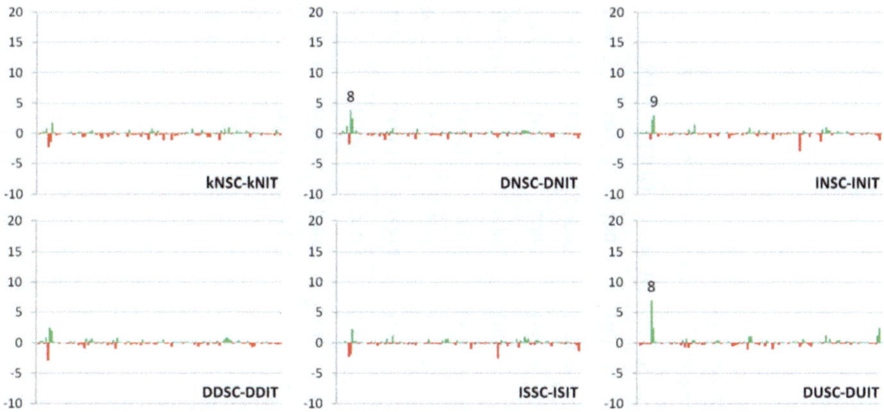

Fig. 6.66 EDR—differences between the classification accuracies of the SC and IT constrained measure (combined with different variants of the kNN classifier) by datasets (labels represent dataset identifiers)

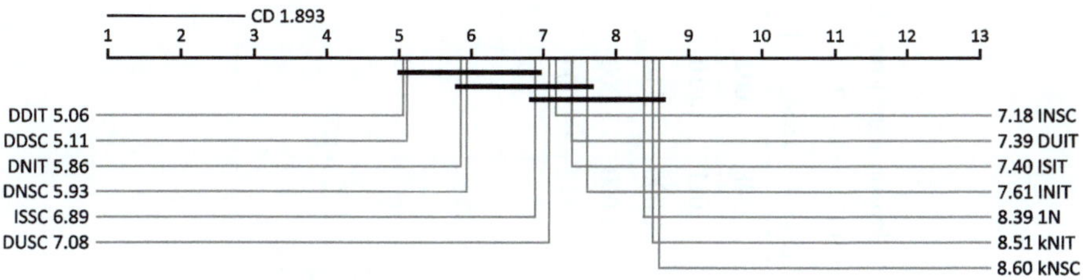

Fig. 6.67 EDR—critical difference diagram comparing average accuracy ranks based on Nemenyi's test: classifiers connected by a solid bar did not differ statistically significantly

pairwise comparisons, in the first step, we used Nemenyi's post-hoc test, the results of which are shown in Fig. 6.67. Based on them, it can be concluded that 1N was significantly outperformed only by the DD and DN weighting schemes (with both SC and IT). Since in the Chap. 5 (Fig. 5.70) in addition to them, IS and DU were also significantly better than 1N, their decline can be attributed to the application of global constraints. In the second step, the more permissive Wilcoxon signed-rank test (adjusted according to the Holm-Bonferroni approach) revealed statistical significance between 1N and all variants of the kNN classifier except for the majority-voting method (Table 6.15) which is in line with the findings in Chap. 5 (see Table 5.20). The inferiority of 1N and kN in relation to DD and DN was also confirmed by the numbers of statistically significant wins and losses obtained using the *t*-test (Table 6.16), again, consistent with the results of the unconstrained EDR in the previous chapter.

Applying the Sakoe-Chiba band, the largest number of datasets in the FP region was generated by the kN method (about 48% of the total of 128) and the smallest by the DD scheme (approximately 26%), as visually presented in Fig. 6.68. By switching to the Itakura parallelogram, their number increased a little for each kNN

Table 6.15 EDR—Holm-Bonferroni-adjusted *p*-values of the pairwise Wilcoxon signed-rank test on classification accuracy

A		79.27	79.24	79.22	79.18	79.17	79.17	79.12	79.08	78.74	78.73	78.67	78.56	77.06
		DDSC	DDIT	DNSC	ISSC	ISIT	DNIT	INSC	INIT	kNIT	kNSC	DUSC	DUIT	1N
79.27	DDSC		1.000	.251	.041	.012	.348	**.002**	**.008**	**< .001**	**< .001**	**< .001**	**< .001**	**< .001**
79.24	DDIT			1.000	.321	.011	.011	**.010**	**.002**	**< .001**	**< .0.01**	**.006**	**< .001**	**< .001**
79.22	DNSC				1.000	.668	1.000	.116	.116	**< .001**	**< .001**	**.007**	**.001**	**< .001**
79.18	ISSC					1.000	1.000	1.000	1.000	**.001**	**.001**	1.000	1.000	**< .001**
79.17	ISIT						.542	1.000	1.000	**.007**	.060	1.000	1.000	**.001**
79.17	DNIT							.364	.051	**< .001**	**< .001**	.099	.015	**< .001**
79.12	INSC								1.000	**.001**	**< .001**	1.000	1.000	**.001**
79.08	INIT									.011	.060	1.000	1.000	**.001**
78.74	kNIT										1.000	1.000	1.000	.323
78.73	kNSC											1.000	1.000	.493
78.67	DUSC												1.000	**< .001**
78.56	DUIT													**< .001**

Note Bolded values indicate significance at the level of $\alpha = 0.01$

Table 6.16 EDR—statistically significant wins (W) and losses (L) reported by the corrected resampled *t*-test on classification accuracy (across 12,800 splits)

	DDSC	DDIT	DNSC	DNIT	ISIT	ISSC	INSC	INIT	DUIT	DUSC	kNIT	kNSC	1N
W	413	391	365	338	275	269	246	245	269	294	152	144	232
L	84	90	123	118	215	215	236	261	350	378	421	436	706
W-L	329	301	242	220	60	54	10	−16	−81	−84	−269	−292	−474

variant (Fig. 6.69), except for the majority-voting rule where it decreased by one. Again, the proportion of wrong estimates of the gain over 1N was the highest for kN and the lowest for DD (about 29%).

EDR was the only measure where limiting the elasticity with either the SC or the IT technique led to a deterioration in both the FP and TP regions of virtually all six kNN variants (only the number of TP datasets increased from 64 to 66 in the case of the kN classifier combined with Itakura parallelogram—see Fig. 5.71 in Chap. 5). The degradation in both the FP and TP regions was most intense with DUIT: the

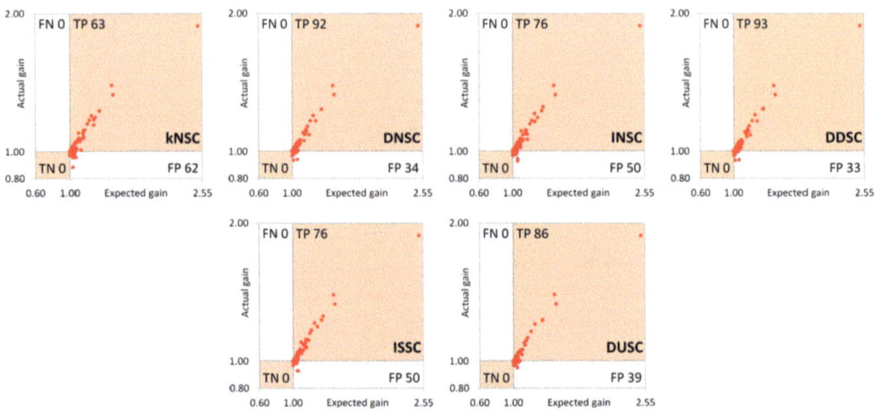

Fig. 6.68 EDR—comparing the actual and expected gain in classification accuracy: kNN variants (combined with the SC constrained measure) versus 1NN (combined with the unconstrained measure)

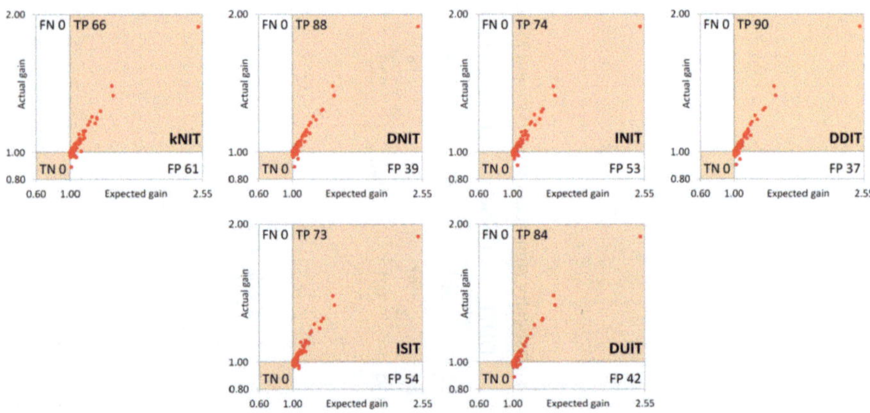

Fig. 6.69 EDR—comparing the actual and expected gain in classification accuracy: kNN variants (combined with the IT constrained measure) versus 1NN (combined with the unconstrained measure)

number of FP datasets increased by about 2.3 times and the number of TP datasets decreased by approximately 1.2 times.

When comparing the average degree of benefit from switching from 1NN to kNN variants (in terms of the number of FP and TP datasets when contrasted with the 1NN classifier paired with the unconstrained measures, see Chap. 4), EDR outperformed all other elastic measures with both the Sakoe-Chiba band and the Itakura parallelogram (in which case LCS achieved an identical average reduction in the number of FP datasets). The greatest progress was achieved by both SC and IT with the DD weighting scheme, both in the FP and in the TP region. In relation to 1NN, by applying the DN method SC reduced the number of FP cases by approximately 2.6 times (from 85 to 33) and IT by approximately 2.3 times (from 85 to 37). The improvement in the TP region was even more striking: SC increased the number of true positive datasets by about 3.2 times (from 29 to 93) and IT by approximately 2.7 times (from 33 to 90).

Compared to the findings with the 1NN classifier in Chap. 4 (see Figs. 6.70 and 4.51c), the success rate in correctly estimating the true advantage of one type of global constraint over the other noticeably worsened in the case of all six kNN variants. Moreover, looking at the averages of the six kNN classifiers, on this issue, EDR produced the worst result among the measures discussed in this chapter. The largest increase in the number of false positive and false negative (from 34 to 52) and the largest decrease in the number of true positive and true negative (from 74 to 63) estimates were detected in combination with the DD weighting scheme.

In the FP region, the error in estimating the advantage of kNN variants paired with the SC constraint over the 1NN classifier combined with the unconstrained EDR was greater than 0.05 for *BeetleFly* (8) (except with the DU scheme), *Lightning7* (68) (except with the DU scheme) and *PickupGestureWiimoteZ* (85), as presented in Fig. 6.71. Besides them, when limiting the width of the warping window with the Itakura parallelogram the difference between the expected and actual gain was

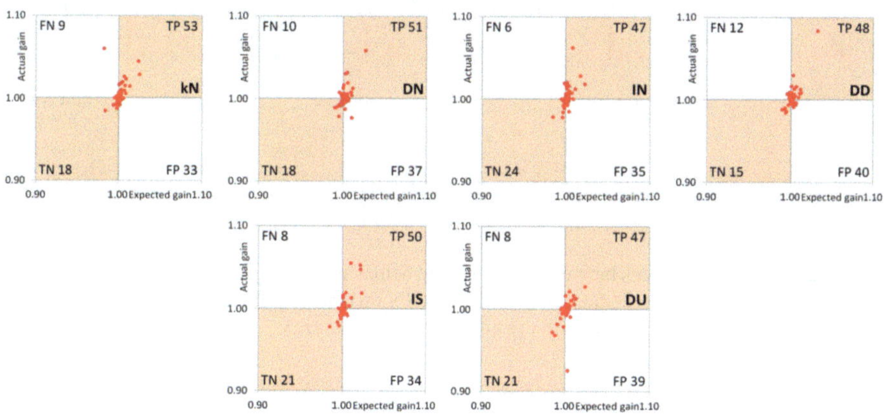

Fig. 6.70 EDR—comparing the actual and expected gain in classification accuracy: IT versus SC (combined with different variants of the kNN classifier)

greater than 0.05 in the FP region also for *Lightning2* (67) and *SemgHandMovementCh2* (99)—DU was an exception in both cases (Fig. 6.72). In addition to the above datasets, there were also some others with this property for each kNN variant with both SC and IT. Datasets numbered 67, 68, 85 and 99, with most of the weighting schemes discussed here, exceeded the 0.03 threshold in Chap. 5 (see Fig. 5.72) as well, and those marked with numbers 8 and 85 exhibited the described trait also in Chap. 4 (Fig. 4.52).

As can be seen from Fig. 6.73, the error in estimating the advantage of IT over SC among the FP datasets was greater than 0.03 only in the case of *BeetleFly* (8) when classifying with the DN and DU weighting schemes.

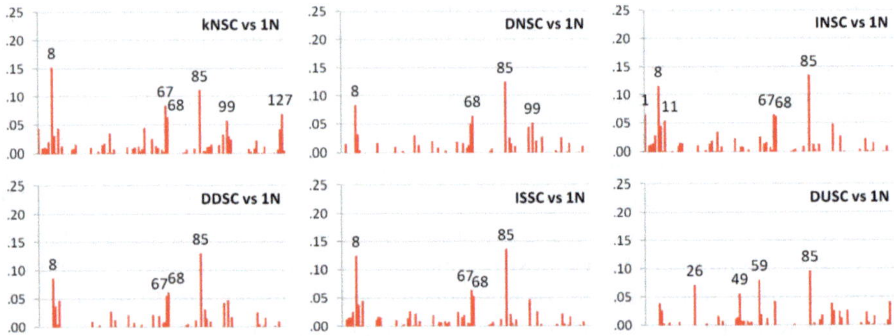

Fig. 6.71 EDR—differences between expected and actual accuracy gains of datasets belonging to the FP region (labels represent dataset identifiers): kNN variants (combined with the SC constrained measure) versus 1NN (combined with the unconstrained measure)

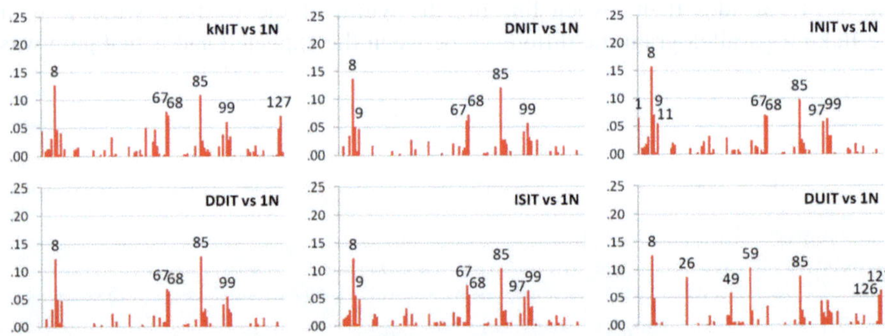

Fig. 6.72 EDR—differences between expected and actual accuracy gains of datasets belonging to the FP region (labels represent dataset identifiers): kNN variants (combined with the IT constrained measure) versus 1NN (combined with the unconstrained measure)

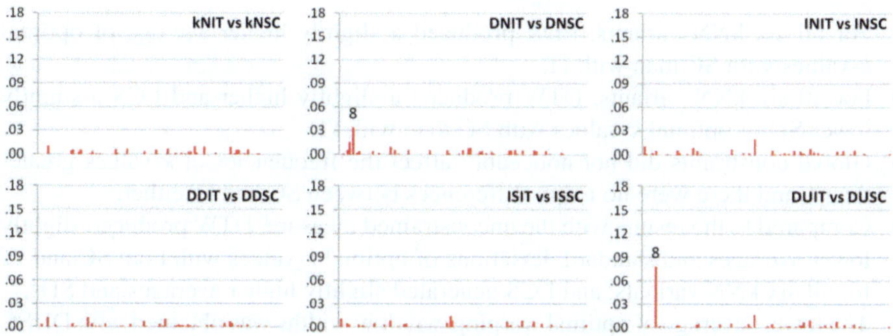

Fig. 6.73 EDR—differences between expected and actual accuracy gains of datasets belonging to the FP region (labels represent dataset identifiers): IT versus SC (combined with different variants of the kNN classifier)

6.5 Highlights

- The average accuracy of 1N (1NN with the corresponding unconstrained measure) was outperformed by all six kNN variants with both types of global constraints.
- On average, the advantage of the kNN variants over 1N (with both types of global constraints), was the smallest when using the ERP, and the largest when applying the DTW similarity measure.
- Except in the case of EDR, the application of global constraints had a positive effect on the average accuracy of the discussed kNN variants (i,e., they increased slightly compared to the results with the corresponding unconstrained measures).
- The positive impact of applying global constraints was the strongest for DTW, followed by ERP, and finally LCS.
- With LCS, ERP and EDR, the highest average accuracy was achieved by DDSC and the lowest by DUIT.
- With DTW, the highest average accuracy was achieved by ISSC (followed by DDSC) and the lowest by kNIT (in front of DUIT).
- The average accuracy of SC was almost always higher than that of IT. The only exceptions were IS with LCS and kN with EDR.
- On average, the standard deviation (SD) of the accuracy was highest for EDR and lowest for ERP, with both types of global constraints.
- With DTW and LCS, all kNN variants produced slightly higher SD of accuracy with SC than with IT.
- For DTW, LCS and ERP, applying both types of global constraints resulted in a slight decrease in the SD of accuracy for all kNN variants compared to the results with the unconstrained measure.
- Regardless of the similarity measure and the type of global constraint, DU generated the highest and kN the lowest average value and standard deviation of the optimal numbers of nearest neighbors.

- For all six kNN variants, EDR produced a slightly higher average of optimal k-values with SC than with IT.
- For all six kNN variants, DTW produced a slightly higher and LCS a slightly lower SD of optimal k-values with SC than with IT.
- Global constraints did not noticeably affect the frequencies of k-values greater than 1 and there were no major differences between SC and IT either.
- Compared to the results with the unconstrained measure, DTW produced slightly lower averages and standard deviations of optimal k-values with both SC and IT for all six kNN variants, and LCS generated slightly higher averages and SDs.
- The lowest average of optimal warping-window widths was obtained with DUSC for all four similarity measures.
- Similar to the results with 1NN, IT generated significantly higher average values and standard deviations of optimal r-values than SC. The difference was the largest with EDR and the smallest with DTW.
- Considering more than one nearest neighbor somewhat increased the average of the optimal warping-window widths for each measure. Whereby, the increase was slightly higher for IT than for SC.
- For ERP and EDR, switching from 1NN to kNN variants led to a slight increase in the SD of optimal r-values for both types of global constraints.
- For DTW, the switch from 1NN to kNN variants affected standard deviation of optimal warping-window widths differently for SC and IT: with IT there was a slight increase and with SC a slight decrease.
- The general characteristics of optimal warping-window widths did not change significantly when switching from the 1NN classifier to kNN variants.
- In terms of the number of datasets for which it produced the best accuracy, 1N was hard to beat. Only DDSC outperformed it in combination with all four elasticity measures.
- The lowest mean rank was produced by the DDSC classifier independent of the distance measure. The highest mean ranks were generated by kNSC (EDR) and 1N (DTW, LCS, ERP).
- The win-loss ratio of each kNN variant was the highest with DTW, and except for DD, it was the lowest with EDR.
- For DTW, applying global constraints led to an improvement in the win-loss ratio for all six weighting schemes.
- For ERP, the application of global constraints led to an improvement in the win-loss ratio for all weighting schemes, except for DU.
- For EDR, the application of global constraints led to a worsening of the win-loss ratio for all six weighting schemes.
- Except for kNSC and ISIT, the positive average difference in accuracy was the highest for all kNN variants when applying the DTW measure, for both types of global constraints.
- The positive average difference in accuracy was the lowest for all kNN variants when applying the ERP measure, for both types of global constraints.

- The employment of global constraints increased the positive average difference of all kNN variants in combination with all four elastic measures. The largest increase for each kNN variant was detected in the case of DTW, both for SC and IT.
- Except for kNSC, the negative average difference in accuracy was the highest for all kNN variants when applying the DTW measure, for both types of global constraints.
- Except for ISSC, the negative average difference in accuracy was the lowest for all kNN variants when applying the ERP measure, for both types of global constraints.
- The employment of global constraints increased the negative average difference of all kNN variants in combination with all four elastic measures.
- Relying on ERP, kNN variants surpassed the average accuracy of 1N by more than 10 percentage points only on two datasets: *FordA* (42) and *InsectWingbeatSound* (64). For the other similarity measures, the number of such datasets varied between 4 (DUIT with DTW) and 11 (DDSC with DTW).
- With DTW, the largest advantage of kNN variants over 1N was achieved on the *InsectWingbeatSound* (64) dataset and ranged between 22 and 26 pp.
- With LCS and EDR, the largest advantage of kNN variants over 1N was achieved on the *ProximalPhalanxTW* (94) dataset and was approximately 25 pp.
- Regardless of the similarity measure used, according to the Friedman test, the difference in the performance of the considered classifiers was statistically significant at the $\alpha = 0.01$ level.
- According to Nemenyi's post-hoc test, DD and DN significantly outperformed 1N regardless of the similarity measure used and the type of global constraints.
- According the Holm-Bonferroni-corrected Wilcoxon signed-rank test, at the $\alpha = 0.01$ level, all six weighting schemes significantly outperformed 1N regardless of the similarity measure used and the type of global constraints. The only exceptions were kNSC and kNIT with EDR.
- For DTW, the application of global constraints led to a decrease in the number of FP and an increase in the number of TP datasets. The only exception was the DU scheme, where the application of the IT constraint resulted in an increase of the number of datasets in the FP region from 25 to 27.
- For EDR, the application of global constraints led to an increase in the number of FP and a decrease in the number of TP datasets. The only exception was the kN scheme, where the application of the IT constraint resulted in an increase of the number of datasets in the TP region from 64 to 66.

References

1. Berndt, D., Clifford, J.: Using dynamic time warping to find patterns in time series. In: Usama, R.U., Fayyad, M. (ed.) knowledge discover databases paper from 1994 AAAI Workshop,

pp. 359–370. AAAI Press, Seattle, Washington (1994). http://dblp.uni-trier.de/rec/bib/conf/kdd/BerndtC94
2. Vlachos, M., Kollios, G., Gunopulos, D.: Discovering similar multidimensional trajectories. In: Proceedings of 18th international conference on data engineering, pp. 673–684. IEEE Computer Society (2002). https://doi.org/10.1109/ICDE.2002.994784
3. Chen, L., Ng, R.: On the marriage of Lp-norms and edit distance. In: Nascimento, M.A., Özsu, M.T., Kossmann, D., Miller, R.J., Blakeley, J.A., Schiefer, K.B. (eds.) Proceedings of 2004 VLDB conference, pp. 792–803. Elsevier (2004). https://doi.org/10.1016/B978-012088469-8.50070-X
4. Chen, L., Özsu, M.T., Oria, V.: Robust and fast similarity search for moving object trajectories. In: Proceedings of 2005 ACM SIGMOD International conference on management Data—SIGMOD'05, pp. 491–502. ACM Press, New York, USA (2005). https://doi.org/10.1145/1066157.1066213
5. Dudani, S.A.: The distance-weighted k-nearest-neighbor rule. IEEE Trans. Syst. Man. Cybern. **SMC-6**, 325–327 (1976). https://doi.org/10.1109/TSMC.1976.5408784
6. Mitchell, T.M.: Machine learning. McGraw-Hill Inc., New York, NY, USA (1997)
7. Gou, J., Du, L., Zhang, Y., Xiong, T.: A new distance-weighted k-nearest neighbor classifier. J. Inf. Comput. Sci. **9**, 1429–1436 (2012)
8. Gou, J., Xiong, T., Kuang, Y.: A novel weighted voting for k-nearest neighbor rule. J. Comput. **6**, 833–840 (2011). https://doi.org/10.4304/jcp.6.5.833-840
9. Sakoe, H., Chiba, S.: Dynamic programming algorithm optimization for spoken word recognition. IEEE Trans. Acoust. **26**, 43–49 (1978). https://doi.org/10.1109/TASSP.1978.1163055
10. Itakura, F.: Minimum prediction residual principle applied to speech recognition. IEEE Trans. Acoust. **23**, 67–72 (1975). https://doi.org/10.1109/TASSP.1975.1162641
11. Anh Dau, H., Keogh, E., Kamgar, K., Michael Yeh, C.-C., Zhu, Y., Gharghabi, S., Ann Ratanamahatana, C., Chen, Y., Hu, B., Begum, N., Bagnall, A., Mueen, A., Batista, G., Hexagon, M.L., The UCR time series classification Archive (2019). https://www.cs.ucr.edu/~eamonn/time_series_data_2018/
12. Paparrizos, J.: 2018 UCR time-series archive: cackward compatibility, missing values, and varying lengths (2019). https://github.com/johnpaparrizos/UCRArchiveFixes
13. Geler, Z.: Role of similarity measures in time series analysis. University of Novi Sad, Serbia (2015). (PhD thesis) https://nardus.mpn.gov.rs/handle/123456789/1703
14. Geler, Z., Kurbalija, V., Ivanovic, M., Radovanovic, M.: Time-series classification with constrained DTW distance and inverse-square weighted k-NN. In: 2020 International Conference on innovation intelligent system application, pp. 1–7. IEEE (2020).https://doi.org/10.1109/INISTA49547.2020.9194639
15. Geler, Z., Kurbalija, V., Ivanović, M., Radovanović, M.: Weighted kNN and constrained elastic distances for time-series classification. Expert Syst. Appl. **162**, 113829 (2020). https://doi.org/10.1016/j.eswa.2020.113829
16. Kurbalija, V., Radovanović, M., Geler, Z., Ivanović, M.: A framework for time-series analysis. In: Dicheva, D., Dochev, D. (eds.) artificial intelligent methodology system application. SE-5, pp. 42–51. Springer, Berlin, Heidelberg (2010). https://doi.org/10.1007/978-3-642-15431-7_5
17. Kurbalija, V., Ivanović, M., Geler, Z., Radovanović, M.: Two faces of the framework for analysis and prediction, part 2—research. Inf. Technol. Control. **47**, 489–502 (2018). https://doi.org/10.5755/j01.itc.47.3.18747
18. Kurbalija, V., Ivanović, M., Geler, Z., Radovanović, M.: Two faces of the framework for analysis and prediction, part 1—education. Inf. Technol. Control. **47**, 249–261 (2018). https://doi.org/10.5755/j01.itc.47.2.18746
19. Witten, I.H., Frank, E., Hall, M.A., Pal, C.J.: Data mining: practical machine learning tools and techniques, 4th ed. Morgan Kaufmann (2017). https://doi.org/10.1016/C2015-0-02071-8
20. Friedman, M.: The use of ranks to avoid the assumption of normality implicit in the analysis of variance. J. Am. Stat. Assoc. **32**, 675 (1937). https://doi.org/10.2307/2279372

21. Friedman, M.: A comparison of alternative tests of significance for the problem of m rankings. Ann. Math. Stat. **11**, 86–92 (1940). https://doi.org/10.1214/aoms/1177731944
22. Demšar, J.: Statistical comparisons of classifiers over multiple data sets. J. Mach. Learn. Res. **7**, 1–30 (2006)
23. Nemenyi, P.B.: Distribution-free multiple comparisons. Princeton University (1963) (PhD thesis)
24. Benavoli, A., Corani, G., Mangili, F.: Should we really use post-hoc tests based on mean-ranks? J. Mach. Learn. Res. **17**, 1–10 (2016). http://jmlr.org/papers/v17/benavoli16a.html
25. Wilcoxon, F.: Individual comparisons by ranking methods. Biometrics Bull. **1**, 80 (1945). https://doi.org/10.2307/3001968
26. Holm, S.: A simple sequentially rejective multiple test procedure. Scand. J. Stat. **6**, 65–70 (1979). http://www.jstor.org/stable/4615733
27. Bouckaert, R.R., Frank, E.: Evaluating the replicability of significance tests for comparing learning algorithms. In: Dai, H., Srikant, R., Zhang, C. (eds.) advanced knowledge discover data mining, pp. 3–12. Springer, Berlin, Heidelberg (2004). https://doi.org/10.1007/978-3-540-24775-3_3
28. Batista, G.E.A.P.A., Wang, X., Keogh, E.J.: A complexity-invariant distance measure for time series. In: Proceedings of 2011 SIAM international conference on data mining, pp. 699–710. Society for Industrial and Applied Mathematics, Philadelphia, PA (2011). https://doi.org/10.1137/1.9781611972818.60

Chapter 7
Conclusion

The focus of this book is a detailed study of characteristics and behavior of elastic similarity measures in distance-based classifiers (1NN and different kNN variants) for time series data. The choice of similarity measure is crucial in various tasks of time-series analysis such as classification, clustering, prediction and anomaly detection. It is extremely important since it should correctly reflect the underlying similarity between the considered time series. This study aims to address several key issues, including (1) the effects of global constraints on similarity measure performance, (2) the influence of constrained elastic similarity measures on classification accuracy of 1NN and several kNN variants, (3) the outcome of applying different weighting schemes on the performance of the kNN classifier and (4) the joint influence of global constraints and weighting schemes on classification accuracy. In addition, an open-source library for time-series analysis and mining is shortly presented.

Euclidean distance is a well-established measure of similarity between time series (mainly thanks to its intuitiveness, simplicity and the fact that it represents a metric), but it has limitations as well: time series must be of the same length and it is sensitive to shifts and distortions along the time axis. To overcome these limitations, various elastic measures based on dynamic programming have been suggested in the research community, such as DTW [1], LCS [2], ERP [3] and EDR [4]. These elastic measures have shown better classification accuracy compared to Euclidean distance but have the drawback of quadratic computation complexity.

Global constraints are introduced to reduce the computation time of elastic measures by narrowing the search path in the warping (editing) matrix. Furthermore, they can prevent some unnatural alignments between time-points, thus contributing to better quality of distance measures.

The reproducibility and traceability of scientific results is crucial in scientific community. Furthermore, the artificial intelligence and data mining findings strongly depend on the data used for training and evaluation. Many newly introduced concepts are tested on a small number of publicly available datasets, or even on non-public, protected datasets. To cope with this problem all experiments in this book are

Z. Gellér et al., *Recent Advances in Time-Series Classification—Methodology and Applications*, Intelligent Systems Reference Library 264,
https://doi.org/10.1007/978-3-031-77527-7_7

performed and evaluated on the largest publicly available repository of labeled time-series data—the UCR Time Series Repository [5].

The UCR repository contains 128 datasets from various domains like: agriculture, sport, astronomy, medicine, robotics, energetic, biology, seismology, food science, physics, etc. Furthermore, the properties of datasets vary significantly within this repository: the length of time series is between 24 and 2844, the number of classes is between 2 and 60 and the size of the datasets varies from 40 to 24,000. The raw data was pre-processed using several pre-processing techniques (missing value imputation, z-normalization) in order that all datasets have uniform characteristics and that various analyzes can be equally applied on all datasets. The repository and the constituting datasets are described in Chap. 2.

Time-series analysis and mining have seen significant growth in research efforts in recent years, resulting in a great diversity of techniques and algorithms being proposed. However, there is a lack of a quality open-source system that can support various aspects of time-series mining and allow for proper comparisons between different methods. To address this need, a universal framework called Framework for Analysis and Prediction (FAP) has been developed [6, 7]. FAP incorporates key concepts such as similarity measures, representations, pre-processing tasks, various classifiers and evaluation methods, making it a valuable tool for researchers to test and compare new ideas in the context of time series data exploration. The overview of FAP is given in Chap. 2.

In its current state, FAP includes implementations of various distance measures such as distances based on L_p norms, DTW, LCS, ERP, EDR, TWED [8] and Swale [9]. All elastic measures are implemented as unconstrained and constrained with the Sakoe-Chiba band (SC) [10] and Itakura parallelogram (IT) [11]. Additionally, the system implements the 1NN and kNN classifiers with different weighting schemes, various evaluation methods like Leave-One-Out, Cross-Validation and Holdout and representations such as Spline [12], SAX [13], PAA [14], PLA [15] and APCA [16]. Pre-processing techniques for scaling, normalization and shifting are also included in FAP. These components have been involved in a wide range of research areas, including psychological time-series mining [17], distributed distance matrix generation [18], medicine [19], emotion recognition [20] and neurology time-series analysis [21], as well as in various educational tasks.

The efficiency of the 1NN classifier is directly dependent on the underlying distance measure, making it suitable for evaluating different time-series similarity measures. Global constraints such as the Sakoe-Chiba band and Itakura parallelogram can be applied to speed up the nearest neighbor search process. In Chap. 4, the impact of global constraints on elastic measures (DTW, LCS, ERP and EDR) was analyzed by comparing NN graphs and classification accuracies with and without constraints. The experiments were conducted on the UCR repository datasets using various warping-window widths in three phases: (1) the quality and quantity of changes in the NN graph generated by constraining the elasticity is explored; (2) the accuracies obtained for individual warping-window width are compared; (3) the performances of the unconstrained measures with their SC and IT constrained counterparts using optimal warping-window widths are compared. The obtained results showed that global constraints can improve the classification performance, with optimal window widths varying depending on the measure used. The matching threshold parameter for LCS and EDR measures was fixed at 0.25 as proposed in the literature.

The results presented in Chap. 4 revealed several important findings. The IT and SC constraints had varying effects on the changes in the nearest neighbor graph, with IT producing more emphasized changes than SC. The amount of changes in the NN graph was largest with DTW and lowest with EDR. Prohibiting elasticity led to lower average classification accuracy compared to the unconstrained measures, with the smallest difference detected for DTW and the largest for LCS. When using the SC constraint, the optimal warping-window width was generally less than 16% of the time series length, while with the IT constraint, wider windows were needed (up to 80% of the time series length). DTW and ERP showed slightly higher average classification accuracies with constraints and EDR performed somewhat better without constraints.

DTW had the highest variability in performance with constraints, while EDR had the lowest variance. Statistically significant differences between the three considered similarity measure variants (unconstrained measure, SC and IT) were observed in the performance of DTW, LCS and EDR according to the Friedman test [22, 23]. As reported by Nemenyi's post-hoc test [24] and the Holm-Bonferroni-corrected [25] Wilcoxon signed-rank test [26], SC and IT were significantly different from the unconstrained variant when using DTW and EDR. LCS showed significant differences between IT and the unconstrained variant. In the case of ERP, no significant differences were revealed by the Nemenyi test, but significant differences were detected with the Wilcoxon signed-rank test between the constrained (both SC and IT) and unconstrained measures. The accuracy gains were most consistent with DTW and least consistent with EDR when choosing between constrained and unconstrained variants. Choosing between SC and IT based on classification accuracy on the training set was most successful with LCS and DTW and least effective with EDR. Finally, the results in Chap. 4 revealed the fact that the choice of constraint had a significant impact on classification accuracy and should be carefully considered when analyzing time series data.

The significance of the nearest neighbor in time-series classification is clearly evident from the relevant literature. Going further, Chap. 5 compared the accuracies of 1NN and kNN classifiers implemented using dynamic programming-based similarity measures and distances based on the L_p norm to better understand the importance of the first neighbor in improving classification accuracy. Different weighting schemes for nearest neighbors (presented and explained in Chap. 3) were discussed and compared in the context of time series data, focusing on unconstrained elastic similarity measures (DTW, LSC, ERP and EDR). The results of Chap. 5 aim to determine if the 1NN may be outperformed by the majority-voting kNN and weighted kNN.

Throughout Chap. 5, the performance of various kNN variants compared to the 1NN classifier was analyzed based on average classification accuracy, optimal number of nearest neighbors and win-loss ratios, supported by statistical tests.

Overall, the kNN variants outperformed 1NN in terms of classification accuracy when combined with different distance measures. The smallest advantage of the kNN variants over 1NN was observed when using the Euclidean distance and the largest when applying the EDR similarity measure. Generally, the DD [27] and DN [28] weighting schemes produced the highest average accuracies and FB [29] performed the worse. The optimal number of nearest neighbors was highest with the DU [30] weighting function, ranging between 14.35 and 16.27. Thereafter, the DD, DN and UN [30] methods produced higher optimal k-values than the other schemes for most distance measures. The lowest optimal value of number of nearest neighbors was observed in the case of the FB weighting approach and the majority-voting kNN. The dominance of the first nearest neighbor was most emphasized with the Chebyshev distance and least with the LCS measure.

The positive average difference in accuracy was lowest with the Manhattan distance, while the negative average difference was highest with DTW. The advantage of the kNN variants over 1NN ranged from below 9.5 percentage points (for several measures) to over 25 percentage points on certain datasets, with significant differences in performance observed for most schemes and distance measures. According to statistical tests, the Friedman test indicated significant differences in classifier performance, while Nemenyi's post-hoc test showed that only DD and DN significantly outperformed 1NN with the Manhattan distance and elastic measures.

The Holm-Bonferroni-corrected Wilcoxon signed-rank test further confirmed that the DD and DU schemes consistently outperformed 1NN for the Euclidean and Chebyshev distances. Additionally, DN, DD, ZA [31], UN and DU significantly outperformed 1NN when combined with the Manhattan distance and the elastic measures. In summary, the impact of different weighting functions and optimal numbers of nearest neighbors on classifier performance turned out to be important. Furthermore, several weighting methods (DN, IN [28], IS [32–34], DD and DU) showed constantly superior performance with different distance measures.

Finally, the last experimental chapter in this book (Chap. 6) examined the joint influence of global constraints, weights and the neighborhood size on classification accuracy. Five promising weighting schemes (selected based on the results of Chap. 5) were combined with SC and IT constrained elastic measures and their performances were compared to that of the 1NN classifier paired with the corresponding unconstrained measure. This chapter investigated the combined effect of two key parameters of NN classification of time series: the width of the warping window (r) and the number of considered neighbors (k). The values of r and k which are used are consistent with previous chapters, providing comparable insights into the performance of the classification models.

Chapter 6 revealed that, on average, all six kNN variants (majority-voting rule, DN, IN, IS, DD, DU) with both types of global constraints (SC and IT) outperformed the 1NN classifier in terms of accuracy. The advantage of the kNN variants was the smallest with the ERP measure and largest with DTW. Global constraints generally had a positive effect on accuracy for the kNN variants, particularly with DTW. The highest accuracy was achieved by DDSC (the combination of the DD weighting scheme and the SC constraint), while the lowest by DUIT (the combination of the

DU weighting function and the IT constraint). DDSC also had the lowest mean rank among all classifiers. The accuracy of SC constraint was almost always higher than of IT.

Considering neighborhood size, the DU weighting method evidently preferred the highest number of nearest neighbors, while the majority-voting kNN favored the lowest number of neighbors. From the perspective of distance measures, DTW selected slightly lower k-values, while LCS chose higher values.

The lowest average optimal warping-window widths were achieved with DUSC for all similarity measures. IT consistently generated higher average of optimal r-values compared to SC, with the largest difference seen with EDR. Using multiple nearest neighbors slightly increased the average optimal warping-window widths, with a somewhat higher increase for IT. In general, the characteristics of optimal warping-window widths did not significantly change when transitioning from 1NN to kNN variants.

In general, all kNN weighting schemes produced the highest win-loss ratio with DTW. Global constraints improved the win-loss ratio of kNN variants for the DTW and ERP measures, while the ratio was reduced for EDR. The performance difference among classifiers was statistically significant in most of the cases, according to all evaluated statistical test: Friedman test, Nemenyi's post-hoc test and Holm-Bonferroni-corrected Wilcoxon signed-rank test.

To conclude, the results of this book highlight the significance of the nearest neighbor in time-series classification and offer insights on how to enhance the performance of kNN classifiers in this domain. In most of the experiments, the kNN variants with global constraints showed superior performance compared to 1NN paired with the corresponding unconstrained measure, with DTW being the most effective similarity measure. We believe that these results will be valuable to the time-series-mining community and that our efforts will inspire other researchers to intensify and expand research activities in this thrilling area.

Future research is diverse and rather motivating. Naturally, future efforts should be extended to more recent elastic similarity measures (like TWED), as well as those that will appear in the future. Moreover, the same research questions are valid to other distance-based classifiers (other than 1NN and kNN), so it would be interesting to examine them also. Recent research actively investigate the hubness phenomenon [35–38] in kNN neighborhoods, hence the development and evaluation of hubs-based weighting schemes is a promising scientific direction. The same stands for the phenomenon of Local Intristic Dimentionality (LID) [39] which also influences the behavior of neighbors in a close proximity of the data point. At the end, the time-series distance measures can find their application in various graph embedding methods [40], which try to embed graphs into tabular-like data. Today, dynamic graphs (graphs which evolve during time) are of special interest and the time-series data are naturally involved in those embeddings.

References

1. Berndt, D., Clifford, J.: Using dynamic time warping to find patterns in time series. In: Usama, R.U., Fayyad, M. (eds.) Knowl. Discov. Databases Pap. from 1994 AAAI Work., pp. 359–370. AAAI Press, Seattle, Washington (1994). http://dblp.uni-trier.de/rec/bib/conf/kdd/BerndtC94
2. Vlachos, M., Kollios, G., Gunopulos, D.: Discovering similar multidimensional trajectories. In: Proc. 18th Int. Conf. Data Eng., pp. 673–684. IEEE Comput. Soc. (2002). https://doi.org/10.1109/ICDE.2002.994784
3. Chen, L., Ng, R.: On the marriage of Lp-norms and edit distance. In: Nascimento, M.A., Özsu, M.T., Kossmann, D., Miller, R.J., Blakeley, J.A., Schiefer, K.B. (eds.) Proc. 2004 VLDB Conf., pp. 792–803. Elsevier (2004). https://doi.org/10.1016/B978-012088469-8.50070-X
4. Chen, L., Özsu, M.T., Oria, V.: Robust and fast similarity search for moving object trajectories. In: Proc. 2005 ACM SIGMOD Int. Conf. Manag. Data—SIGMOD '05, pp. 491–502. ACM Press, New York, NY, USA (2005). https://doi.org/10.1145/1066157.1066213
5. Anh Dau, H., Keogh, E., Kamgar, K., Michael Yeh, C.-C., Zhu, Y., Gharghabi, S., Ann Ratanamahatana, C., Chen, Y., Hu, B., Begum, N., Bagnall, A., Mueen, A., Batista, G.: Hexagon-ML, the UCR Time Series Classification Archive (2019). https://www.cs.ucr.edu/~eamonn/time_series_data_2018/
6. Kurbalija, V., Ivanović, M., Geler, Z., Radovanović, M.: Two faces of the framework for analysis and prediction, Part 1—Education. Inf. Technol. Control **47**, 249–261 (2018). https://doi.org/10.5755/j01.itc.47.2.18746
7. Kurbalija, V., Ivanović, M., Geler, Z., Radovanović, M.: Two faces of the framework for analysis and prediction, Part 2—Research. Inf. Technol. Control **47**, 489–502 (2018). https://doi.org/10.5755/j01.itc.47.3.18747
8. Marteau, P.-F.: Time warp edit distance with stiffness adjustment for time series matching. IEEE Trans. Pattern Anal. Mach. Intell.Intell. **31**, 306–318 (2009). https://doi.org/10.1109/TPAMI.2008.76
9. Morse, M.D., Patel, J.M.: An efficient and accurate method for evaluating time series similarity. In: Proc. 2007 ACM SIGMOD Int. Conf. Manag. Data—SIGMOD '07, pp. 569–580. ACM Press, New York, NY, USA (2007). https://doi.org/10.1145/1247480.1247544
10. Sakoe, H., Chiba, S.: Dynamic programming algorithm optimization for spoken word recognition. IEEE Trans. Acoust. **26**, 43–49 (1978). https://doi.org/10.1109/TASSP.1978.1163055
11. Itakura, F.: Minimum prediction residual principle applied to speech recognition. IEEE Trans. Acoust. **23**, 67–72 (1975). https://doi.org/10.1109/TASSP.1975.1162641
12. Kurbalija, V., Ivanović, M., Budimac, Z.: Case-based curve behaviour prediction. Softw. Pract. Exp. **39**, 81–103 (2009). https://doi.org/10.1002/spe.891
13. Lin, J., Keogh, E., Wei, L., Lonardi, S.: Experiencing SAX: a novel symbolic representation of time series. Data Min. Knowl. Discov. **15**, 107–144 (2007). https://doi.org/10.1007/s10618-007-0064-z
14. Keogh, E., Chakrabarti, K., Pazzani, M., Mehrotra, S.: Dimensionality reduction for fast similarity search in large time series databases. Knowl. Inf. Syst.. Inf. Syst. **3**, 263–286 (2001). https://doi.org/10.1007/PL00011669
15. Keogh, E.J., Chu, S., Hart, D., Pazzani, M.: Segmenting time series: a survey and novel approach. In: Last, M., Kandel, A., Bunke, H. (eds.) Data Min. Time Ser. Databases, pp. 1–22. World Scientific Publishing Company (2004)
16. Keogh, E., Chakrabarti, K., Pazzani, M., Mehrotra, S.: Locally adaptive dimensionality reduction for indexing large time series databases. In: Proc. 2001 ACM SIGMOD Int. Conf. Manag. Data, pp. 151–162. ACM, New York, NY, USA (2001). https://doi.org/10.1145/375663.375680

17. Kurbalija, V., von Bernstorff, C., Burkhard, H.-D., Nachtwei, J., Ivanović, M., Fodor, L.: Time-series mining in a psychological domain. In: Proc. Fifth Balk. Conf. Informatics—BCI '12, pp. 58–63. ACM Press, New York, NY, USA (2012). https://doi.org/10.1145/2371316.2371328
18. Mitrovic, D., Ivanović, M., Geler, Z.: Agent-based distributed computing for dynamic networks. Inf. Technol. Control **43**, 88–97 (2014). https://doi.org/10.5755/j01.itc.43.1.4588
19. Kurbalija, V., Radovanović, M., Ivanović, M., Schmidt, D., von Trzebiatowski, G.L., Burkhard, H.-D., Hinrichs, C.: Time-series analysis in the medical domain: a study of tacrolimus administration and influence on kidney graft function. Comput. Biol. Med.. Biol. Med. **50**, 19–31 (2014). https://doi.org/10.1016/j.compbiomed.2014.04.007
20. Kurbalija, V., Ivanović, M., Radovanović, M., Geler, Z., Dai, W., Zhao, W.: Emotion perception and recognition: an exploration of cultural differences and similarities. Cogn. Syst. Res.. Syst. Res. **52**, 103–116 (2018). https://doi.org/10.1016/j.cogsys.2018.06.009
21. Kurbalija, V., Geler, Z., Stankov, T.V., Petrušić, I., Ivanović, M., Kononenko, I., Semnic, M., Daković, M., Semnic, R., Bosnić, Z.: Analysis of neuropsychological and neuroradiological features for diagnosis of Alzheimer's disease and mild cognitive impairment. Int. J. Med. Inform. **178**, 105195 (2023). https://doi.org/10.1016/j.ijmedinf.2023.105195
22. Friedman, M.: The use of ranks to avoid the assumption of normality implicit in the analysis of variance. J. Am. Stat. Assoc. **32**, 675 (1937). https://doi.org/10.2307/2279372
23. Friedman, M.: A comparison of alternative tests of significance for the problem of m rankings. Ann. Math. Stat. **11**, 86–92 (1940). https://doi.org/10.1214/aoms/1177731944
24. Nemenyi, P.B.: Distribution-free multiple comparisons. Princeton University (1963) (PhD thesis)
25. Holm, S.: A simple sequentially rejective multiple test procedure. Scand. J. Stat. **6**, 65–70 (1979). http://www.jstor.org/stable/4615733
26. Wilcoxon, F.: Individual comparisons by ranking methods. Biometrics Bull. **1**, 80 (1945). https://doi.org/10.2307/3001968
27. Gou, J., Du, L., Zhang, Y., Xiong, T.: A new distance-weighted k-nearest neighbor classifier. J. Inf. Comput. Sci. **9**, 1429–1436 (2012)
28. Dudani, S.A.: The distance-weighted k-nearest-neighbor rule. IEEE Trans. Syst. Man. Cybern. SMC-6, 325–327 (1976). https://doi.org/10.1109/TSMC.1976.5408784
29. Pao, T.-L., Chen, Y.-T., Yeh, J.-H., Cheng, Y.-M., Lin, Y.-Y.: A comparative study of different weighting schemes on KNN-based emotion recognition in mandarin speech. In: Huang, D.-S., Heutte, L., Loog, M. (eds.) Adv. Intell. Comput. Theor. Appl. With Asp. Theor. Methodol. Issues, pp. 997–1005. Springer, Berlin, Heidelberg (2007). https://doi.org/10.1007/978-3-540-74171-8_101
30. Gou, J., Xiong, T., Kuang, Y.: A novel weighted voting for K-nearest neighbor rule. J. Comput.Comput. **6**, 833–840 (2011). https://doi.org/10.4304/jcp.6.5.833-840
31. Zavrel, J.: An empirical re-examination of weighted voting for k-NN. In: Proc. 7th Belgian-Dutch Conf. Mach. Learn., pp. 139–148 (1997)
32. Wu, X., Kumar, V., Ross Quinlan, J., Ghosh, J., Yang, Q., Motoda, H., McLachlan, G.J., Ng, A., Liu, B., Yu, P.S., Zhou, Z.-H., Steinbach, M., Hand, D.J., Steinberg, D.: Top 10 algorithms in data mining. Knowl. Inf. Syst. **14**, 1–37 (2008). https://doi.org/10.1007/s10115-007-0114-2
33. Larose, D.T., Larose, C.D.: Discovering knowledge in data, 2nd edn. Wiley (2014). https://doi.org/10.1002/9781118874059
34. Mitchell, T.M.: Machine learning. McGraw-Hill Inc., New York, NY, USA (1997)
35. Radovanović, M., Nanopoulos, A., Ivanović, M.: Hubs in space: popular nearest neighbors in high-dimensional data. J. Mach. Learn. Res. **11**, 2487–2531 (2010). http://dl.acm.org/citation.cfm?id=1756006.1953015
36. Radovanovic, M., Nanopoulos, A., Ivanovic, M.: Reverse nearest neighbors in unsupervised distance-based outlier detection. IEEE Trans. Knowl. Data Eng.Knowl. Data Eng. **27**, 1369–1382 (2015). https://doi.org/10.1109/TKDE.2014.2365790
37. Tomašev, N., Radovanović, M., Mladenić, D., Ivanović, M.: Hubness-based fuzzy measures for high-dimensional k-nearest neighbor classification. Int. J. Mach. Learn. Cybern.Cybern. **5**, 445–458 (2014). https://doi.org/10.1007/s13042-012-0137-1

38. Bratić, B., Houle, M.E., Kurbalija, V., Oria, V., Radovanović, M.: The influence of hubness on NN-descent. Int. J. Artif. Intell. ToolsArtif. Intell. Tools **28**, 1960002 (2019). https://doi.org/10.1142/S0218213019600029
39. Houle, M.E.: Dimensionality, discriminability, density and distance distributions. In: 2013 IEEE 13th Int. Conf. Data Min. Work., pp. 468–473. IEEE (2013). https://doi.org/10.1109/ICDMW.2013.139
40. Savić, M., Kurbalija, V., Radovanović, M.: Local intrinsic dimensionality measures for graphs, with applications to graph embeddings. Inf. Syst. **119**, 102272 (2023). https://doi.org/10.1016/j.is.2023.102272

Appendix A
Similarity Measures

In the formulas below, A and B denote time series of length n and m, i.e. $A = (a_1, a_2, \ldots, a_n)$ and $B = (b_1, b_2, \ldots, b_m)$ (in the case of L_p norms, it must be $n = m$).

L_p norms

$$d_p(A, B) = ||A - B||_p = \sqrt[p]{\sum_{i=1}^{n} |a_i - b_i|^p}$$

where $p \geq 1$ is an arbitrary real number.

- **L_2—Euclidean distance**

$$d_2(A, B) = \sqrt{\sum_{i=1}^{n} (a_i - b_i)^2}$$

- **L_1—Manhattan distance**

$$d_1(A, B) = ||A - B||_1 = \sum_{i=1}^{n} |a_i - b_i|$$

- **L_∞—Chebyshev distance**

$$d_\infty(A, B) = ||A - B||_\infty = \max_{i=1,\ldots,n} |a_i - b_i|$$

DTW—Dynamic Time Warping

$$d_{DTW}(A, B) = DTW[n, m]$$

Z. Gellér et al., *Recent Advances in Time-Series Classification—Methodology and Applications*, Intelligent Systems Reference Library 264,
https://doi.org/10.1007/978-3-031-77527-7

$$DTW[i,j] = \begin{cases} 0 & i = j = 0 \\ \infty & (i = 0 \text{ and } j > 0) \text{ or } (i > 0 \text{ and } j = 0) \\ d(a_i, b_j) + min\begin{cases} DTW[i, j-1] \\ DTW[i-1, j-1] \\ DTW[i-1, j] \end{cases} & i, j > 0 \end{cases}$$

(*in this book*: $d(a_i, b_j) = (a_i - b_j)^2$).

LCS—Longest Common Subsequence

$$d_{LCS}(A, B) = \frac{n + m - 2LCS[n, m]}{n + m}$$

$$LCS[i,j] = \begin{cases} 0 & i = 0 \text{ or } j = 0 \\ LCS[i-1, j-1] + 1 & i, j > 0 \text{ and } |a_i - b_j| \le \varepsilon \\ max\{LCS[i, j-1], LCS[i-1, j]\} & i, j > 0 \text{ and } |a_i - b_j| > \varepsilon \end{cases}$$

where $\varepsilon \ge 0$ is the matching threshold (*in this book*: $\varepsilon = 0.25$, i.e. a quarter of the maximum standard deviation of the time series belonging to the given dataset).

ERP—Edit Distance with Real Penalty

$$d_{ERP}(A, B) = ERP[n, m]$$

$$ERP[i,j] = \begin{cases} 0 & i = j = 0 \\ ERP[i, j-1] + |b_j - g| & i = 0 \text{ and } j > 0 \\ ERP[i-1, j] + |a_i - g| & i > 0 \text{ and } j = 0 \\ min\begin{cases} ERP[i-1, j-1] + |a_i - b_j| \\ ERP[i, j-1] + |b_j - g| \\ ERP[i-1, j] + |a_i - g| \end{cases} & i, j > 0 \end{cases}$$

where g is a constant value (*in this book*: $g = 0$).

EDR—Edit Distance on Real sequence

$$d_{EDR}(A, B) = EDR[n, m]$$

$$EDR[i,j] = \begin{cases} 0 & i = j = 0 \\ EDR[i,j-1] + 1 & i = 0 \text{ and } j > 0 \\ EDR[i-1,j] + 1 & i > 0 \text{ and } j = 0 \\ min\begin{cases} EDR[i-1,j-1] + cost_{EDR}(a_i \to b_j) \\ EDR[i,j-1] + 1 \\ EDR[i-1,j] + 1 \end{cases} & i,j > 0 \end{cases}$$

where

$$cost_{EDR}(a \to b) = \begin{cases} 0 & |a - b| \leq \varepsilon \text{ and } a, b \neq \emptyset \\ 1 & \text{otherwise} \end{cases}$$

and $\varepsilon \geq 0$ is the matching threshold (*in this book*: $\varepsilon = 0.25$, i.e. a quarter of the maximum standard deviation of the time series belonging to the given dataset).

Appendix B
Weighting Functions

In the formulas below, k is the number of nearest neighbors, w_i denotes the weight of the i-th nearest neighbor and d_i is the distance between the query time series Q and its i-th nearest neighbor.

The weighting functions are listed here in the order they appear in Chap. 5.

MC—Macleod's Weighting Function

$$w_i = \begin{cases} 1 & d_s = d_1 \\ \frac{(d_s - d_i) + \alpha \times (d_s - d_1)}{(1+\alpha) \times (d_s - d_1)} & d_s \neq d_1 \end{cases}$$

where $s \geq k$ and $\alpha \geq 0$ (*in this book*: $s = k$ and $\alpha = 1$).

RK—Rank Weight

$$w_i = k - i + 1.$$

DN—Dudani's Weighting Scheme

$$w_i = \begin{cases} 1 & d_k = d_1 \\ \frac{d_k - d_i}{d_k - d_1} & d_k \neq d_1 \end{cases}$$

IN—Inverse Distance Weight

$$w_i = \frac{1}{d_i}$$

(*in this book*: to avoid division by zero, 0.001 is added to the denominator)

Z. Gellér et al., *Recent Advances in Time-Series Classification—Methodology and Applications*, Intelligent Systems Reference Library 264,
https://doi.org/10.1007/978-3-031-77527-7

DD—Dual Distance-Weighted Function

$$w_i = \begin{cases} 1 & d_k = d_1 \\ \frac{d_k - d_i}{d_k - d_1} \times \frac{d_k + d_1}{d_k + d_i} & d_k \neq d_1 \end{cases}$$

IS—Inverse of the Squared Distance Weight

$$w_i = \frac{1}{d_i^2}$$

(*in this book*: to avoid division by zero, 0.001 is added to the denominator).

ZA—Zavrel's Weighting Function

$$w_i = e^{-\alpha d_i^{\beta}}$$

where α and β represent constant values (*in this book*: $\alpha = \beta = 1$).

UN—Uniform Weighting Function

$$w_i = \frac{1}{i}$$

FB—Fibonacci Weighting Function

$$w_i = w_{i+1} + w_{i+2}$$

$$w_{k-1} = w_k = 1$$

DU—Dual-Uniform Weighting Function

$$w_i = \begin{cases} 1 & d_k = d_1 \\ \frac{d_k - d_i}{d_k - d_1} \times \frac{1}{i} & d_k \neq d_1 \end{cases}$$

Appendix C
The Relationship Between the Average Difference in Performance and the Average Distance from the Diagonal

Let y_i denote the classification accuracy of method A over the dataset S_i and x_i the classification accuracy of method B over the same dataset and let $y_i > x_i$ for each $i \leq n$. Based on Fig. C.1, the average distance of the points (x_i, y_i) from the diagonal is equal to the average difference between y_i and x_i divided by the square root of 2:

$$d^2 + d^2 = (y - x)^2$$

$$d^2 = \frac{(y - x)^2}{2}$$

$$d = \frac{y - x}{\sqrt{2}}$$

It follows:

$$\frac{1}{n}\sum_{i=1}^{n} d_i = \frac{1}{2}\sum_{i=1}^{n} \frac{y_i - x_i}{\sqrt{2}} = \frac{1}{\sqrt{2}}\frac{1}{n}\sum_{i=1}^{n}(y_i - x_i)$$

Z. Gellér et al., *Recent Advances in Time-Series Classification—Methodology and Applications*, Intelligent Systems Reference Library 264,
https://doi.org/10.1007/978-3-031-77527-7

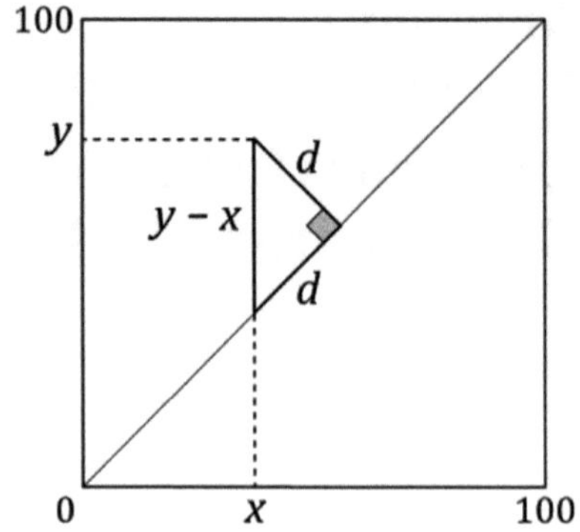

Fig. C.1 The relationship between the difference in performance and the distance from the diagonal

Appendix D
Explanation of the Behavior of kNN Classifier Variants with $k = 2$ and $k = 3$

Regardless of the distance measure, $k = 2$ cannot produce a better result than $k = 1$, because the weight of the second nearest neighbor cannot be greater than the weight of the first one in the case of any considered scheme. The two weights can be equal, but the implementation of the NN algorithm always selects the first class with the highest weight (i.e., in this case, the class of the (first) nearest neighbor).

When it comes to the RK, UN and FB weighting schemes, even $k = 3$ can never be selected as the optimal solution by the SCV1×9 algorithm, regardless of the distance measure used. The reason for this phenomenon is that the sum of the weights of the second and third neighbors cannot be greater than the weight of the first one, as presented in Table D.1. Therefore, the classification accuracy with $k = 3$ does not differ from the accuracy with $k = 1$ (since the corresponding NN classifier will always return the class of the first nearest neighbor even with $k = 3$). Furthermore, as the smallest value of the parameter k that gives the best result is always chosen, the optimal number of nearest neighbors will always be 1.

Table D.1 Nearest neighbor weights when using the RK, UN and FB schemes with $k = 3$

	RK	UN	FB
	$w_i = k - i + 1$	$w_i = \frac{1}{i}$	$w_i = w_{i+1} + w_{i+2}$ $w_{k-1} = w_k = 1$
w_1	3	1	2
w_2	2	1/2	1
w_3	1	1/3	1

Z. Gellér et al., *Recent Advances in Time-Series Classification—Methodology and Applications*, Intelligent Systems Reference Library 264,
https://doi.org/10.1007/978-3-031-77527-7

MIX
Papier aus verantwortungsvollen Quellen
Paper from responsible sources
FSC® C105338

If you have any concerns about our products,
you can contact us on
ProductSafety@springernature.com

In case Publisher is established outside the EU,
the EU authorized representative is:
Springer Nature Customer Service Center GmbH
Europaplatz 3, 69115 Heidelberg, Germany

Printed by Libri Plureos GmbH
in Hamburg, Germany